TAILORED FUNCTIONAL MATERIALS FOR CLEAN AND SUSTAINABLE DEVELOPMENT

TAILORED FUNCTIONAL MATERIALS FOR CLEAN AND SUSTAINABLE DEVELOPMENT

Edited by

Tamara Tatrishvili, PhD

Neha Kanwar Rawat, PhD

Ann Rose Abraham, PhD

A. K. Haghi, PhD

AAP | APPLE ACADEMIC PRESS

First edition published 2024

Apple Academic Press Inc.
1265 Goldenrod Circle, NE,
Palm Bay, FL 32905 USA

760 Laurentian Drive, Unit 19,
Burlington, ON L7N 0A4, CANADA

CRC Press
2385 NW Executive Center Drive,
Suite 320, Boca Raton FL 33431

4 Park Square, Milton Park,
Abingdon, Oxon, OX14 4RN UK

Library and Archives Canada Cataloguing in Publication

Title: Tailored functional materials for clean and sustainable development / edited by Tamara Tatrishvili, PhD, Neha Kanwar Rawat, PhD, Ann Rose Abraham, PhD, A.K. Haghi, PhD.
Names: Tatrishvili, Tamara, editor. | Rawat, Neha Kanwar, editor. | Abraham, Ann Rose, editor. | Haghi, A. K., editor.
Description: First edition. | Includes bibliographical references and index.
Identifiers: Canadiana (print) 20230495893 | Canadiana (ebook) 20230495931 | ISBN 9781774913659 (hardcover) | ISBN 9781774913666 (softcover) | ISBN 9781003394761 (ebook)
Subjects: LCSH: Nanostructured materials. | LCSH: Nanotechnology. | LCSH: Biopolymers. | LCSH: Polymers.
Classification: LCC TA418.9.N35 T35 2024 | DDC 620.1/15—dc23

Library of Congress Cataloging-in-Publication Data

CIP data on file with US Library of Congress

ISBN: 978-1-77491-365-9 (hbk)
ISBN: 978-1-77491-366-6 (pbk)
ISBN: 978-1-00339-476-1 (ebk)

About the Editors

Tamara Tatrishvili, PhD
Assistant Professor, Main Specialist, Office of the Academic Process Management, Ivane Javakhishvili Tbilisi State University (TSU); Director of the Institute of Macromolecular Chemistry and Polymeric Materials at TSU, Tbilisi, Georgia

Tamara Tatrishvili, PhD, is Assistant Professor and the main specialist at the Office of Academic Process Management (Faculty of Exact and Natural Sciences) at Ivane Javakhishvili Tbilisi State University as well as the Director of the Institute of Macromolecular Chemistry and Polymeric Materials at TSU, Tbilisi, Georgia; DAAD alumni, and a member of Georgian chemical society.

Her research interests include polymer chemistry, polymeric materials, and chemistry of silicon-organic compounds. She is the Executive Editor of *Journal of the Georgian Chemical Society*.

Dr. Tatrishvili is the author of more than 190 scientific publications, 12 books, and several monographs.

Neha Kanwar Rawat, PhD
Researcher, Department of Materials Science and Engineering, Indian Institute of Science, Bengaluru, India

Neha Kanwar Rawat, PhD, is a recipient of a prestigious DS-Kothari Postdoctoral Fellowship and DST Young Scientist Postdoctoral Fellowship and is presently a researcher in the Department of Materials Science and Engineering, Indian Institute of Science, Bengaluru, India. She received her PhD in Chemistry from Jamia Millia Islamia (a Central University), India. Her main interests include nanotechnology and nanostructured materials synthesis and characterization, with her main focus on green chemistry along with novel sustainable chemical processing of nano-conducting polymers/nanocomposites; conducting films, ceramics, silicones, and matrices: epoxies, alkyds, polyurethanes, and others. She is also pursuing her interest in fusing new technology in areas that include electrochemistry and organic–inorganic hybrid nanocomposites for biomedical applications, which also includes protective surface coatings for corrosion inhibition and MW shielding materials. She has published numerous peer-reviewed research articles in journals of high repute. Her contributions have led to many books and chapters in international books published with the Royal Society of Chemistry, Wiley, Elsevier, Apple Academic Press, Nova US, and many others in progress. She is a peer reviewer for many international books and is a member of many groups, including the Royal Society of Chemistry, and the American Chemical Society (USA) and a life member of the Asian Polymer Association.

Ann Rose Abraham, PhD
Assistant Professor, Sacred Heart College (Autonomous), Thevara, Kochi, Kerala, India

Ann Rose Abraham, PhD, is currently an Assistant Professor at the Department of Physics, Sacred Heart College (Autonomous), Thevara, Kochi, Kerala, India. She has expertise in the field of condensed matter physics, nanomagnetism, multiferroics and polymeric nanocomposites, etc. She has research experience at various reputed national institutes, including Bose Institute, Kolkata, India; SAHA Institute of Nuclear Physics, Kolkata, India; UGC-DAE CSR Centre, Kolkata, India. She has also collaborated with various international laboratories. She is a recipient of a Young Researcher Award in physics and Best Paper Awards–2020, 2021. She served as assistant professor and examiner at the Department of Basic Sciences, Amal Jyothi College of Engineering, under APJ Abdul Kalam Technological University, Kerala, India. Dr. Abraham is a frequent speaker at national and international conferences. She has a good number of publications to her credit in many peer-reviewed high impact journals of international repute. She has authored many book chapters and edited more than 10 books with Taylor and Francis, Elsevier, etc. Dr. Abraham received her MSc, MPhil, and PhD degrees in Physics from the School of Pure and Applied Physics, Mahatma Gandhi University, Kerala, India.

A. K. Haghi, PhD
Professor Emeritus of Engineering Sciences, Former Editor-in-Chief, International Journal of Chemoinformatics and Chemical Engineering; Member, Canadian Research and Development Center of Sciences and Culture

A. K. Haghi, PhD, has published over 250 academic research-oriented books as well as over 1000 research papers published in various journals and conference proceedings. He has received several grants, consulted for several major corporations, and is a frequent speaker to national and international audiences. He is Founder and former Editor-In-Chief of the *International Journal of Chemoinformatics and Chemical Engineering*, published by IGI Global (USA) as well as *Polymers Research Journal*, published by Nova Science Publishers (USA). Professor Haghi has acted as an editorial board member for many international journals. He has served as a member of the Canadian Research and Development Center of Sciences and Cultures (CRDCSC) and the Research Chemistry Centre, Coimbra, Portugal. Dr. Haghi holds a BSc degree in urban and environmental engineering from the University of North Carolina (USA), an MSc degree in mechanical engineering from North Carolina A&T State University (USA), and an MSc degree in applied mechanics, acoustics, and materials from the Université de Technologie de Compiègne (France). He holds a PhD in engineering sciences from Université de Franche-Comté (France).

Contents

Contributors

Abhishek Anand
Non-Conventional Energy Laboratory, Rajiv Gandhi Institute of Petroleum Technology, Jais, Amethi, Uttar Pradesh, India

Mohammad Ashfaq
Department of Biotechnology, University Centre for Research & Development, Chandigarh University, Mohali, Punjab, India

Andreea Irina Barzic
"Petru Poni" Institute of Macromolecular Chemistry, Laboratory of Physical Chemistry of Polymers, Iasi, Romania

Aarthi S. Bhatt
Department of Chemistry, NMAM Institute of Technology (Deemed to be University), Nitte, Karkala Taluk, Udupi, Karnataka, India

Elguja Chagelishvili
Institute of Mining and Technology, Tbilisi, Georgia

Tanmoy Chakraborty
Department of Chemistry and Biochemistry, School of Basic Sciences and Research, Sharda University, Geater Noida, India

Divya Chauhan
Department of Chemical and Biomedical Engineering, University of South Florida, Tampa, FL, USA

Ivan Chodák
Polymer Institute, Slovak Academy of Sciences, Bratislava, Slovakia

Shayeri Das
Department of Mechatronics Engineering, Manipal University Jaipur, Jaipur, Rajasthan, India
Department of Electrical Engineering, Ideal Institute of Engineering, Kalyani, Nadia, West Bengal, India

N. G. Divya
Bharata Mata College, Thrikkakara, Edappally, Cochin, Kerala, India

Chkhaidze Ekaterina
Department of Chemical and Biological Technologies, Georgian Technical University, Tbilisi, Georgia

R. A. Gakhokidze
Department of Chemistry, Ivane Javakhishvili State University, Tbilisi, Georgia

Rinky Ghosh
Department of Materials Science and Engineering, Indian Institute of Technology Kanpur, Kanpur, Uttar Pradesh, India

Antia Giorgi
Faculty of Medicine, Tbilisi State Medical University, Tbilisi, Georgia

Bagrat Godibadze
Ferdinad Tavadze Institute of Metallurgy and Materials Science, Tbilisi, Georgia

Divya Neravathu Gopi
Bharata Mata College, Thrikkakara, Edappally, Cochin, Kerala, India

Berdzenishvili Irine
Department of Chemical and Biological Technologies, Georgian Technical University, Tbilisi, Georgia

Harshitha M Jathanna
Department of Biotechnology, NMAM Institute of Technology (Deemed to be University), Nitte, Karnataka, India

Peter Jurkovič
VIPO, a.s., gen. Svobodu Partizánske, Slovakia

Laszlo Kecskes
John Hopkins University, Baltimore, MD, USA

Angela Kleinová
Polymer Institute, Slovak Academy of Sciences, Slovakia

V. I. Kodolov
Department of Chemistry and Chemical Engineering of Kalashnikov Izhevsk State Technical University, Basic Research-Educational Center of Chemical Physics and Mesoscopy, UdSC, UrD, RAS, Russia

G. A. Korablev
Department of Physics at Izhevsk State Agricultural Academy, Izhevsk State Agricultural Academy, Izhevsk, Udmurt Republic, Russia

Mohammed Saju M.
Bharata Mata College, Thrikkakara, Edappally, Kerala, India

Siradze Manana
Department of Chemical and Biological Technologies, Georgian Technical University, Tbilisi, Georgia

Jacob Mathew
Department of Chemical Engineering, National Taiwan University of Science and Technology, Taiwan, Republic of China

Ján Matyašovský
VIPO, a.s., gen. Svobodu Partizánske, Slovakia

Matej Mičušík
Polymer Institute, Slovak Academy of Sciences, Slovakia

Igor Novák
Polymer Institute, Slovak Academy of Sciences, Slovakia

M. O. Nutsubidze
Department of Chemistry, Ivane Javakhishvili State University, Tbilisi, Georgia

Saurabh Pandey
Non-Conventional Energy Laboratory, Rajiv Gandhi Institute of Petroleum Technology, Jais, Amethi, Uttar Pradesh, India

Akaki Peikrishvili
Ferdinad Tavadze Institute of Metallurgy and Materials Science, Tbilisi, Georgia

Deepak Poddar
Netaji Subhas University of Technology, New Delhi, India

Anton Popelka
Center for Advanced Materials, Qatar University, Doha, Qatar

Kiran Rana
Department of Plastics Engineering, Central Institute of Plastics, Engineering and Technology, GIDC Estate, Vatva, Ahmedabad, Gujarat, India

Prabhat Ranjan
Department of Mechatronics Engineering, Manipal University Jaipur, Dehmi Kalan, Jaipur, Rajasthan, India

Neha Kanwar Rawat
Materials Research Laboratory, Department of Chemistry, Jamia Millia Islamia, New Delhi, India; Department of Materials Engineering, Indian Institute of Science, Bangalore, Karnataka, India

Ann Roshna
Cochin University of Science and Technology, South Kalamassery, Cochin, Kerala, India

G. Santhosh
Department of Mechanical Engineering, NMAM Institute of Technology (Deemed to be University), Udupi, Karnataka, India

Mohammod Saju
Bharata Mata College, Thrikkakara, Edappally, Cochin, Kerala, India

Ján Sedliačik
Technical University in Zvolen, Slovakia

Atul Sharma
Non-Conventional Energy Laboratory, Rajiv Gandhi Institute of Petroleum Technology, Jais, Amethi, Uttar Pradesh, India

Nidhi Sharma
1Department of Physics, Dhanauri P.G. College, Roorkee Haridwar, Uttarakhand, India

Amritanshu Shukla
Non-Conventional Energy Laboratory, Rajiv Gandhi Institute of Petroleum Technology, Jais, Amethi, Uttar Pradesh, India
Department of Physics, Lucknow University, Lucknow, Uttar Pradesh, India

N. N. Sidamonidze
Department of Chemistry, Ivane Javakhishvili State University, Tbilisi, Georgia

Ladislav Šoltés
Centre of Experimental Medicine, Slovak Academy of Sciences, Slovakia

Neetu Talreja
Department of Science, Faculty of Science and Technology, Alliance University, Anekal, Bengaluru, Karnataka, India

Darieo Thankachan
Department of Materials and Science Engineering, National Taiwan University of Science and Technology, Taiwan, Republic of China

Riju K. Thomas
Bharata Mata College, Thrikkakara, Edappally, Cochin, Kerala, India

Katarína Valachová
Centre of Experimental Medicine, Slovak Academy of Sciences, Slovakia

R. O. Vardiashvili
Department of Chemistry, Ivane Javakhishvili State University, Tbilisi, Georgia

G. E. Zaikov
N. M. Emmanuel Institute of Biochemical Physics, RAS, Russia

Abbreviations

AFM	atomic force microscopy
AG	alginate
ATR-FTIR	attenuated total reflectance Fourier-transform infrared
ATP	adenosine triphosphate
CA	capric acid
CAc	cellulose acetate
CDs	cyclodextrin
CFU	colony-forming units
CMC	carboxymethyl cellulose
CNFs	carbon nanofibers
CNTs	carbon nanotubes
CS	chitosan
CT	computed tomography
CVD	chemical vapor deposition
DAN	1,5-diaminonaphthalene
DC	dispersive component
DEP	dielectrophoresis
DFT	density functional theory
DHLA	dihydrolipoic acid
DSSCs	dye-sensitized solar cell
DTPA	diethylenetriaminepentaacetic acid
ECH	epichlorohydrin
EDLC	electrochemical double layer type capacitor
ESP	electrospinning
ESPR	elucidate the transition from electrospraying
ESR	electron spin resonance
eV	electron-volts
FeMG	Fe-based metallic glasses
FEP	fluorinated ethylene propylene
FN	fibronectin
FXI	coagulation factors XI
FXII	coagulation factors XII
GCE	glassy carbon electrode
$g\text{-}C_3N_4$	graphitic carbon nitride

GE	gelatin
HA	hyaluronic acid
HAP	hydroxyapatite
HBCA	4-hydroxy-4′-biphenylcarboxylic acid
HEC	hot explosive consolidation
HMWK	high-molecular-weight-kininogen
HOMO	highest occupied molecular orbital
HRP	horseradish peroxidase
ITO	indium tin oxide
LA	lauric acid
LbL	layer-by-layer
LHC	latent heat of crystallization
LHF	latent heat of fusion
LHH	long-term high-temperature heating
LUMO	lowest unoccupied molecular orbital
MA	myristic acid
MEMS	microelectromechanical systems
MMT	montmorillonite
M-NPs	metal nanoparticles
MGs	metallic glasses
MW	molecular weight
MWCNTs	multiwalled carbon nanotubes
NEMS	nanoelectromechanical systems
NGF	nerve growth factor
OE	open ended
OWRK	Owens–Wendt–Rabel–Kaelble
PA	polyamide
PA	palmitic acid
PAN	polyacrylonitrile
PA6	polyamide 6
PC	polar component
PCL	polycaprolactone
PCMs	phase change material
PCR	polymerase chain reaction
PDMS	polydimethylsiloxane
PEC	polyelectrolyte complex
PEDOT	poly(3,4-ethylenedioxythiophene)
PEG	polyethylene glycol
PEM	polyelectrolyte multilayer

PET	polyethylene terephthalate
PGA	poly(L-glutamic acid)
PHA	polyhydroxyalkanoates
PHB	polyhydroxybutyrate
PI	polyimide
PLA	polylactide
PLA	polylactic acid
PLL	poly(L-lysine)
PMIA	poly(meta-phenylene isophthalamide)
PS	polystyrene
PTCA	percutaneous transluminal coronary angioplasty
PTFE	polytetrafluoroethylene
PU	polyurethane
PV	peroxide values
PVA	polyvinyl alcohol
PVC	polyvinyl chloride
PVCs	photovoltaic cells
PVdF	polyvinylidene fluoride
P4VP	poly (4-vinylpyridine)
QDs	quantum dots
RF	radio frequency
RGIPT	Rajiv Gandhi Institute of Petroleum Technology
RSF	regenerated silk fibroin
RSM	response surface methodology
SA	stearic acid
SAN	poly(styrene-coacrylonitrile)
SCM	self-complementary multiple
SEM	scanning electron microscopy
SEP	spatial-energy parameter
SF	silk fibroin
SMDD	substrate-mediated drug delivery
SPR	surface plasmon resonance
SPUs	segmented polyurethanes
ssDNA	single-stranded DNA
SWCNTs	single-walled carbon nanotubes
TENG	triboelectric nanogenerators
T_f	freezing temperature
T_m	melting temperature
TN	tribonegative

TP	tribopositive
TNT	titanium dioxide nanotubes
W-MG	tungsten-metallic glass
WVP	water vapor permeability
XPS	X-ray photoelectron spectroscopy
1-D	one dimensional
2-D	two dimensional
3-D	three dimensional

Preface

Tailored functional materials have grown enormously in recent years due to their wide-ranging applications in both science and engineering. This new volume presents tailoring and characterization of modern materials with functional applications and clean technologies in different sectors.

It covers new topics in nanotechnology, nanoscience, biopolymers, energy, carbon nanotubes, solid materials, biomaterials, biomedical, and healthcare for green environment.

This unique research-oriented volume can be used as a reference book in science and applications of tailored functional materials in an inclusive manner for postgraduate students in materials science and chemical engineering.

The book is divided into four main parts:

In the first part, we explain why the unique features of tailored nanomaterials make them an ideal candidate for applications in biosensors technology, pathogenic microbe detection, and modification of polyester-based polymers.

In the second part of this book, the chapters summarize polymer functionalities and their applications in medical science and engineering.

The third part is devoted to understanding tailored custom-designed solids along with computational chemistry within the scope of the book showing how fast growth in solid materials can influence different engineering sectors.

By covering case studies in materials science, the fourth part of this volume explains how tailored materials can be prepared and utilized as application-oriented composites.

The last part of this new title provides comprehensive studies on major innovations in tailored energy materials, which focuses on their technocommercial applications.

PART I

Nanotechnology and Nanomaterials

CHAPTER 1

Carbon Nanotubes and Fullerenes: Promising Materials for Biosensor Technology

NIDHI SHARMA[1], NEETU TALREJA[2], DIVYA CHAUHAN[3], and MOHAMMAD ASHFAQ[4]

[1]*Department of Physics, Dhanauri P.G. College, Dhanauri, Roorkee Haridwar, Uttarakhand, India*

[2]*Department of Science, Alliance University, Anekal, Bengaluru, Karnataka, India*

[3]*Department of Chemical and Biomedical Engineering, University of South Florida, Tampa, FL, USA*

[4]*Department of Biotechnology, University Centre for Research & Development, Chandigarh University, Gharuan, Mohali, Punjab, India*

ABSTRACT

Biosensor technology is one of the leading technologies to detect environmental pollutants, drug molecules, proteins, enzymes, and diseases. Conventional technology is used to detect diseases with high precision. However, these technologies are ineffective in early-stage detection of diseases. Therefore, newer biosensor technology is required for early detection of various diseases, including cancer. With the help of nanomaterials/nanotechnology (NMs) and materials engineering, researchers continue to focus on developing novel biosensor technologies. Carbon-based materials like carbon nanotubes (CNTs) and fullerenes are the most promising materials

Tailored Functional Materials for Clean and Sustainable Development. Tamara Tatrishvili, PhD, Neha Kanwar Rawat, PhD, Ann Rose Abraham, PhD & A. K. Haghi, PhD (Eds.)

for developing biosensors. The extraordinary characteristics of the CNTs and fullerenes, such as high surface area, high electrical and mechanical properties, and easy functionalization, make them an appropriate candidate for the development of biosensors. In this chapter, our focus is mainly on synthesizing CNTs and fullerenes, surface functionalization, and its application in cancer and microorganism detection.

1.1 INTRODUCTION

Sensor technology is one the leading technologies that can detect environmental pollutants, drug molecules, proteins, enzymes, and diseases. The biosensor technology in medicine and life sciences has been efficiently used to detect biological marker or diseases even at an early stage.[37,55,82] The early-stage diagnosis and treatment of diseases with a minimally invasive approach is important to improve the quality of the treatment. Usually, traditional treatment/diagnosis approaches, mainly DNA array, enzyme-linked immunosorbent assay (ELISA), immunoassay, computed tomography (CT) scan, X-ray, and polymerase chain reaction (PCR), have been used for the diagnosis of diseases. However, expensive chemicals/reagents, sophisticated instruments, and less sensitive and skilled personnel are required. Moreover, early-stage diseases, especially cancer, are difficult to diagnose using such technologies.[7,17,53,58,64,73] Therefore, there is a requirement for such a biosensor technology that can detect/diagnose numerous diseases even at the early stages of infections.

Significant research has been done so far to develop novel biosensor technologies because of the requirement of early detection of diseases with a minimally invasive approach. With the help of nanomaterials/nanotechnology (NMs) and materials engineering, researchers continue to focus on developing novel biosensor technologies. The unique characteristics of the NMs like size, shape, chemical, and electrical properties might help in improving the treatment of patients, or patients care by minimally invasive, accurate, and highly sensitive biosensors.[57,60,86] NMs such as metal nanoparticles (M-NPs), carbon nanotubes (CNTs), carbon nanofibers (CNFs), graphene, graphene oxide, and fullerenes have been effusively used in numerous applications, including environmental remediation/water treatment technologies (removal of heavy metal ions, pharmaceutical compounds (vitamins and antibiotics), and antibacterials),[18,19,44,69,70,76,77] energy,[78] antibiotic materials,[12] wound dressing materials,[13] agriculture,[14,43] drug delivery,[11] and sensors[15,41] due to extraordinary characteristics like high surface area, easy to functionalize, high electrical, and high mechanical property.[3,5,16,26] Among all of them, CNTs and fullerenes have the potential ability to develop highly sensitive and accurate biosensors for the diagnosis and treatment of various diseases, including cancer. The extraordinary diversity of unconventional, magnetic, and electrical properties of CNTs makes them suitable candidates for nanoelectronics. The CNTs easily adsorb sensing molecules onto the surface of CNTs and the unparalleled surface unit providing high sensitivity that makes them promising materials for the development of chemical and biosensors. Despite several advantages of CNTs-based sensor, there are some limitations like CNTs cannot be applied directly in biological applications due to metallic properties and insoluble solvents. Moreover, the size of CNTs is not uniform, which might affect the sensitivity or reproducibility. This book chapter focuses on the synthesis and characteristics of CNTs and fullerenes, and their application in biosensor technology. The future aspects of CNTs and fullerenes are also discussed.

1.2 SYNTHESIS OF THE CARBON-BASED MATERIALS

The CNTs are synthesized using numerous processes that mainly involve the gas-phase process. Usually, there are three types of synthesis procedures for the synthesis of CNTs: (1) chemical vapor deposition (CVD) process, (2) laser ablation process, and (3) carbon arc-discharge process.

1.2.1 CVD PROCESS

The CVD process is the utmost popular method for synthesizing the CNTs. Usually in this process, thermal decomposition of the vapor of the hydrocarbon is carried out in the presence of the metal catalyst, and thereby CVD process is also known as thermal CVD or catalytic CVD process. The metal foils or silicon wafers were used as substrates and placed into the CVD reactor. Initially, metal foils or silicon wafers were exposed to precursor at higher temperatures depending on the metals in an inert atmosphere. The precursor reacts or decomposes onto the surface of metal foils or silicon wafers to produce nanomaterials, especially CNTs. Figure 1.1 shows the schematic representation of the CVD process. Several types of CNTs have been developed using the CVD process. For example, Sankararamakrishnan et al. synthesized functionalized CNTs by using floating catalysts. The data suggested that the floating catalyst-based CNTs were successfully prepared without any substrate, even at a lower temperature.[67] Magrez et al. synthesized CNTS using catalytic CVD process for high yield of CNTs at low temperature. The data suggested that the catalytic CVD process increases the yield of CNTs even at low temperature.[50] In general, the CVD process is a simple and economical technique for synthesizing CNTs.

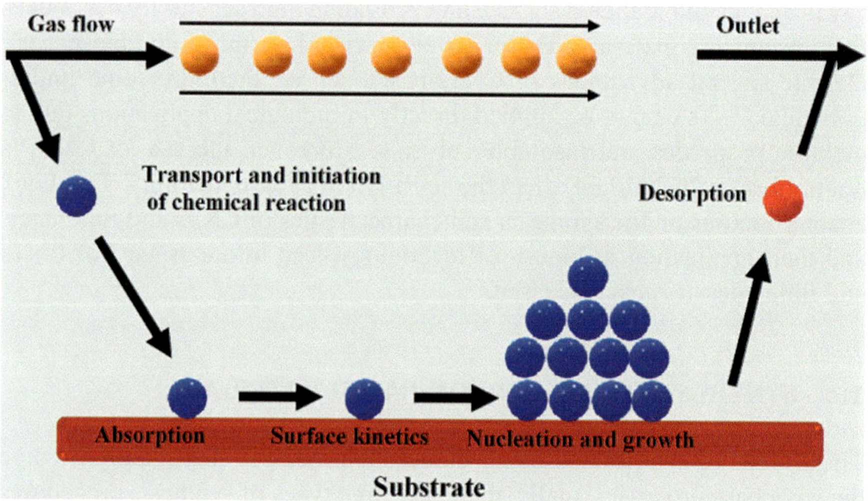

FIGURE 1.1 Schematic representation of the CVD process.

Source: Reprinted with permission from Ref. [17] © 2022. Elsevier.

1.2.2 LASER ABLATION PROCESS

In the laser ablation process, graphite is used for numerous varieties of carbon-based materials, including CNTs and fullerene. The synthesis of carbon-based materials using the laser ablation process has a high purity grade of around 90%, and its structure is better (high graphitization) than the arc process. Usually, the laser ablation process provides a high temperature which converts graphite and metallic catalyst into carbon–metal vapor. The evaporated carbon–metal vapor is further heated by laser pulses (up to several tons of Kilokelvin). After cooling the temperature, metal and carbon condense into a larger structure. However, relatively low yield limits their applicability on a large scale.[8,20,29,36,45]

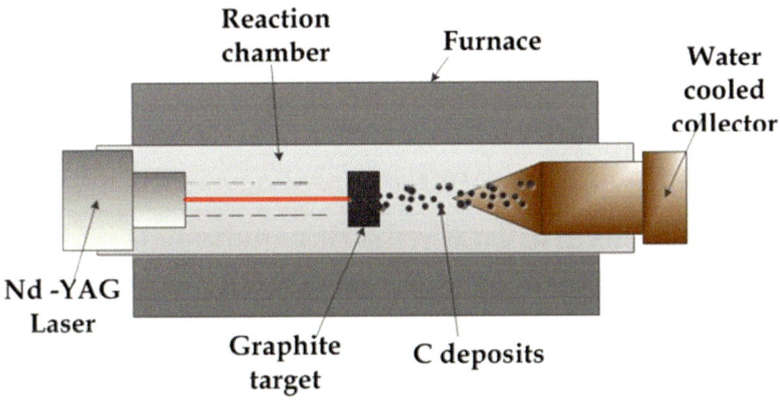

FIGURE 1.2 Schematic representation of the laser ablation process.
Source: Reprinted with permission from Ref. [20]. https://creativecommons.org/licenses/by/3.0/

1.2.3 ARC-DISCHARGE PROCESS

The arc-discharge process is a well-known process for the synthesis of carbon-based nanomaterials, including CNTs and fullerene. In this process, helium (as buffer gas) is introduced into a reactor that contains graphite anode, cathode, metals (like nickel, cobalt, and iron), and carbon molecules. The reactor/chamber is heated up to 4000 K, and a direct current is passed under pressure. Approximately, 50% of the carbon vapor is solidified or deposited in the cathode shoot in this process. The cathode shoot contains carbon-based nanomaterials (like CNTs and fullerene).[38,46,61,71,74] Figure 1.3 shows the schematic representation of the arc-discharge process.

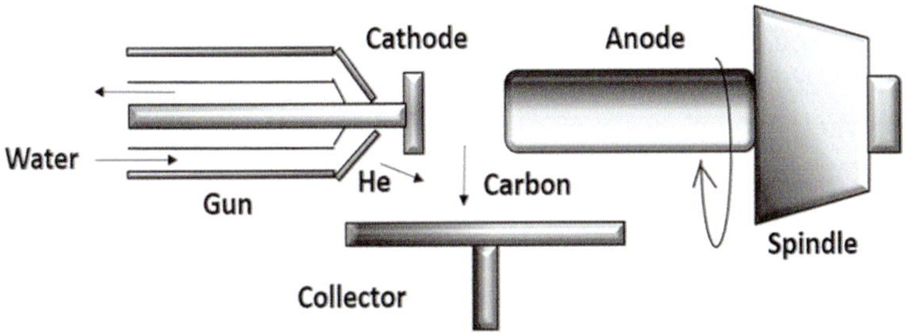

FIGURE 1.3 Schematic representation of the arc-discharge process.
Source: Reprinted with permission from Ref. [62]. https://creativecommons.org/licenses/by/4.0/

Literature suggests that CVD is a simple and cost-effective technique in comparison with the laser ablation and arc-discharge process. Moreover, the crystallinity of the laser ablation and arc-discharge process is better compared to the CVD process. However, in terms of yield and percent purity CVD process is superior.

1.3 SURFACE FUNCTIONALIZATION OF CARBON-BASED NANOMATERIALS

The surface functionalization or modification of surface chemistry or incorporation of a surface functional group onto the surface of carbon-based materials like CNTs and fullerenes enhances the specific characteristic of the materials such as biocompatibility, sensitivity, and selectivity in different applications. The surface functionalization of the materials results in modifications of the surface energy. This surface modification facilitates the better coupling with other substances at the molecular level or bonding of drugs, genomes, luminescent molecules, and catalysts. Moreover, the incorporation of polar or non-polar onto the surface of materials enhances the surface wettability, which improves the applicability of the materials. Numerous reports suggested that the surface modification of the CNTs and fullerenes improved the applicability of the materials.[5,21,30,47,51,52,75,83] Figure 1.4 shows the schematic representation of the numerous types of surface functionalization.

Usually, CNTs tend to agglomerate uncontrollably due to higher surface energy and stabilization. The weak dispersion ability and insolubility in solvents limit their application. The extensive use of CNTs requires surface functionalization to make them processable and alter their characteristics. To overcome such shortcoming, mainly two different approaches: (1) chemical and (2) mechanical process (high shear, high impact, grinding, and rubbing), have been used. However, these processes break CNTs, are time-consuming, and are ineffective approaches. The functionalization of CNTs might increase their electrical, mechanical, and optical properties.[24] Therefore, surface functionalization of carbon-based materials is an attractive area of research nowadays.

1.4 BIOCOMPATIBILITY OF THE CARBON-BASED MATERIALS

The biocompatibility of carbon-based materials is the first prerequisite of the materials used for biological applications. With the increasing applicability of the CNTs and fullerenes in biological applications, researchers continue to focus on the potential toxicity of carbon-based materials. The toxicity of the carbon-based materials might be determined using an exposure of different cell lines according to the applicability in real time. The cytotoxicity of the CNTs is mainly due to the presence of metal ions. The metal ions present in the CNTs increase the production of reactive oxygen species, which lead to cell death.[6,10,25,39,42,51,62,65,85] Figure 1.5 shows the cellular toxicity of the CNTs. As observed from the figure, the different lengths of the CNTs caused various degrees of toxicity. The larger CNTs induced a higher inflammation, whereas smaller CNTs showed a minor degree of inflammation. Moreover, different lengths of CNTs cause a different degree of granuloma (Fig. 1.5).[48] It is important to mention here that the surface functionalization of carbon-based nanomaterials might decrease or increase the biocompatibility of the materials that improves the applicability of the materials towards biological applications.

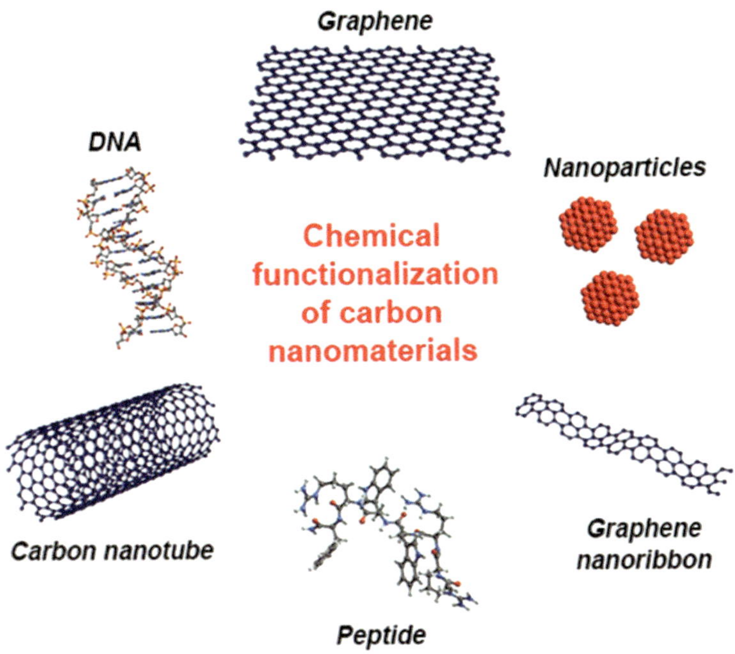

FIGURE 1.4 Schematic illustration of the various types of surface functionalization of carbon-based materials.
Source: Reprinted with permission from Ref. [24]. https://creativecommons.org/licenses/by/4.0/

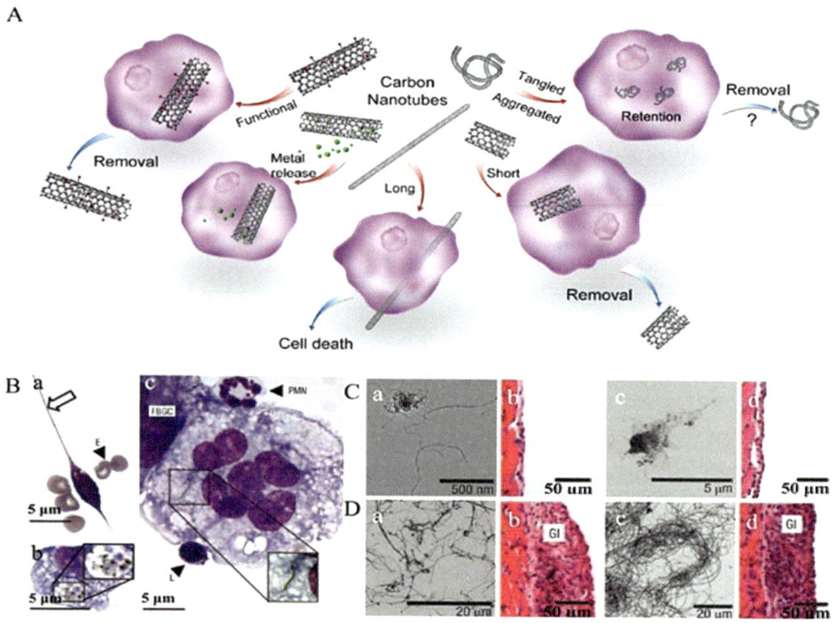

FIGURE 1.5 Cellular toxicity of the CNTs: (A) varying types of CNTs affect the cellular toxicity of the CNTs, (B) (a–c) CNTs leading to phagocytosis, and (C) granulomas response. *Source*: Reprinted with permission from Ref. [49] © 2013. American Chemical Society.

1.5 CARBON-BASED NANOMATERIALS FOR BIOSENSORS

1.5.1 DETECTION OF CANCER

Cancer is a disease in which there is uncontrolled growth of the cells due to the changes in the genes (mutation of genes), and is one of the leading causes of morbidity and mortality globally. The development of the disease is a several-step process comprising physiocellular process, cellular signaling, and apoptosis; therefore it is an extremely complicated disease. Moreover, early diagnosis of cancer might improve the treatment and decrease the mortality rate. Therefore, it is necessary to progress the early diagnosis approach and help to improve the effective treatment strategies for controlling cancer disease.[2,23,27,28,31,40,79,87] Figure 1.6 shows the overview of the uses of CNTs in the detection of cancer. In this context, carbon-based nanomaterials, mainly CNTs and fullerenes-based biosensors, might become alternative tools for early cancer detection.

The CNTs and fullerenes are effectively used as a biosensor in detecting cancer disease. For example, Xiao et al. synthesized the HER2-IgY antibody functionalized CNTs (HER2-CNTs) to detect cancer cells (SK-BR3 and MCF-7 cells). The prepared HER2-CNTs are effectively used for detecting and selective photothermal ablation of breast cancer. The extraordinary characteristics of the CNTs with HER2 antibodies show high sensitivity and specificity that might provide newer strategies for the detection of cancer.[84]

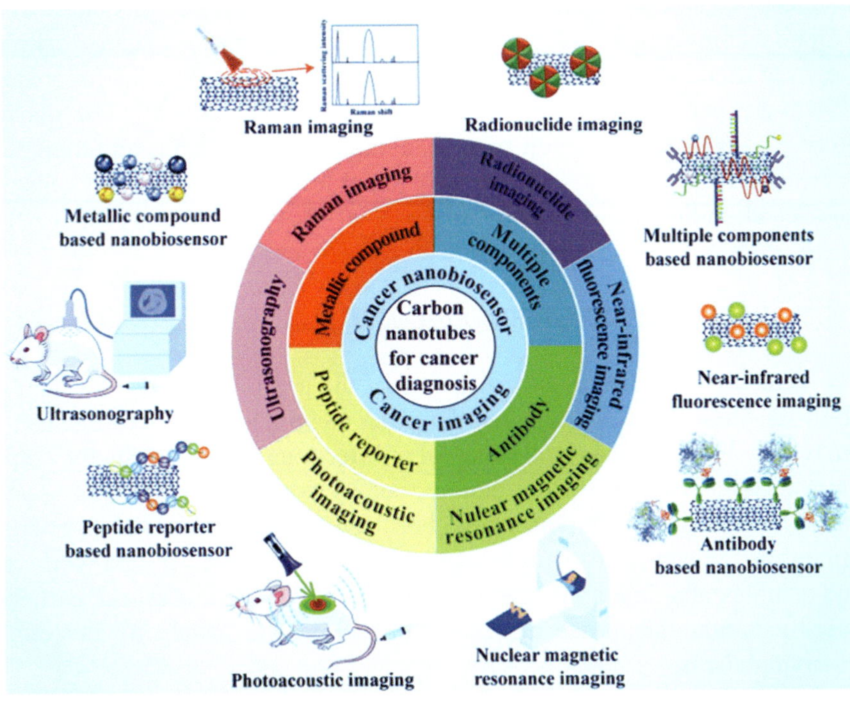

FIGURE 1.6 An illustration of CNTs for the diagnosis of cancer.
Source: Reprinted with permission from Ref. [80] © 2005. Elsevier Ltd.

Abdolahad et al. fabricated the vertically aligned CNTs-based impedance sensing biosensor (VA-CNTs-ISB) to detect cancer (SW48). The prepared VA-CNTs-ISB effectively detects cancer cells with the detection limit of 4000 cells/cm^2. The prepared VA-CNTs-ISB first ensnares cancer cells on VA-CNTs, which leads to mechanical and electrical interaction, thereby changing the impedance.[1] Zheng et al. synthesized the polydopamine-CNTs-folate (P-CNTs-F)-based nanoprobe to detect cancer. The data suggested that the P-CNTs-F-based nanoprobes have high biocompatibility, stability, reactivity, and conductivity. The prepared P-CNTs-F efficiently detect HeLa cells with the detection limit of 5×10^2 cells/mL.[87] Liu et al. synthesized PEI functionalized Multiwalled CNTs (MW-CNTs) incorporated with epithelial cell adhesion molecules (EpCAM) (PEI-MWCNTs-EpCAM)-based nanoprobes for the detection of cancer cells ($HepG_2$ and HeLa). The data suggest that the prepared PEI-MWCNTs-EpCAM-based nanoprobes effectively detect the cancer cells with the detection limit of 10 cells/mL. The process of PEI-MWCNTs-EpCAM-based biosensor is simple and can precisely control the electrode surface, thereby is highly sensitive.[49] Raoof et al. synthesized malonodiserinolamide-fullerenes (M-fullerene) for the localization within the liver cancer cell line. The data suggest that the prepared M-fullerenes effectively permeate through the vasculature of the tumor.[63] The above literature suggests that carbon-based nanomaterials like CNTs and fullerenes have effectively detected cancer cells even at low concentrations of cells or low detection limits.

1.5.2 DETECTION OF MICROORGANISM

The detection of microorganisms is one of the great challenges nowadays. High-sensitive, selective, and fast detection techniques for microorganisms are required to control or treat bacterial or infectious diseases. Usually, polymerase chain reaction (PCR), culture, colony counting, fluorescent assay, and immunological techniques have been used to identify or detect microorganisms. However, it requires sophisticated instruments, skilled workers, and complex and time-consuming processes.[33,34,66,72,81] In this aspect, carbon-based nanomaterials-based biosensors provide newer insight for detecting microorganisms.

Several nanomaterials, including the CNTs and fullerenes-based biosensors have been used to detect microorganisms.[22,32,56,68,80] For example, Garcia Aljaro et al. synthesized antibody immobilized CNTs (Ab-CNTs) to detect bacteria. The data suggested that they effectively detect *E. coli* bacteria with the detection limit of 10^1–10^3 CFU/mL. Moreover, the prepared Ab-CNTs show a faster response with a 5-min detection time.[35] Andrade et al. synthesized the peptide clavanin A immobilized onto the CNTs (CA-CNTs) for the detection of bacteria. The data suggested that the developed CA-CNTs-based biosensor effectively detects bacterial strains (*Klebsiella pneumoniae (K. pneumoniae), Enterococcus faecalis (E. faecalis), Escherichia coli (E. coli),* and *Bacillus subtilis (B. subtilis)) with the detection limit of* 10^2–10^6 *CFU/*

mL. The AFM confirms the functionality of biosensor-aided advantages for detecting bacteria.[9] Nilbler et al. synthesized peptide-functionalized CNTs (f-CNTs) incorporated with hydrogel-based infrared fluorescent sensor for detecting bacteria. The data suggested that the surface functionalization of CNTs quickly detected the metabolites like polysaccharides, DNase, and protease. The f-CNTs-based hydrogel effectively detects microorganisms like *P. aeruginosa* and *S.aureus*.[59] Matos Pires and Dong synthesized poly-carbazole incorporated fullerenes (PC-fullerenes)-based photodiode to detect pathogens. The data suggested that the PC-CNTs-based chemiluminescent sensor effectively detects *C. jejuni* and adenovirus with the detection limit of 10^5 cells/mL and 10^{-8} cells/mL, respectively.[54] Table 1.1 summarizes CNTs and fullerene-based biosensor to detect various diseases, especially infectious diseases and cancer. The literature reports suggested that the surface functionalization of the CNTs/fullerenes has numerous advantages, such as selectivity and sensitivity. The biosensor's selectivity and sensitivity are very important aspects or one of the leading challenges nowadays. The surface functionalization of the CNTs and fullerenes might resolve such associated issues like selectivity and sensitivity.

TABLE 1.1 CNTs- and Fullerenes-Based Biosensors for the Detection of Diseases.

S. No.	Nanoprobes	Technique	Cancer/ bacterial cells	Detection limit	References
1.	HER2-CNTs	NIR irradiation	SK-BR3 and MCF-7	–	[84]
2.	P-CNTs-F	Electrochemical	HeLa	5×10^2 cells/mL	[87]
3.	M-fullerene	Imaging	Liver cell lines	–	[63]
4.	VA-CNTs-ISB	Electrochemical	SW48	4000 cells/cm^2	[1]
5.	PEI-MWCNTs-EpCAM	Electrochemical	$HepG_2$ and HeLa	10 cells/mL	[49]
6.	f-CNTs-based hydrogel	NIR	*P. aeruginosa* and *S.aureus*	–	[59]
7.	CA-CNTs	Electrochemical	*K. pnuemoniae*, *E. faecalis*, *E. coli* and *B. subtilis*	*10^2–10^6 CFU/mL*	[9]
8.	Ab-CNTs	Chemiresistive	*E. coli*	10^3–10^1 CFU/mL	[35]
10.	PC-fullerenes	Chemiluminescent	*C. jejuni and adenovirus*	10^5 cells/mL and 10^{-8} cells/mL	[54]

1.6 CONCLUSIONS

Biosensor technology in biological sciences, especially biomedical, has been effectively used to detect various biomolecules or biomarkers of diseases even early. In this aspect, carbon-based nanomaterials like CNTs and fullerenes have a potential ability due to their unique characteristic properties (high surface area, easy functionalization, and high electrical and mechanical property). The CNTs easily adsorb sensing molecules/biomolecules onto the surface of CNTs and the unparalleled surface unit providing high sensitivity that makes them promising materials for the development of chemical and biosensors. These inherent properties of CNTs and fullerenes make them a suitable candidate for developing precise, sensitive, and rapid response biosensors for the detection of various diseases. Despite several advantages of CNTs-based biosensor, there is some limitation like CNTs cannot be applied directly in biological applications due to metallic property and insoluble solvents.

Moreover, the size of CNTs is not uniform, which might affect the sensitivity or reproducibility. Researchers continue to focus on developing the most accurate biosensors based on the CNTs and fullerenes. With the help of surface functionalization (chemical/polymers), the sensitivity and selectivity of the biosensor is significantly improved. Therefore, carbon-based materials, especially CNTs and fullerenes, become promising materials that accurately diagnose various diseases or cancer even at an early stage.

KEYWORDS

- **CNTs**
- **fullerenes**
- **biosensor**
- **cancer**
- **infectious diseases**

REFERENCES

1. Abdolahad, M.; Taghinejad, M.; Taghinejad, H.; Janmaleki, M.; Mohajerzadeh, S. A Vertically Aligned Carbon Nanotube-Based Impedance Sensing Biosensor for Rapid and High Sensitive Detection of Cancer Cells. *Lab Chip* **2012,** *12*, 1183–1190.

2. Afreen, S.; Muthoosamy, K.; Manickam, S.; Hashim, U. Functionalized Fullerene (C60) as a Potential Nanomediator in the Fabrication of Highly Sensitive Biosensors. *Biosens. Bioelectron.* **2015,** *63*, 354–364.
3. Afreen, S.; Omar, R. A.; Talreja, N.; Chauhan, D.; Ashfaq, M. Carbon-Based Nanostructured Materials for Energy and Environmental Remediation Applications. In *Approaches in Bioremediation: The New Era of Environmental Microbiology and Nanobiotechnology*; Prasad, R., Aranda, E., Eds.; Springer International Publishing: Cham, 2018; pp 369–392.
4. Afreen, S.; Talreja, N.; Chauhan, D.; Ashfaq, M. Chapter 15——Polymer/Metal/Carbon-Based Hybrid Materials for the Detection of Heavy Metal Ions. In *Multifunctional Hybrid Nanomaterials for Sustainable Agri-Food and Ecosystems*; Abd-Elsalam, K. A., Ed.; Elsevier, 2020; pp 335–353.
5. Afreen, S.; Talreja, N.; Ashfaq, M.; Chauhan, D. Chapter 11—Carbon Nanostructure-Based Sensor: A Promising Tools for Monitoring Crops. In *Nanotechnology-Based Sustainable Alternatives for the Management of Plant Diseases*; Balestra, G. M, Fortunati, E., Eds.; Elsevier, 2022; pp 287–300.
6. Albert, K.; Hsu H-Y. Carbon-Based Materials for Photo-Triggered Theranostic Applications. *Molecules* **2016,** *21*.
7. Alimardani, V.; Abolmaali, S. S.; Tamaddon, A. M.; Ashfaq, M. Recent Advances on Microneedle Arrays-Mediated Technology in Cancer Diagnosis and Therapy. *Drug Deli. Transl. Res.* **2021,** *11*, 788–816.
8. AlMalki, F. A.; Khashan, K. S.; Jabir, M. S.; Hadi, A. A.; Sulaiman, G. M.; Abdulameer, F. A.; Albukhaty, S.; Al-Karagoly, H.; Albaqami, J. Eco-Friendly Synthesis of Carbon Nanoparticles by Laser Ablation in Water and Evaluation of Their Antibacterial Activity. *J. Nanomater.* **2022,** *2022*, 7927447.
9. Andrade, C. A. S.; Nascimento, J. M.; Oliveira, I. S.; de Oliveira, C. V. J.; de Melo, C. P.; Franco, O. L.; Oliveira, M. D. L. Nanostructured Sensor Based on Carbon Nanotubes and Clavanin A for Bacterial Detection. *Colloids Surfaces B Biointerfaces* **2015,** *135*, 833–839.
10. Ashfaq, M.; Singh, S.; Sharma, A.; Verma, N. Cytotoxic Evaluation of the Hierarchical Web of Carbon Micronanofibers. *Ind. Eng. Chem. Re.* **2013,** *52*, 4672–4682.
11. Ashfaq, M.; Khan, S.; Verma, N. Synthesis of PVA-CAP-Based Biomaterial In Situ Dispersed with Cu Nanoparticles and Carbon Micro-Nanofibers for Antibiotic Drug Delivery Applications. *Biochem. Eng. J.* **2014,** *90*, 79–89.
12. Ashfaq, M.; Verma, N.; Khan, S. Copper/Zinc Bimetal Nanoparticles-Dispersed Carbon Nanofibers: A Novel Potential Antibiotic Material. *Mater. Sci. Eng. C* **2016,** *59*, 938–947.
13. Ashfaq, M.; Verma, N.; Khan, S. Carbon Nanofibers as a Micronutrient Carrier in Plants: Efficient Translocation and Controlled Release of Cu Nanoparticles. *Environ. Sci. Nano* **2017a,** *4*, 138–148.
14. Ashfaq, M.; Verma, N.; Khan, S. Highly Effective Cu/Zn-Carbon Micro/Nanofiber-Polymer Nanocomposite-Based Wound Dressing Biomaterial Against the *P. aeruginosa* Multi- and Extensively Drug-Resistant Strains. *Mater. Sci. Eng. C* **2017b,** *77*, 630–641.
15. Ashfaq, M.; Verma, N.; Khan, S. Novel Polymeric Composite Grafted with Metal Nanoparticle-Dispersed CNFs as a Chemiresistive Non-Destructive Fruit Sensor Material. *Mater. Chem. Phys.* **2018,** *217*, 216–227.
16. Ashfaq, M.; Talreja, N.; Chuahan, D.; Srituravanich, W. Carbon Nanostructure-Based Materials: A Novel Tool for Detection of Alzheimer's Disease. In *Biological, Diagnostic*

and Therapeutic Advances in Alzheimer's Disease: Non-Pharmacological Therapies for Alzheimer's Disease; Ashraf, G. M., Alexiou, A., Eds.; Springer Singapore: Singapore, 2019; pp 71–89.
17. Ashfaq, M.; Talreja, N.; Chauhan, D.; Afreen, S.; Sultana, A.; Srituravanich, W. Two-Dimensional (2D) Hybrid Nanomaterials for Diagnosis and Treatment of Cancer. *J. Drug Deliv. Sci. Technol.* **2022a,** *70*, 103268.
18. Ashfaq, M.; Talreja, N.; Chauhan, D.; Rodríguez, C. A.; Mera, A. C.; Ramalinga Viswanathan, M. Synthesis of Reduced Graphene Oxide Incorporated Bimetallic (Cu/Bi) Nanorods Based Photocatalyst Materials for the Degradation of Gallic Acid and Bacteria. *J. Ind. Eng. Chem.* **2022b.**
19. Ashfaq, M.; Talreja, N.; Chauhan, D.; Viswanathan, M. R. Synthesis of Cu-doped 2D-WS2 Nanosheet-Based Nano-Antibiotic Materials for Inhibiting *E. Coli* and *S. aureus* Bacterial Strains. *New J. Chem.* **2022c,** *46*, 5581–5587.
20. Awais Rouf, S.; Usman, Z.; Tariq Masood, H.; Mannan Majeed, A.; Sarwar, M.; Abbas, W. Synthesis and Purification of Carbon Nanotubes, 2021.
21. Baig, N.; Kammakakam, I.; Falath, W. Nanomaterials: A Review of Synthesis Methods, Properties, Recent Progress, and Challenges. *Mater. Adv.* **2021,** *2*, 1821–1871.
22. Bardhan, N. M.; Ghosh, D.; Belcher, A. M. Carbon Nanotubes as In Vivo Bacterial Probes. *Nat. Commun.* **2014,** *5*, 4918.
23. Bohunicky, B.; Mousa, S. A. Biosensors: The New Wave in Cancer Diagnosis. *Nanotechnol Sci. Appl.* **2010,** *4*, 1–10.
24. Burdanova, M. G.; Kharlamova, M. V.; Kramberger, C.; Nikitin, M. P. Applications of Pristine and Functionalized Carbon Nanotubes, Graphene, and Graphene Nanoribbons in Biomedicine. *Nanomaterials* **2021,** *11*.
25. Cha, C.; Shin, S. R.; Annabi, N.; Dokmeci, M. R.; Khademhosseini, A. Carbon-Based Nanomaterials: Multifunctional Materials for Biomedical Engineering. *ACS Nano* **2013,** *7*, 2891–2897.
26. Chauhan, D.; Talreja, N.; Ashfaq, M. Chapter 13—Nanoadsorbents for Wastewater Remediation. In *Aquananotechnology*; Abd-Elsalam, K. A., Zahid, M., Eds.; Elsevier, 2021; pp 273–290.
27. Chen, Z.; Zhang, A.; Wang, X.; Zhu, J.; Fan, Y.; Yu, H.; Yang, Z. The Advances of Carbon Nanotubes in Cancer Diagnostics and Therapeutics. *J. Nanomater.* **2017,** *2017*, 3418932.
28. Chikkaveeraiah, B. V.; Bhirde, A. A.; Morgan, N. Y.; Eden, H. S.; Chen, X. Electrochemical Immunosensors for Detection of Cancer Protein Biomarkers. *ACS Nano* **2012,** *6*, 6546–6561.
29. Das, R.; Shahnavaz, Z.; Ali, M. E.; Islam, M. M.; Abd Hamid, S. B. Can We Optimize Arc Discharge and Laser Ablation for Well-Controlled Carbon Nanotube Synthesis? *Nanoscale Res. Lett.* **2016,** *11*, 510.
30. Datta, D.; Das, K. P.; Deepak, K. S.; Das, B. Chapter 3—Candidates of Functionalized Nanomaterial-Based Membranes. In *Membranes with Functionalized Nanomaterials*; Dutta, S., Hussain, C. M., Eds.; Elsevier, 2022; pp 81–127.
31. Dizaji, B. F.; Farboudi, A.; Rahbar, A.; Azarbaijan, M. H.; Asgary, M. R. The Role of Single- and Multi-Walled Carbon Nanotube in Breast Cancer Treatment. *Therap. Deliv.* **2020,** *11*, 653–672.
32. Emelyantsev, S.; Prazdnova, E.; Chistyakov, V.; Alperovich, I. Biological Effects of C60 Fullerene Revealed with Bacterial Biosensor—Toxic or Rather Antioxidant? *Biosensors* **2019,** *9*.

33. Ferone, M.; Gowen, A.; Fanning, S.; Scannell, A. G. M. Microbial Detection and Identification Methods: Bench top assays to omics approaches. *Compreh. Rev. Food Sci. Food Safety* **2020,** *19*, 3106–3129.
34. Franco-Duarte, R.; Černáková, L.; Kadam, S.; Kaushik, K. S.; Salehi, B.; Bevilacqua, A.; Corbo, M. R.; Antolak, H.; Dybka-Stępień, K.; Leszczewicz, M.; Relison Tintino, S.; Alexandrino de Souza, V. C.; Sharifi-Rad, J.; Coutinho, H. D. M.; Martins, N.; Rodrigues, C. F. Advances in Chemical and Biological Methods to Identify Microorganisms-from Past to Present. *Microorganisms* **2019,** *7*, 130.
35. García-Aljaro, C.; Cella, L. N.; Shirale, D. J.; Park, M.; Muñoz, F. J.; Yates, M. V.; Mulchandani, A. Carbon Nanotubes-Based Chemiresistive Biosensors for Detection of Microorganisms. *Biosens. Bioelectron.* **2010,** *26*, 1437–1441.
36. Guo, T.; Nikolaev, P.; Rinzler, A. G.; Tomanek, D.; Colbert, D. T.; Smalley, R. E. Self-Assembly of Tubular Fullerenes. *J. Phys. Chem.* **1995,** *99*, 10694–10697.
37. Haleem, A.; Javaid, M.; Singh, R. P.; Suman, R.; Rab, S. Biosensors Applications in Medical Field: A Brief Review. *Sens. Int.* **2021,** *2*, 100100.
38. Hutchison, J. L.; Kiselev, N. A.; Krinichnaya, E. P.; Krestinin, A. V.; Loutfy, R. O.; Morawsky, A. P.; Muradyan, V. E.; Obraztsova, E. D.; Sloan, J.; Terekhov, S. V.; Zakharov, D. N. Double-Walled Carbon Nanotubes Fabricated by a Hydrogen Arc Discharge Method. *Carbon* **2001,** *39*, 761–770.
39. Islam, M.; Lantada, A. D.; Mager, D.; Korvink, J. G. Carbon-Based Materials for Articular Tissue Engineering: From Innovative Scaffolding Materials Toward Engineered Living Carbon. *Adv. Healthcare Mater.* **2022,** *11*, 2101834.
40. Ji, S-r.; Liu, C.; Zhang, B.; Yang, F.; Xu, J.; Long, J.; Jin, C.; Fu, D-l.; Ni, Q-x.; Yu, X-j. Carbon Nanotubes in Cancer Diagnosis and Therapy. *Biochim. Biophys. Acta (BBA)—Rev. Cancer* **2010,** *1806*, 29–35.
41. Kumar, D.; Talreja, N. Nickel Nanoparticles-Doped Rhodamine Grafted Carbon Nanofibers as Colorimetric Probe: Naked Eye Detection and Highly Sensitive Measurement of Aqueous Cr3+ and Pb2+. *Korean J. Chem. Eng.* **2019,** *36*, 126–135.
42. Kumar, N.; Chamoli, P.; Misra, M.; Manoj, M. K.; Sharma, A. Advanced Metal and Carbon Nanostructures for Medical, Drug Delivery and Bio-Imaging Applications. *Nanoscale* **2022,** *14*, 3987–4017.
43. Kumar, R.; Ashfaq, M.; Verma, N. Synthesis of Novel PVA–Starch Formulation-Supported Cu–Zn Nanoparticle Carrying Carbon Nanofibers as a Nanofertilizer: Controlled Release of Micronutrients. *J. Mater. Sci.* **2018,** *53*, 7150–7164.
44. Kumar, V.; Talreja, N.; Deva, D.; Sankararamakrishnan, N.; Sharma, A.; Verma, N. Development of Bi-Metal Doped Micro- and Nano Multi-Functional Polymeric Adsorbents for the Removal of Fluoride and Arsenic(V) from Wastewater. *Desalination* **2011,** *282*, 27–38.
45. Kuo, T. F.; Chi, C. C.; Lin, I. N. Synthesis of Carbon Nanotubes by Laser Ablation of Graphites at Room Temperature. *Jpn. J. Appl. Phys.* **2001,** *40*, 7147–7150.
46. Li, N.; Wang, Z.; Zhao, K.; Shi, Z.; Gu, Z.; Xu S. Large Scale Synthesis of N-Doped Multi-Layered Graphene Sheets by Simple Arc-Discharge Method. *Carbon* **2010,** *48*, 255–259.
47. Liu, W.; Speranza, G. Functionalization of Carbon Nanomaterials for Biomedical Applications. **2019,** C 5.
48. Liu, Y.; Zhao, Y.; Sun, B.; Chen, C. Understanding the Toxicity of Carbon Nanotubes. *Acc. Chem. Res.* **2013,** *46*, 702–713.

49. Liu, Y.; Zhu, F.; Dan, W.; Fu, Y.; Liu, S. Construction of Carbon Nanotube Based Nanoarchitectures for Selective Impedimetric Detection of Cancer Cells in Whole Blood. *Analyst* **2014,** *139*, 5086–5092.
50. Magrez, A.; Seo, J. W.; Smajda, R.; Mionić, M.; Forró, L. Catalytic CVD Synthesis of Carbon Nanotubes: Towards High Yield and Low Temperature Growth. *Materials* **2010,** *3*.
51. Maiti, D.; Tong, X.; Mou, X.; Yang, K. Carbon-Based Nanomaterials for Biomedical Applications: A Recent Study. *Front Pharmacol*. **2019,** *9*, 1401–1401.
52. Makvandi, P.; Iftekhar, S.; Pizzetti, F.; Zarepour, A.; Zare, E. N.; Ashrafizadeh, M.; Agarwal, T.; Padil, V. V. T.; Mohammadinejad, R.; Sillanpaa, M.; Maiti, T. K.; Perale, G.; Zarrabi, A.; Rossi, F. Functionalization of Polymers and Nanomaterials for Water Treatment, Food Packaging, Textile and Biomedical Applications: A Review. *Environ. Chem. Lett.* **2021,** *19*, 583–611.
53. Mardian, Y.; Kosasih, H.; Karyana, M.; Neal, A.; Lau C-Y. Review of Current COVID-19 Diagnostics and Opportunities for Further Development. *Front. Med.* **2021,** *8*.
54. Matos Pires, N. M.; Dong, T. Microfluidic Biosensor Array with Integrated Poly(2,7-Carbazole)/Fullerene-Based Photodiodes for Rapid Multiplexed Detection of Pathogens. *Sensors* **2013,** *13*.
55. Mehrotra, P. Biosensors and Their Applications—A Review. *J. Oral. Biol. Craniofac. Res.* **2016,** *6*, 153–159.
56. Münzer, A. M.; Michael, Z. P.; Star, A. Carbon Nanotubes for the Label-Free Detection of Biomarkers. *ACS Nano* **2013,** *7*, 7448–7453.
57. Nagraik, R.; Sharma, A.; Kumar, D.; Mukherjee, S.; Sen, F.; Kumar, A. P. Amalgamation of Biosensors and Nanotechnology in Disease Diagnosis: Mini-Review. *Sens. Int.* **2021,** *2*, 100089.
58. Naresh, V.; Lee, N. A Review on Biosensors and Recent Development of Nanostructured Materials-Enabled Biosensors. *Sensors* **2021,** *21*.
59. Nißler, R.; Bader, O.; Dohmen, M.; Walter, S. G.; Noll, C.; Selvaggio, G.; Groß, U.; Kruss, S. Remote Near Infrared Identification of Pathogens with Multiplexed Nanosensors. *Nat. Commun.* **2020,** *11*, 5995.
60. Purohit, B.; Vernekar, P. R.; Shetti, N. P.; Chandra, P. Biosensor Nanoengineering: Design, Operation, and Implementation for Biomolecular Analysis. *Sens. Int.* **2020,** *1*, 100040.
61. Rahman, G.; Najaf, Z.; Mehmood, A.; Bilal, S.; Shah, A. U.; Mian, S. A.; Ali, G. An Overview of the Recent Progress in the Synthesis and Applications of Carbon Nanotubes. **2019,** C 5.
62. Rahmati, M.; Mozafari, M. Biological Response to Carbon-Family Nanomaterials: Interactions at the Nano-Bio Interface. *Front. Bioeng. Biotechnol.* **2019,** *7*.
63. Raoof, M.; Mackeyev, Y.; Cheney, M. A.; Wilson, L. J.; Curley, S. A. Internalization of C60 Fullerenes into Cancer Cells with Accumulation in the Nucleus via the Nuclear Pore Complex. *Biomaterials* **2012,** *33*, 2952–2960.
64. Rastogi, A.; Yadav, K.; Mishra, A.; Singh, M. S.; Chaudhary, S.; Manohar, R.; Parmar, A. S. Early Diagnosis of Lung Cancer Using Magnetic Nanoparticles-Integrated Systems. *Nanotechnol. Rev.* **2022,** *11*, 544–574.
65. Rauti, R.; Musto, M.; Bosi, S.; Prato, M.; Ballerini, L. Properties and Behavior of Carbon Nanomaterials When Interfacing Neuronal Cells: How Far Have We Come? *Carbon* **2019,** *143*, 430–446.

66. Reta, D. H.; Tessema, T. S.; Ashenef, A. S.; Desta, A. F.; Labisso, W. L.; Gizaw, S. T.; Abay, S. M.; Melka, D. S.; Reta, F. A. Molecular and Immunological Diagnostic Techniques of Medical Viruses. *Int. J. Microbiol.* **2020,** *2020*, 8832728.
67. Sankararamakrishnan, N.; Chauhan, D.; Dwivedi, J. Synthesis of Functionalized Carbon Nanotubes by Floating Catalytic Chemical Vapor Deposition Method and Their Sorption Behavior Toward Arsenic. *Chem. Eng. J.* **2016,** *284*, 599–608.
68. Sanvicens, N.; Pastells, C.; Pascual, N.; Marco, M. P. Nanoparticle-Based Biosensors for Detection of Pathogenic Bacteria. *TrAC Trends Anal. Chem.* **2009,** *28*, 1243–1252.
69. Saraswat, R.; Talreja, N.; Deva, D.; Sankararamakrishnan, N.; Sharma, A.; Verma, N. Development of Novel In Situ Nickel-Doped, Phenolic Resin-Based Micro–Nano-Activated Carbon Adsorbents for the Removal of Vitamin B-12. *Chem. Eng. J.* **2012,** *197*, 250–260.
70. Sasidharan, V.; Sachan, D.; Chauhan, D.; Talreja, N.; Ashfaq, M. Three-Dimensional (3D) Polymer—Metal–Carbon Framework for Efficient Removal of Chemical and Biological Contaminants. *Sci. Rep.* **2021,** *11*, 7708.
71. Shi, Z.; Lian, Y.; Zhou, X.; Gu, Z.; Zhang, Y.; Iijima, S.; Zhou, L.; Yue, K. T.; Zhang, S. Mass-Production of Single-Wall Carbon Nanotubes by Arc Discharge Method 11 This Work Was Supported by the National Natural Science Foundation of China, No. 29671030. *Carbon* **1999,** *37*, 1449–1453.
72. Sibley, C. D.; Peirano, G.; Church, D. L. Molecular Methods for Pathogen and Microbial Community Detection and Characterization: Current and Potential Application in Diagnostic Microbiology. *Infect. Genet. Evol.* **2012,** *12*, 505–521.
73. Song, M.; Yang, M.; Hao, J. Pathogenic Virus Detection by Optical Nanobiosensors. *Cell Rep. Phys. Sci.* **2021,** *2*, 100288.
74. Subrahmanyam, K. S.; Panchakarla, L. S.; Govindaraj, A.; Rao, C. N. R. Simple Method of Preparing Graphene Flakes by an Arc-Discharge Method. *J. Phys. Chem. C* **2009,** *113*, 4257–4259.
75. Sultana, A.; Omar, R. A.; Talreja, N.; Chauhan, D.; Mangalaraja, R. V.; Ashfaq, M. Chapter 28—Copper-Based Metal-Organic Framework for Environmental Applications. In *Copper Nanostructures: Next-Generation of Agrochemicals for Sustainable Agroecosystems*; Abd-Elsalam, K. A., Ed.; Elsevier, 2022; pp 701–717.
76. Talreja, N.; Kumar, D.; Verma, N. Removal of Hexavalent Chromium from Water Using Fe-Grown Carbon Nanofibers Containing Porous Carbon Microbeads. *J. Water Process Eng.* **2014,** *3*, 34–45.
77. Talreja, N.; Verma, N.; Kumar, D. Carbon Bead-Supported Ethylene Diamine-Functionalized Carbon Nanofibers: An Efficient Adsorbent for Salicylic Acid. *CLEAN—Soil, Air, Water* **2016,** *44*, 1461–1470.
78. Talreja, N.; Jung, S.; Yen, L. T. H.; Kim, T. Phenol-Formaldehyde-Resin-Based Activated Carbons with Controlled Pore Size Distribution for High-Performance Supercapacitors. *Chem. Eng. J.* **2020,** *379*, 122332.
79. Tang, L.; Xiao, Q.; Mei, Y.; He, S.; Zhang, Z.; Wang, R.; Wang, W. Insights on Functionalized Carbon Nanotubes for Cancer Theranostics. *J. Nanobiotechnol.* **2021,** *19*, 423.
80. Tegos, G. P.; Demidova, T. N.; Arcila-Lopez, D.; Lee, H.; Wharton, T.; Gali, H.; Hamblin, M. R. Cationic Fullerenes Are Effective and Selective Antimicrobial Photosensitizers. *Chem. Biol.* **2005,** *12*, 1127–1135.

81. Vasala, A.; Hytönen, V. P.; Laitinen, O. H. Modern Tools for Rapid Diagnostics of Antimicrobial Resistance. *Front. Cell. Infect. Microbiol.* **2020,** *10.*
82. Vigneshvar, S.; Sudhakumari, C. C.; Senthilkumaran, B.; Prakash, H. Recent Advances in Biosensor Technology for Potential Applications—An Overview. *Front. Bioeng. Biotechnol.* **2016,** *4.*
83. Wang, X.; Liu L-H, Ramström, O.; Yan, M. Engineering Nanomaterial Surfaces for Biomedical Applications. *Exp. Biol. Med. (Maywood)* **2009,** *234*, 1128–1139.
84. Xiao, Y.; Gao, X.; Taratula, O.; Treado, S.; Urbas, A.; Holbrook, R. D.; Cavicchi, R. E.; Avedisian, C. T.; Mitra, S.; Savla, R.; Wagner, P. D.; Srivastava, S.; He H. Anti-HER2 IgY Antibody-Functionalized Single-Walled Carbon Nanotubes for Detection and Selective Destruction of Breast Cancer Cells. *BMC Cancer* 9, 351.
85. Zain, M.; Yasmeen, H.; Yadav, S. S.; Amir, S.; Bilal, M.; Shahid, A.; Khurshid, M. Chapter 10—Applications of Nanotechnology in Biological Systems and Medicine. In *Nanotechnology for Hematology, Blood Transfusion, and Artificial Blood*; Denizli, A., Nguyen, T. A., Rajan, M., Alam, M. F., Rahman, K., Eds.; Elsevier, 2022; pp 215–235.
86. Zhang, X.; Guo, Q.; Cui, D. Recent Advances in Nanotechnology Applied to Biosensors. *Sensors (Basel)* **2009,** *9*, 1033–1053.
87. Zheng T-T, Zhang, R.; Zou, L.; Zhu, J-J. A Label-Free Cytosensor for the Enhanced Electrochemical Detection of Cancer Cells Using Polydopamine-Coated Carbon Nanotubes. *Analyst* **2012,** *137*, 1316–1318.

CHAPTER 2

Applications of Carbon Nanotubes in Pathogenic Microbe Detection

DIVYA NERAVATHU GOPI, MOHAMMED SAJU M., and RIJU K. THOMAS

Bharata Mata College, Thrikkakara, Edappally, Kochi, Kerala, India

ABSTRACT

The presence of pathogenic microbes around us, especially the presence of these microbes in food that we consume, causes severe health issues. Thus, the identification and precise detection of these pathogenic microbes and toxins are crucial in controlling many of these health issues. In this perspective, the detection of pathogenic microbes is very significant, as also the development of pathogenic microbe detectors employing nanomaterials such as carbon nanotubes, graphene nanomaterials, quantum dots, etc. These nanomaterials are predominantly used to make the detectors more efficient, highly sensitive, and selective due to their inherent optical, chemical, electrochemical, mechanical, and physical properties. Especially carbon nanomaterials such as single-walled carbon nanotubes (SWCNTs), multiwalled carbon nanotubes (MWCNTs), and their hybrid composites are really promising materials to improve the sensitivity and selectivity of pathogenic microbe detectors. The development of nanotechnology and nanoengineering leads to a platform for rapid, simple, and more advanced electronic pathogenic microbe detectors for the transfer of pathogenic signals into measurable electronic signals.

Tailored Functional Materials for Clean and Sustainable Development. Tamara Tatrishvili, PhD, Neha Kanwar Rawat, PhD, Ann Rose Abraham, PhD & A. K. Haghi, PhD (Eds.)

2.1 INTRODUCTION

In 1991, Japanese scientist Sumio Iijima, discovered fullerene-related carbon nanotubes accidentally.[28] Carbon nanotubes (CNTs) are cylindrical structures consisting of hexagonal rings of carbon atoms. Besides its low-dimensional physical form, it shows interesting chemical–physical properties, such as high electrical and heat conductivity, and has considerable tensile strength.[4] Hence, CNTs are considered potential candidates in nanopowered systems and their research applications. The CNTs are categorized into two broad groups: single-walled carbon nanotubes (SWCNTs) and multiwalled carbon nanotubes (MWCNTs). The SWCNTs contain a single cylindrical thin layer of carbon atoms with their ends either open or capped with hemispherical-shaped fullerene structures. In contrast, MWCNTs comprise many concentric cylinders of graphene sheets.[10] The outermost diameter of SWCNTs usually ranges from 1 to 3 nm, and their innermost diameter ranges from 0.6 to 2.4 nm. The MWCNTs have a much longer diameter, with an outer diameter ranging from 2.5 to 100 nm and an inner diameter ranging from 1.5 to 15 nm.[24]

The CNTs first reported were MWCNTs by Iijima in 1991. Following this discovery, SWCNTs were discovered in 1993.[12] Since the CNTs have unique properties compared to conventional materials, these discoveries led to many research breakthroughs in nanotechnology material science, condensed matter physics, and microelectronics such as electron emitter source with high current density, highly conductive electrical wire, the high thermal conductor for the heat radiator, the probe needle for scanning probe microscopes, molecular sieves, gas adsorbers, carriers for drug delivery carrier systems, etc. Because of their unique physicochemical features, CNTs are highly adaptable and ultralight-weighted.[14]

The cylindrically shaped CNTs have a nanometer-sized diameter and a few micrometer length dimensions. These carbon sheets comprise planar sp^2 bonding and a small amount of sp^3 bonding. The graphene layers are not a perfect stack but a layered structure with strong bonding within the planar layers and weak van der Waals bonding between the adjacent layers. The adjacent graphene layers are rotationally disaligned, and the relative position of these layers is randomly oriented.[27] The structure containing these 2D graphene layers having no mutual relationship is called turbostratic graphite.[17] Like in the turbostratic graphite, graphene sheets in carbon nanotubes are positioned. As a result, ideal nanotube cylinders separated by a considerable spatial distance should be able to slide past one another.

Raman spectroscopy is one of the extensively used characterization tools to characterize CNTs and graphene. Compared to other spectroscopic techniques, it is much cheaper, nondestructive, and able to give enough information about the sample.[29] Raman spectra of SWCNT are much more intense than graphene. The spectra provide information about morphological and structural properties, bonding types, defects, diameter, and chirality of the nanotubes, even though their precise interpretation is challenging.[9] Other characterization tools commonly employed are neutron diffraction, X-ray diffraction, STM, TEM, X-ray photoelectron spectroscopy, and IR spectroscopy. The relation of electronic dispersion with diameter along the structures of SWCNT and MWCNT is studied by their respective STM and TEM images.[19] Furthermore, the tubes' stability and nanotubes' growth processes can also be observed through TEM. These studies confirm that the lattice constant of nanotube structures is comparable to that of graphene and graphite.[23] Moreover, the TEM images of SWCNT show that the largest inflated is approximately 2.5 nm in diameter at ambient conditions. The X-ray diffraction technique has a lesser contribution to the studies of carbon nanotubes because of the low X-ray scattering cross-section of carbon and the Scherrer broadening phenomenon. Nevertheless, X-ray diffraction can give important information regarding graphene's negative thermal expansion coefficient [Prato et al., 2003]. Another exciting feature of X-ray diffraction is its high penetrating power. X-rays are successfully used in probing the sample when electron beams are not appropriate. On the other hand, AFM is used to examine the bundles of carbon nanotubes to study and measure their mechanical properties and find Young's modulus of bundles of SWCNTs in the order of terapascal range.

There are different synthesis methods to prepare and grow carbon nanostructures such as graphene and carbon nanotubes. The growth mechanisms should aid in the control and optimization of nanotube synthesis as nanotubes' top prospects for applications will be achieved. Many methods achieve nanotubes' growth, albeit many challenges remain, such as the production of defect-free nanotubes continuously to macroscopic lengths and the lack of an efficient growth process to obtain structurally perfect nanotubes at large scales. Various methods for CNT growth are laser ablation, arc-discharge, and chemical vapor deposition (CVD) on a transition metal substrate. The CVD growth methods include catalytic CVD and plasma-enhanced CVD; both are reliable in making defect-free CNT structures at large scales. Moreover, methods like laser ablation and arc discharge cannot control the growth of aligned and ordered CNT structures, and both these methods require solid graphite as starting material, whereas, controlling nanotube development using CVD techniques has resulted in the growth of well-organized nanowires.[7]

As mentioned earlier, carbon nanotubes find many applications in wide branches of science and industry. Some of the biological applications of CNTs are schematically represented in Figure 2.1. Potential applications of carbon nanotubes are numerous, especially in electronics. The distinctive electronic properties of SWCNTs are beneficial in fabricating carbon-made circuits. On contrary, its chiral variability is a significant challenge in applying SWCNTs.[2] The sensitivity of SWCNTs to small chemical, mechanical, and electronic variations present on substrates is low. Another concern is placing SWCNTs in the preferred site with desired orientation on a wafer scale. Bottom-up assembly by dielectrophoresis (DEP) is one of the best methodologies to overcome many of the above challenges.[30] The DEP is when a polarizable particle is placed in nonuniform electric fields, the particle motion is affected by the ambient electric fields. The deposition using DEP techniques can achieve individual metallic SWCNTs in the desired locations.[25]

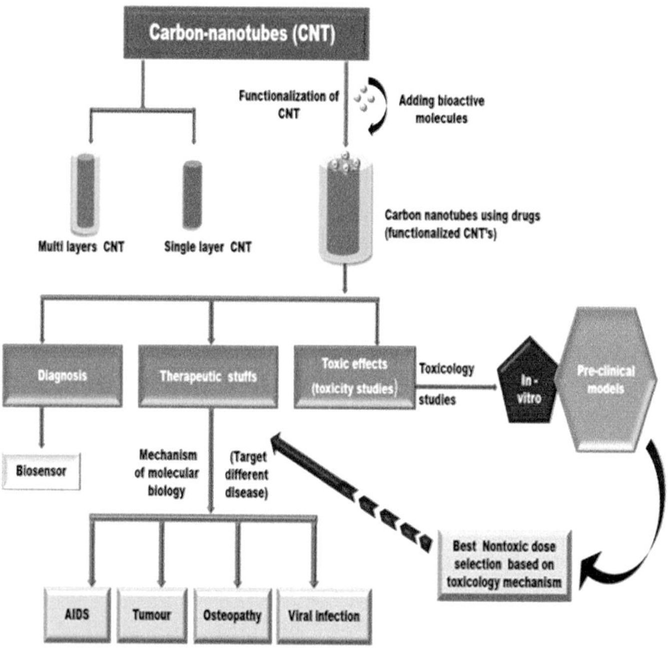

FIGURE 2.1 Schematic representation of relevant biological applications of CNTs.
Source: Reproduced with permission from Ref. [22], ©2020. Elsevier.

One of the earliest commercial applications is implanting nanotubes into a polymer matrix and electrically conductive plastics. Because nanotubes can form conductive percolation networks at tiny volume fractions, at higher densities, the nanotubes can start to tailor the matrix's mechanical properties and maintain the natural properties of the matrix, which improve transparency, flexibility, and strength.[5]

The conductive nanotube films exhibit large surface areas and good mechanical flexibility. So, these CNTs are used as electrodes in electrochemical double-layer charge injection devices. The property mentioned above of CNTs was primarily explored in the fabrication of supercapacitors that exhibit very large capacitance for their size. An electrode made by CNTs can have an effective area more extensive than its macroscopic area. As the capacitance of an electrochemical device depends on the distance between the electrodes and the charge in the electrolyte, the nanometer separation in the nanotube electrodes can result in substantial capacitance values.[3]

Moreover, these nanotube supercapacitors' charging and discharging power rates are enough to meet the applications requiring higher power and storage capacities than any other ordinary devices. The power rates are typically in the range of 20–30 W/g [Du et al., 2005]. These power rates are suitable for applications in hybrid vehicles.

Another vital application of CNTs in computing and electronics is that it is possible to tailor using SWCNTs and MWCNTs as field emission electron sources.[8] These CNTs exhibit the properties of a good emitter as they provide stable emission, long lifetime, and low emission threshold potentials.[6] The field emission can be achieved even at low threshold electric fields of 1–2 V/μm, and because of the tolerance of nanotubes, operation at high power densities is possible. The manufacture of nanotube field-emitting surfaces is comparatively simple compared to field-emitting surfaces made up of tungsten and molybdenum.[3]

Another significant advance of carbon nanotubes is their applicability in understanding and employing biological specimens at their fundamental levels. They can do better medication and easy drug delivery because of their sizeable inner volume, and they can fill and accommodate biomolecules of the size up to proteins. The CNTs have potential applications in pathogenic detection and electronic biosensing.[20] They are better than conventional culturing techniques used to detect pathogens. Because of their unique mechanical, electrical, and chemical properties, SWCNTs can be used to build nanosensor platforms. Due to their sensitivity and less response time, SWCNTs can amplify the electrochemical reactivity of biomolecules, and even a minute variation in the surrounding environment can be sensed.[1,31]

2.2 WHY CNTS FOR PATHOGENIC DETECTION

Globally, toxic food is responsible for the death of around 400,000 people in a year and moreover, around 30% of death occurring among children below 5 years of age is merely due to foodborne infections.[13] The presence of pathogenic microbes in food and water is a major concern as a consequence of globalized food supply, large-scale food production, and global transportation and distribution of food products. Many attempts have been made by manufacturers and regulatory agencies day by day to minimize the health issues that arise from contaminated food. Yet, controlling the presence of microbial contamination remains a challenge.

Conventional culture techniques include nucleic acid-based polymerase chain reaction (PCR), enzyme-linked immunosorbent assay (ELISA) technique, etc. But these methods are not ideal for fast or in vivo detection of pathogenic microbes as they require laboratory settings, and are time-consuming and costly. The commonly used microbial culture test may detect the specific bacterium, which may take several days for obtaining the results. However, by employing nanotechnology, pathogenic microbes can be detected with high accuracy as compared to conventional methods. The nanotechnology-based pathogenic microbe detectors employ nanomaterials that are highly sensitive and specific in their nature when compared with the traditional sensors. The applications of dispersed CNTs for pathogenic microbe detection have undergone in-depth investigations recently, and as a result, the CNTs have become a promising functional nanomaterial for the development of advanced pathogenic microbe detectors with novel features, which could enhance the electron-transfer kinetics with different redox-active biomolecules of pathogenic bacteria.

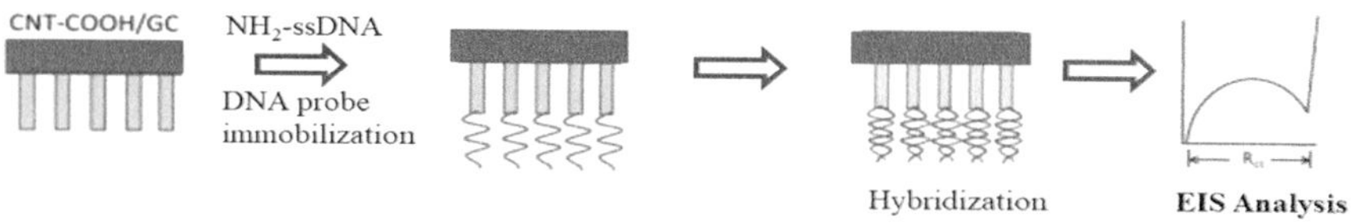

FIGURE 2.2 Schematic representation of DNA and CNT mutual attachment and DNA hybridization measurement.

Source: Reproduced with permission from Ref. [32], ©2022. Elsevier.

Several scientific researches indicate that SWCNTs or MWCNTs can be dispersed in suitable solvents in the presence of surfactants resulting in the formation of individual tubes.[15,16,18,26,] These CNT dispersions are ideal to perform several applications such as electrochemical glucose sensors, gas sensors, biosensors, etc., in which the dispersion parameters such as pH of the dispersion, sonication time, target DNA concentration, etc. are the major parameters on which the efficiency of these sensors depends.[26,28] The schematic representation of biosensor applications of CNTs for the pathogenic virus is shown in Figure 2.3. In pathogenic microbe detection, the CNTs get hybridized with the DNAs of the pathogenic microbes that can be subsequently converted into a measurable signal by an electrochemical sensing process (Fig. 2.2).[13]

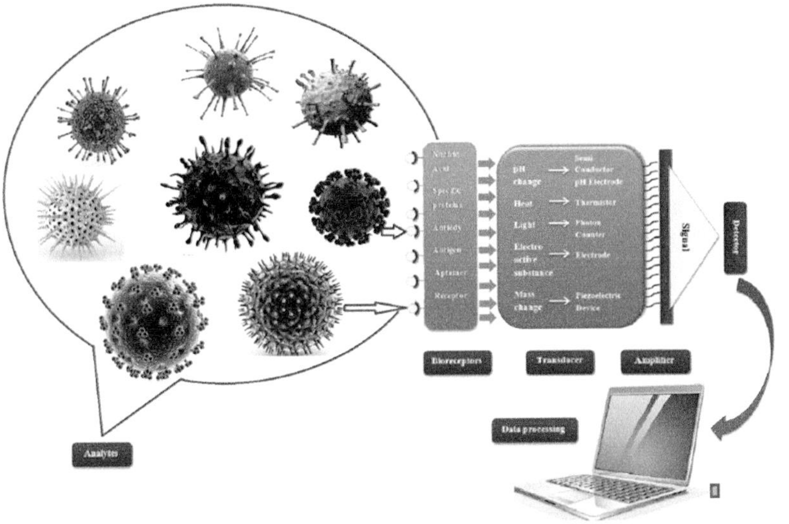

FIGURE 2.3 Schematic representation of biosensor applications of CNTs for the pathogenic virus.
Source: Reproduced with permission from Ref. [21], ©2017. Elsevier.

2.3 NANOMATERIALS FOR PATHOGENIC MICROBE DETECTION

Pathogenic microbe detection involves the identification of biological elements of microbes such as DNA, antibodies, receptors, etc. of any specific microbes that can interact with analyte molecules of the detector, and the resultant signals thus produced will get converted into measurable signals.

The development of pathogenic microbe detectors or sensors using nanomaterials empowers the in vivo detection of bacteria or microbes.

The characteristics of nanomaterials such as good reactivity, high surface permeability, and high surface-to-volume ratio, etc. improve the efficiency in physical and chemical reactivity during sensing applications. Thus, rapid and cost-effective microbial detection devices can be developed by empowering nanomaterials.[13] However, the selection of nanomaterials for pathogenic microbe detectors is dependent on their properties.

The most commonly used nanomaterials in pathogenic microbe detectors include gold nanoparticles, carbon nanoparticles, magnetic nanomaterials, silicon nanoparticles, etc. Among these materials, the most promising nanomaterials for the production of cost-effective pathogenic microbe detectors are carbon nanoparticles due to their added benefits such as equilibrium structure, density, interlayer spacing, electrical conductivity, thermal properties, and elastic behavior. Because of these unique properties, they also find promising applications in biomedical fields such as artificial implants, cancer cell identification, tissue engineering, and drug delivery. Electrochemical applications of these materials include supercapacitors, electromechanical actuators, hybrid electric vehicles, etc. The energy applications are hydrogen storage, fuel cells, laptops, lamps, gas discharge tubes, microwave generators, etc. The sensing applications are pathogenic microbe detector probes, gas detectors, etc.

Nanoelectromechanical system (NEMS) and microelectrochemical system (MEMS) devices are devices with nanoscale and microscale functions. These technologies provide unique properties in energy harvesting, portable power generation, display, imaging, drug delivery, etc. They can also get attached to biological systems which lead to improvement of bioadhesion properties and response stimuli. This technology enables the development of biomarkers for the rapid diagnosis of diseases. A component of biosensors called fluorophores absorbs and emits light within the excitation wavelength and emission spectra, which utilizes total internal reflection. The nanomaterials used for sensing applications must have a sharp and precise optical response with surface plasmon resonance (SPR) of incident light. The SPR allows the detection of biological interactions with these materials. However, the safety of the earth and humans is of prime concern and many reports have established the toxicity of widespread use of nanomaterials.

2.4 PATHOGENIC MICROBE DETECTION USING CNTS

Among the different nanomaterials, nanocomposites, hybrid nanomaterials, and carbon nanostructures such as SWCNTs and MWCNTs have emerged as the building blocks for pathogenic microbe detection due to their unique properties. The CNTs are the quintessential electrode material due to their mechanical strength, and electrical and thermal conductivity. In the last decade, CNTs have come out as one of the most widely investigated nanomaterials for pathogenic microbe detection. The major requirement of a pathogenic microbe detector is its selectivity, sensitivity, detection limit, and response time to assess the presence of pathogenic microbes. The proper modification of electrodes with CNTs enhances the electrochemical performance as a pathogenic microbe detector. In particular, the CNTs-modified electrodes have enhanced electrochemical properties, catalytic activity, etc.

Recent investigations indicate that CNTs can improve the electrochemical reactivity of biomolecules and can enhance the electron-transfer kinetics of protein molecules. Moreover, their high length-to-diameter aspect ratio and high surface-area-to-weight ratio provides enhanced surface functionalization. However, the lack of solubility of CNTs in several mediums remains a challenge in some situations, which can be addressed by covalent modification on the nanotube surface by a treatment process such as treatment with sulfuric and nitric acid, that can open the ends of CNTs to introduce carboxylic acid functional groups. This can effectively address the issue of solubility of CNTs and improve the functional group attachments on the surface as well. As a result, molecules like proteins can covalently get attached to the ends of CNTs, which allows the electrical communication between the electrodes and redox protein during an electrochemical process.

The SWCNT-based microbe detectors have been developed based on the field-effect transistor design, in which SWCNTs provide an efficient path for the transfer of electrons between source and drain.[33] The SWCNT-FET microbe detectors have sensitivity for the detection of pathogenic microbes in food due to their ability to sense changes at the interface. The enhanced sensing of SWCNTs-integrated microbe detectors is attributable to its size, low capacitance, axial electrical conductivity, biocompatibility, structural flexibility, etc. Thus, the SWCNTs can enhance the electrochemical reactivity of microbes, as it is sensitive to variations in their environment. However, some studies indicate that it lacks the essential reproducibility of a pathogenic microbe detector. Still, SWCNTs-integrated GCE electrodes have the potential to improve the electron transfer kinetics and working surface area to enhance the efficiency as a pathogenic microbe detector.[33]

The chemically modified SWCNTs on a glassy carbon electrode (GCE) enable the detection of *Salmonella enterica* serovar Typhimurium. The hybridization of SWCNTs with the corresponding complementary single-stranded

DNA (ssDNA) produces a shift in the electrochemical impedance spectra due to a higher charge transfer kinetic in ssDNA. The SWCNT-COOH deposited on the glassy carbon electrode provides an active and high surface area matrix for binding with ssDNA. The detector has excellent selectivity and sensitivity for the targeted analyte DNA strand from *Salmonella enterica* serovar Typhimurium.[32]

An amino-modified aptasensor using MWCNTs-deposited indium tin oxide (ITO) electrode has good sensitivity and selectivity for the detection of pathogenic *Salmonella* bacteria in food samples, which are found to be more effective as compared to the PCR method in terms of its detection limit. An amino-modified aptamer ssDNA binds to *Salmonella* cells to develop ssDNA/MWCNT/ITO electrode. The impedance measurement using ssDNA/MWCNT/ITO electrode shows variations on exposure to *Salmonella* cells due to the binding of these cells on the aptamer-functionalized surface of the glassy carbon electrode.[11] Integration of SWCNTs and immobilized antibodies has sensitivity towards *Escherichia coli* K-12 microbes. The electrical current shows a linear relationship between current and concentrations of *Escherichia coli* K-12 microbe suspension in which the electrical current decreases as the cell concentration increases in the sample due to an increased bacterial resistance on the bionano-modified electrode surface.[33]

2.5 CONCLUSIONS

Pathogenic detection using the CNTs is a simple, convenient, and inexpensive technique for practical pathogenic detection applications. The use of pathogenic microbe detectors in food industries leads to identifying the presence of contaminants in food with high sensitivity and accuracy. Different diagnostics require the development of portable devices for the precise detection of pathogenic microbes in food industries. The pathogenic detectors using CNTs have the potential to develop low-cost, rapid, sensitive, miniaturized, and analytical devices for practical applications. The detection efficiency relies on the unique properties of CNTs as it can increase the sensing surface area for microbe recognition, sensitivity to surface modification, etc. Thus, the incorporation of CNTs into the sensing platform is a promising technique to enhance the sensing signal response of pathogenic detectors.

KEYWORDS

- **carbon nanotube**
- **pathogenic detection**
- **pathogenic microbes**
- **biosensors**
- **microbe detectors**

REFERENCES

1. Allen, B. L.; Kichambare, P. D.; Star, A. Carbon Nanotube Field-Effect-Transistor-Based Biosensors. *Adv. Mater.* **2007,** *19*, 1439–1451.
2. Banerjee, S.; Hemraj-Benny, T.; Wong, S. S. *Adv. Mater.* **2005,** *17*, 17.
3. Baughman, R. H.; Zakhidov, A. A.; de-Heer, W. A. *Science* **2002,** *297*, 787.
4. Bianco, A.; Kostarelos, K.; Partidos, C. D.; Prato, M. Biomedical, Applications of Functionalized Carbon Nanotubes. *Chem. Commun.* **2005,** *1*, 571–577.
5. Biercuk, M. J.; Llaguno, M. C.; Radosavljevic, M.; Hyun, J. K.; Johnson, A. T. *Appl. Phys. Lett.* **2002,** *80*, 2767.
6. Bonard, J. M.; Kind, H.; Stockli, T.; Nilsson, L. A. *Solid-State Electron.* **2001,** *45*, 893.
7. Dai, H.; Kong, J.; Zhou, C.; Franklin, N.; Tombler, T.; Cassell, A.; Fan, S.; Chapline, M. *J. Phys. Chem.* **1999,** *103*, 11246–11255.
8. de-Heer, W. A.; Chaˆtelain, A.; Ugarte, D. *Science* **1995,** *270*, 1179.
9. Ferrari, A. C.; Basko, D. M. *Nat. Nanotechnol.* **2013,** *8*, 235.
10. Foldvari, M.; Bagonluri, M. Carbon Nanotubes as Functional Excipients for Nanomedicines: I. Pharmaceutical Properties. *Nanomed. Nanotechnol. Biol. Med.* **2008,** *4*, 173–182.
11. Hasan, M. R.; Pulingam, T.; Appaturi, J. N.; Zifruddin, A. N.; The, S. J.; Lim, T. W.; Ibrahim, F.; Leo, B. F.; Thong, K. L. Carbon Nanotube-Based Aptasensor for Sensitive Electrochemical Detection of Whole-Cell Salmonella. *Anal. Biochem.* **2018,** *554*, 34–43.
12. Iijima, S.; Ichihashi, T. Single-Shell Carbon Nanotubes of 1-nm Diameter. *Nature* **1993,** *363*, 603–605.
13. Kumar, H.; Kuča, K.; Bhatia, S. K.; Saini, K.; Kaushal, A. K.; Verma, R.; Bhalla, T. C.; Kumar, D. Applications of Nanotechnology in Sensor-Based Detection of Foodborne Pathogens. *Sensors* **2020,** *20* (7), 1966.
14. Lacerda, L.; Bianco, A.; Prato, M.; Kostarelos, K. Carbon Nanotubes as Nanomedicines: from Toxicology to Pharmacology. *Adv. Drug Deliv. Rev.* **2006,** *58*, 1460–1470.
15. Li, Z.; Wu, Z.; Li, K. The High Dispersion of DNA-Multi-walled Carbon Nanotubes and Their Properties. *Anal. Biochem.* **2009,** *387*, 267–270.
16. Madni, I.; Hwang, C. Y.; Park, S. D.; Choa, Y. H.; Kim, H. T. Mixed Surfactant System for Stable Suspension of Multi-walled Carbon Nanotubes. *Colloids Surf. A: Physicochem. Eng. Asp.* **2010,** *358*, 101–107.
17. Maire, J.; M′ering, J. Proceedings of the First Conference of the Society of Chemical and Industrial Conference on Carbon and Graphite, London, 1958.

18. Malik, S.; Vogel, S.; Rosner, H.; Arnold, K.; Hennrich, F.; Köhler, A. K.; Richert, C.; Kappes, M. M. Physical-Chemical Characterization of DNA-SWNT Suspensions and Associated Composites. *Compos. Sci. Technol.* **2007,** *67*, 916–921.
19. Maultzsch, J.; Reich, S.; Thomsen, C. *Phys. Rev. B* **2001,** *64*, 121470.
20. Mello, L. D.; Kubota, L. T. Review of the Use of Biosensors as Analytical Tools in the Food and Drink Industries. *Food Chem.* **2002,** *77*, 237–256.
21. Mokhtarzadeh, A.; Keihanc, E. R.; Pashazadeh, P.; Hejazie, M.; Gharaatifar, N.; Hasanzdeh, M.; Baradaran, B.; Guardia, M. *Trends Analyt Chem* **2017,** *97*, 445–457.
22. Nigar, Z.; Hasan, R.; Tyagi, M.; Yadav, N.; Narang, J. Carbon Nanotube—A Review on Synthesis, Properties and Plethora of Applications in the Field of Biomedical Science. *Sens. Int.* **2020,** *1*, 100003.
23. Odom, T. W.; Huang, J. L.; Kim, P.; Lieber, C. M. *Nature* **1998,** *39*, 62.
24. Prato, M.; Kostarelos, K.; Bianco, A. Functionalized Carbon Nanotubes in Drug Design and Discovery. *Acc. Chem. Res.* **2007,** *41*, 60–68.
25. Proctor, J. E.; Armada, D. A. M.; Vijayaraghavan, A. An Introduction to Graphene and Carbon Nanotubes. *MRS Bull.* **2018,** *43*, 148–149.
26. Rubianes, M. D.; Rivas, G. A. Dispersion of Multi-Wall Carbon Nanotubes in Polyethylenimine: A New Alternative for Preparing Electrochemical Sensors. *Electrochem. Commun.* **2007,** *9*, 480–484.
27. Saito, R.; Dresselhaus, G.; Dresselhaus, M. S. *Physical Properties of Carbon Nanotubes*. Imperial College Press: London, 1998.
28. Thuy, N. T.; Tam, P. D.; Tuan, M. A.; Le, A.; Tam, L. T.; Thu, V. V.; Hieu, N. V.; Chien, N. D. Detection of Pathogenic Microorganisms Using Biosensor Based on Multi-Walled Carbon Nanotubes Dispersed in DNA Solution. *Curr. Appl. Phys.* **2012,** *12* (6), 1553–1560.
29. Tuinstra, F.; Koenig, J. L. *J. Chem. Phys.* **1970,** *53*, 1126.
30. Vijayaraghavan, A. *Phys. Stat. Sol. B* **2013,** *250*, 2505.
31. Wang, J. Carbon-Nanotube Based Electrochemical Biosensors: A Review. *Electroanalysis* **2005,** *17*, 7–14.
32. Weber, J. E.; Pillai, S.; Rama, M. K.; Kumar, A.; Singh, S. R. Electrochemical Impedance-Based DNA Sensor Using a Modified Single-Walled Carbon Nanotube Electrode. *Mater. Sci. Eng. C.* **2011,** *31*, 821–825.
33. Yamada, K.; Kim, C. T.; Kim, J. H.; Chung, J. H.; Lee, H. G.; Jun, S. Single-Walled Carbon Nanotube-Based Junction Biosensor for Detection of *Escherichia coli*. *PLoS ONE* **2014,** *9*, 105767.

CHAPTER 3

Cold Plasma—An Efficient Tool for Nanotechnological Modification of Polyester-Based Polymers

IGOR NOVÁK[1], ANTON POPELKA[2], IVAN CHODÁK[1], JÁN SEDLIAČIK[3], ANGELA KLEINOVÁ[1], MATEJ MIČUŠÍK[1], JÁN MATYAŠOVSKÝ[4], PETER JURKOVIČ[4], KATARÍNA VALACHOVÁ[5], and LADISLAV ŠOLTÉS[5]

[1]*Polymer Institute, Slovak Academy of Sciences, Bratislava, Slovakia*

[2]*Center for Advanced Materials, Qatar University, Doha, Qatar*

[3]*Technical University in Zvolen, Zvolen, Slovakia*

[4]*VIPO, a.s., gen. Svobodu Partizánske, Slovakia*

[5]*Centre of Experimental Medicine, Slovak Academy of Sciences, Bratislava, Slovakia*

ABSTRACT

Polyester surface was pretreated by cold plasma to improve its adhesive properties. The changes in the chemical structure of the pretreated polymer were analyzed by attenuated total reflectance Fourier-transform infrared spectroscopy and X-ray photoelectron spectroscopy. The changes in the polyester surface morphology were studied using scanning electron microscopy and atomic force microscopy. The surface energy, and its polar contribution, as well as the peel strengths of the adhesive joints of polyester, modified by

Tailored Functional Materials for Clean and Sustainable Development. Tamara Tatrishvili, PhD, Neha Kanwar Rawat, PhD, Ann Rose Abraham, PhD & A. K. Haghi, PhD (Eds.)

cold plasma to polyacrylate, increased significantly. The efficiency of the material modification depends on a type of discharge, the used gases, the power of plasma source, as well as on the time of modification treatment. Aging of the polymer modified by plasma was also investigated. A correlation between adhesion of polyester modified by cold plasma to polyacrylate and its surface energy was found.

3.1 INTRODUCTION

The plasma-based nanotechnology, typically less than 100 nm in a single dimension, represents a rapidly developing area of research ranging from physics of gaseous plasma to nanofabrication. In general, plasma-modified biomaterial possesses great potential for the applications in macromolecular chemistry, tissue engineering, and drug delivery.[1] The cold plasma finds numerous applications in growth and processing of various nanomaterials, for example, carbon nanotubes, inorganic nanowires, biopolymeric materials, and others. Cold plasma is widely used as a modification tool for biomaterial surfaces, such as polymers, metals, alloys, ceramics, and nanocomposites.[2,3] The effect of nonpolymer-forming plasma (e.g., plasma of argon, oxygen, and nitrogen) can be viewed as the reaction of active species with polymer as well as the formation of free radicals in polymer, which is mainly due to the ultraviolet emitted by the plasma. The incorporation of nitrogen into the polymer surface by nitrogen plasma and the surface oxidation by oxygen plasma are typical processes, which are carried out in nano scale and results in plasma-treated polymer. The plasma-treated chemical processes lead to the incorporation of oxygen in the form of carbonyl and hydroxyl and to some degree of cross-linking depending on the type of polymer.[4,5].

Polyesters represent important polymers that are frequently used in many industrial applications, for example, in the automotive industry for cars' construction, due to its excellent properties. The surface energy of polyesters is insufficient in some applications, for example, bonding, printing, etc. This problem can be solved by using various surface modification methods. Several of these methods have been used to improve the surface and adhesive properties of polymers.[6,7] The application of cold plasmas for pretreatment of polymeric surfaces[8,9] is a dry, ecological method of modification, which can tailor polymers to modify their surface energy and adhesion to other materials. The most important feature of the plasma treatment technique is the modification of surface properties of a polymer without changing their intrinsic bulk properties.[10–16] It is well known that the modification of polymers by cold plasma leads to changes in surface and adhesive properties. Depending on the processing gases used, different functional groups are formed in a very thin surface layer of the polymer.[17–19] Poly(ethylene terephthalate) (PET), with a substantially higher surface energy than polyethylene (surface energy of PET is in the range of 40–50 $mJ.m^{-2}$, and surface energy of low-density polyethylene is in the range of 30–32 $mJ.m^{-2}$), usually does not need to be pretreated by plasma for printing, because the surface energy in the untreated state is higher than the value needed for printing.[20–27] However, for the preparation of strong adhesive joints, modification of PET by cold plasma is essential. A significant problem with discharge plasma modification of PET is that the effect of surface modification decays relatively fast, and its surface and adhesive properties are deteriorated.

In this paper, the surface and adhesive properties of polyester/PET modified by cold plasma have been investigated using contact angle measurements, peel tests, as well as Attenuated Total Reflectance Fourier-Transform Infrared (ATR-FTIR) spectroscopy, X-ray Photoelectron Spectroscopy (XPS), Scanning Electron Microscopy (SEM), Atomic Force Microscopy (AFM) and Electron Spin Resonance (ESR) spectroscopy analyses. The process of hydrophobic recovery of modified PET during aging has been studied. The correlation between the adhesive properties of PET modified by cold plasma and surface energy was also investigated.

3.2 EXPERIMENTAL PART

3.2.1 MATERIALS

In the PET polymer structure (Scheme 1) the aromatic ring and its associated C-C bonds give rigidity to macromolecules. The dissociation of chemical bonds on the surface of the PET macromolecule might occur at C-H or C-C when the polymer chains receive energy from activated species during modification by low-temperature plasma. In our experiments we used oriented PET foils (Tenolan OAN, Technoplast, Czech Republic) with thickness 0.12 mm for modification by discharged plasma, and poly(2-ethylhexyl acrylate) (PEHA) (Polysciences, USA) as a polar adhesive in PET adhesive joints. The PET foils were treated in acetone to eliminate the additives influencing their surface properties. The adhesive joints of modified PET were prepared by using a solution of PEHA in ethyl acetate. The layer with a thickness of 0.12

mm was deposited on supported biaxially oriented isotactic polypropylene with the aid of a coating ruler (Dioptra, Czech Republic).

3.3 COLD PLASMA MODIFICATION

The modification of the PET foils by cold plasma was performed in a diffuse coplanar surface barrier discharge (DCSBD) plasma source at atmospheric pressure in medium of N_2 or O_2 processing gases of technical purity. The DCSBD plasma generator consists of electrodes separated by an alumina dielectric plate. The discharge electrodes having an area of 80 × 80 mm, and consisting of 1-mm-wide and 80-mm-long tungsten strips, are fixed to the upper surface of the aluminum plate. The electrodes are located inside a glass cover allowing passing of the medium gases. The voltage of the DCSBD source was 100 V, current intensity 1 A, and frequency 6 kHz. The power used for modification by DCSBD plasma in N_2 or O_2 was 100 W.

The radiofrequency (RF) plasma source, working at a reduced air pressure of 40 Pa, consists of a stainless steel locked-up vacuum vessel with two circular brass electrodes having diameters of 240 mm. The voltage of the RF plasma source was 2 kV, current density 0.6 A, and frequency 13.56 MHz, and the power applied for the modification of the polymer by the RF plasma in air was 200 W.

3.4 CHARACTERIZATION METHODS

3.4.1 SEM

The morphology of the PET before and after plasma irradiation was investigated by scanning electron microscope JSM-6400 (Jeol, Japan). The foils were sputter-coated (SCD 050, BALTEC) with a Pt layer (ca 4 nm).

3.4.2 AFM

The surface morphology and local surface heterogeneities of the modified polymer were measured by AFM. All measurements were performed under ambient conditions using a commercial atomic force microscope (NanoScope™ Dimension IIIa, MultiMode Digital Instr., USA) equipped with a PPP-NCLR tapping-mode probe (Nanosensors™ Switzerland; spring constant 39 N.m^{-1}, resonance frequency ≈ 160 kHz). The surface properties of all the films were measured in x- and y-axis sizes in the range of 2–25 μm on different positions of the films to find characteristic and significant surface features. The AFM analyses were performed in tapping mode for all the images. This technique allows obtaining two- or three-dimensional information of both height and material heterogeneity contrast with high resolution when recording height and phase shifts simultaneously.

3.4.3 EPR

Samples of PET foils were inserted into silica quartz tubes and then placed into the resonator of an EPR spectrometer. The EPR spectra were subsequently obtained in a Bruker–Biospin, ElexSys E-540 spectrometer equipped with a Bruker high-sensitivity cavity ER 4119HS-W1 operating at 9.75 GHz (X-band) and room temperature. The EPR spectrometer settings were as follows: modulation frequency 100 kHz, modulation amplitude 1.0 G, microwave power 20 mW, receiver gain 60 dB, time constant 1.28 ms, conversion time 5.12 ms, magnetic field scan 6 kG (center at 3480 G), and number of scans 30. A Xepr (Bruker) Linux-based package was used for acquisition of data, and the Origin (OriginLab) software was used for data analysis and presentation. The EPR measurements of PET were performed at room temperature.

3.4.4 ATR-FTIR

The ATR-FTIR spectroscopy measurements of the PET foils[28,29] were performed with a Nicolet Impact 400 FTIR spectrometer (Nicolet, USA) having a resolution of 4 cm^{-1}, a scan range was 4000–400 cm^{-1}, and a total of 1024 scans per analysis. The ATR contained a KRS-5 (thallium-bromide-iodide) crystal.

3.4.5 XPS

The XPS spectra[30–33] were recorded using VG Scientific ESCALAB 250 system equipped with a micro-focused, monochromatic Al K_α X-ray source (1486.6 eV) and a magnetic lens that increases the electron acceptance angle and hence the sensitivity. An X-ray beam of 650 nm size was used at power of 20 mA × 15 kV. The spectra were acquired in the constant analyzer energy mode, with pass energies of 150 and 20 eV for the survey and narrow regions, respectively. Charge compensation was achieved with an electron flood gun coupled to an argon gun. The energy and emission current of the electrons were 4 eV and 0.35 mA, respectively. The argon partial pressure in the chamber was set at 2×10^{-8} mbar. Under these conditions, the surface charge was negative but uniform. The Avantage software, version 2.2, was used for digital acquisition and data processing. Spectral calibration was performed by setting the main C1s peak at 285 eV. The O/C atomic ratios were determined by considering the integrated peak areas of C1s and O1s, and their respective Scoffield sensitivity factors corrected for the analyzer transmission function.

3.5 SURFACE ENERGY

The surface energy of PET was determined *via* measurements of the contact angles[34–39] of a set of testing liquids: redistilled water, ethylene glycol, formamide, methylene iodide, and α-bromo naphthalene) with an SEE (Surface Energy Evaluation) system (Advex, Czech Republic). The drops of the testing liquid (volume 3 μL) were placed on the PET foil surface with a micropipette (Biohit, Finland), and the dependence = f(t) was extrapolated to t = 0. The surface energy of the polymer as well as its polar and dispersive components were evaluated by the Owens-Wendt-Rable-Kaelble (OWRK) method modified by a least-squares method.[37–39]

$$(1 + \cos \theta)\gamma_{LV}/2 = (\gamma_{LV}{}^{d}\gamma s^{d})^{1/2} + (\gamma_{LV}{}^{p}\gamma s^{p})^{1/2} \quad (3.1)$$

$$\gamma_{LV} = \gamma_{LV}{}^{p} + \gamma_{LV}{}^{d} \quad (3.2)$$

$$\gamma_{s} = \gamma_{s}{}^{p} + \gamma_{s}{}^{d} \quad (3.3)$$

where θ = contact angle (deg), γ_{LV} = surface free energy (SFE) of the testing liquid (mJ.m^{-2}), $_{LV}{}^{d}$, $\gamma_{LV}{}^{p}$ = dispersive component (DC), and polar component (PC) of the SFE of the testing liquid (mJ.m^{-2}), and $\gamma s^{d} + \gamma s^{p}$ = DC and PC of the SFE of the polymer (mJ.m^{-2}). The values of the surface energies of the testing liquids at 23°C used for the OWRK determination of the polymer surface energy are shown in Table 3.1.

TABLE 3.1 Surface Energy and Polar Ratio of the Testing Liquids at 23°C.

Testing liquid	γ_L (mJ.m^{-2})	x_L^p
Water	72.8	0.701
Ethylene glycol	48.3	0.393
Formamide	58.2	0.321
Methylene iodide	50.8	0.045
α-Bromo naphthalene	45.3	0

Note: γ_L is the surface tension, $x_L^p = \gamma_L^p/\gamma_L$ is the polar ratio of the testing liquid.

3.6 PEEL STRENGTH OF THE ADHESIVE JOINT

The peel strength of the adhesive joint (P_{peel}) of the plasma-modified PET foil to polyacrylate was determined by peeling of the adhesive joint (peel tests) at a 90° angle using a 5 kN universal testing machine Instron 4301 (Instron, England). The adhesive joints were fixed in the aluminum peeling circle. The width of the adhesive joints was 20 mm, and its length was 140 mm. The adhesive joints were prepared from modified PET foils and biaxially oriented isotactic polypropylene impregnated with polyacrylate. The speed of peeling of the adhesive joint was 1.5 mm.min^{-1}. The relation for calculating of P_{peel} was introduced in previous papers.[37–39].

3.7 RESULTS AND DISCUSSION

3.7.1 SEM

The changes in the surface topography of the polymer modified by low-temperature plasma were investigated by SEM.[22] The SEM microphotographs for the original PET film (A) and the film after DCSBD plasma irradiation in N_2 (B) and in O_2 (C) are presented in Figure 3.1. The surface of unmodified PET film is relatively smooth, but some traces as a consequence of foil production can be observed in a microphotograph (Fig. 3.1A). Numerous small blisters can be seen on the surfaces of PET modified by the DCSBD plasma in N_2 (Fig. 1B) or O_2 (Fig. 1C). The surface of the PET foil modified by the RF plasma in air (Fig. 1D) shows a rough surface consisting of many smaller blisters.

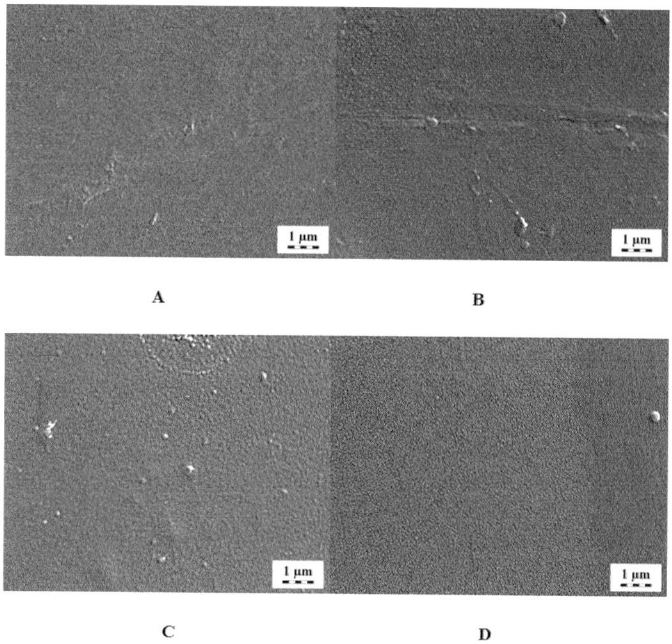

FIGURE 3.1 SEM micrographs of PET foil surfaces: unmodified PET (A), PET modified by DCSBD plasma in N_2, 10s (B), PET modified by DCSBD plasma in O_2, 10s (C), PET modified by RF plasma in air, 120s (D).

3.7.2 AFM

The changes in the surface topography as well as in the roughness of the plasma-modified polymer were investigated by AFM.[23–26] The AFM results of the PET surface topography and local heterogeneities (phase image) after DCSBD and RF plasma modification are shown in Figure 3.2. The surface properties of each PET foil were measured in three different places, but no extensive differences were found between these places. The phase image of the unmodified polymer film (Fig. 3.2A) shows no significant differences in the phase profiles (local heterogeneities) for untreated PET, except for localized melting marks on the surface. The phase image of unmodified PET is rather rough, but regular. After plasma treatment very fine heterogeneous regions were formed, and the polymeric surface became more irregular. The local heterogeneities after modification by the DCSBD plasma are higher in O_2 (Fig. 3.2B) than in N_2 atmosphere (Fig 3.2C). The phase image of the PET surface modified by the RF plasma in air is similar (Fig. 3.2D). The presence of the heterogeneities in Figure 3.2D can be explained by local melting of the polymer during modification by the RF plasma.

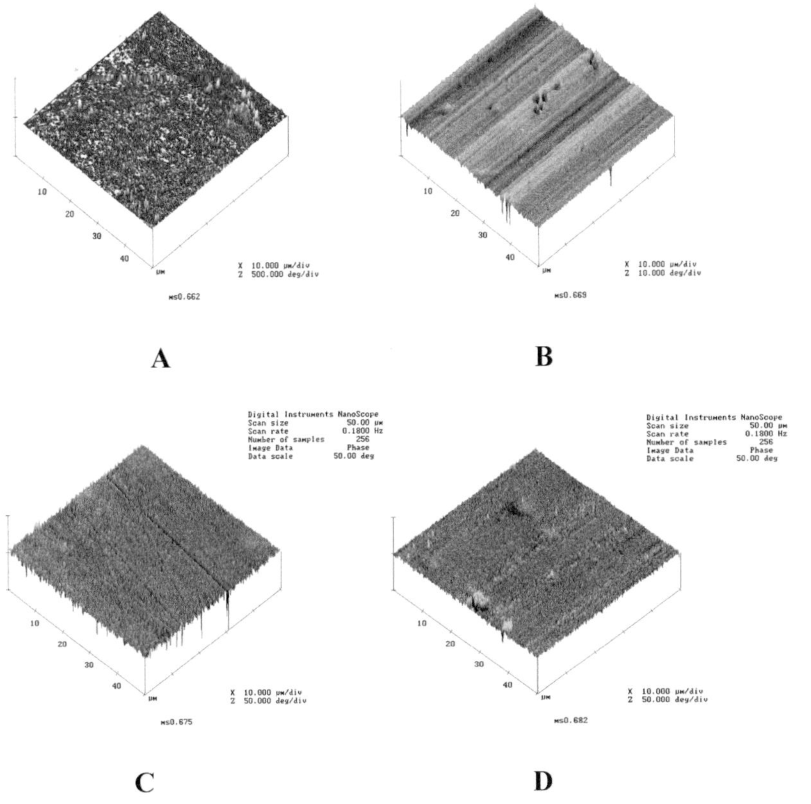

FIGURE 3.2 AFM phase images of PET foil: untreated PET (A), PET treated by DCSBD plasma in N_2, 10s (B), PET modified by DCSBD plasma in O_2, 10s (C), PET treated by RF plasma in air, 120s (D).

3.7.3 ESR

The irradiation of PET by DCSBD plasma led to a breakdown of the C-H or C-C bonds creating free radicals on the PET surface. The relative free-radical intensity of the plasma-treated PET was measured by means of ESR spectroscopy.[27] The ESR results of the PET modified by the DCSBD plasma in N_2 and in O_2 are shown in Figure 3.3.

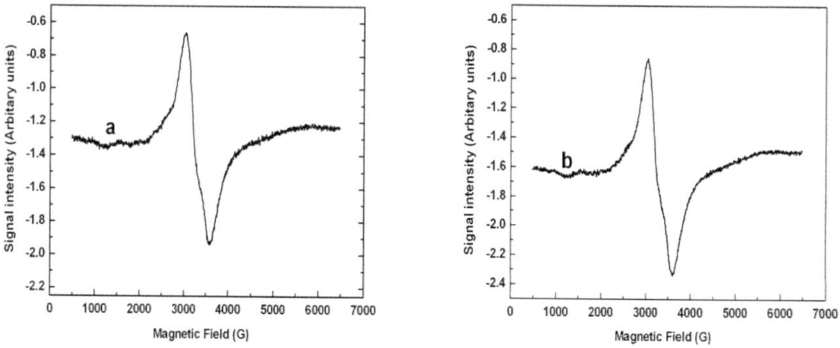

FIGURE 3.3 ESR spectra of PET modified by DCSBD plasma in (a) N_2 and (b) O_2.

The PET foil treated by plasma was exposed to excited and unstable particles, nitrogen ions (Fig. 3.3a), and oxygen ions (Fig. 3.3b), which could transfer their energy to the polymeric surface. The ions could cause bond breakage and formation of free radicals, which could lead to an increased reaction probability with the nitrogen or oxygen ions. The oxygen concentration depends on the removal of the surface contamination of the PET foil, or on the presence of oxidizing species (oxygen, water) in the plasma generator. When subsequently exposed to air, these radicals react with oxygen in air to form peroxides and hydroperoxides.

The ESR spectrum in Figure 3.4 consists of two main singlet signals with parameters: first signal: g-factor (g) = 2.2017, the width of the signal (ΔH_{pp}) = 238.5 G; second signal: g = 2.0004, ΔH_{pp} = 259.5 G.

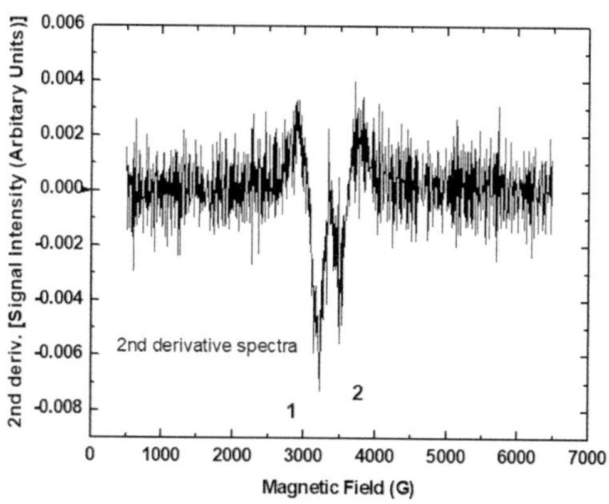

FIGURE 3.4 The second derivative of signal intensity in PET modified by DCSBD plasma in O_2 versus magnetic field.

Both signals result from products of the radical breakdown of the polymer. This spectrum also shows two other signals: first with g = 6.25, which probably belongs to a high-spin iron(III) complex present in traces only; second with g = 4.13 belongs to iron(III) in a tetrahedral environment, also present in traces only. It is interesting that the relation between the two main signal intensities changed during aging of the polymer. In ESR spectra shown in Figure 3.4 the evidence of two singlet signals is inspectional.

TABLE 3.2 ESR Measurements of Free Radicals Concentration of PET Modified by DCSBD Plasma in N_2 or O_2 During Aging.

Aging Time (days)	Processing gas	A_{pp} (arbitrary units) (signal intensity)	ΔH_{pp} (Gauss) (signal width)	Free radicals concentration (relative units)
0	N_2	1.457	589.48	5.062×10^5
	O_2	1.4612	597.68	5.219×10^5
4	N_2	1.2677	557.828	3.944×10^5
	O_2	1.4184	548.78	4.271×10^5
6	N_2	1.4922	568.382	4.820×10^5
	O_2	1.4753	557.828	4.590×10^5
14	N_2	1.40263	578.935	4.701×10^5
	O_2	1.5502	538.2295	4.490×10^5
41	N_2	0.70075	370.2	0.960×10^5
	O_2	0.7456	388.6	1.126×10^5

The free radical concentrations of the PET, modified by a DCSBD plasma in N_2 and in O_2 (Fig. 3.5), decreased by a factor 5 over a period of 41 days (Table 2), as determined from the intensities of the two main singlet signals.

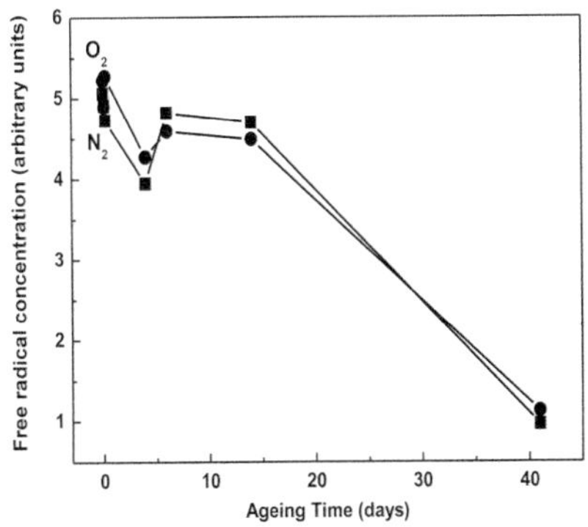

FIGURE 3.5 Free radicals content in PET modified by DCSBD plasma in O_2 and N_2 versus aging time.

3.7.4 ATR-FTIR

Figure 3.6 shows the ATR-FTIR spectra of unmodified PET (Fig. 3.6c), as well as the PET modified by a DCSBD plasma in N_2 (Fig. 3.6b) and O_2 (Fig. 3.6a) atmospheres after treatment for 10 s. For unmodified PET foil (Fig. 3.6c) , the characteristic IR bands at 1710, 1505, and 1173 cm^{-1} were observed for the C-H, C=O, and benzene-ring -C-C- stretching vibrations, as well as the ring C-H in plane bending. The IR bands at 1358 cm^{-1} of the wagging, benzene-ring in-plane C-H bending, and the -C-C- stretching vibration bands appeared near 1173 and 1037 cm^{-1}. The spectra of the DCSBD plasma-treated PET in Figure 3.6b (N_2 plasma) and Figure 3.6a (O_2 plasma) show important differences from that of unmodified PET. The C=O stretch at 1710 cm^{-1} broadened due to the creation of oxygen-containing sites. The C=O content is higher for the PET modified by the O_2 DCSBD plasma (Fig. 3.6a) than for the PET modified by the N_2 plasma (Fig. 3.6b), and consequently a broadening of the peak at 1710 cm^{-1} was observed for the O_2 plasma-modified PET.

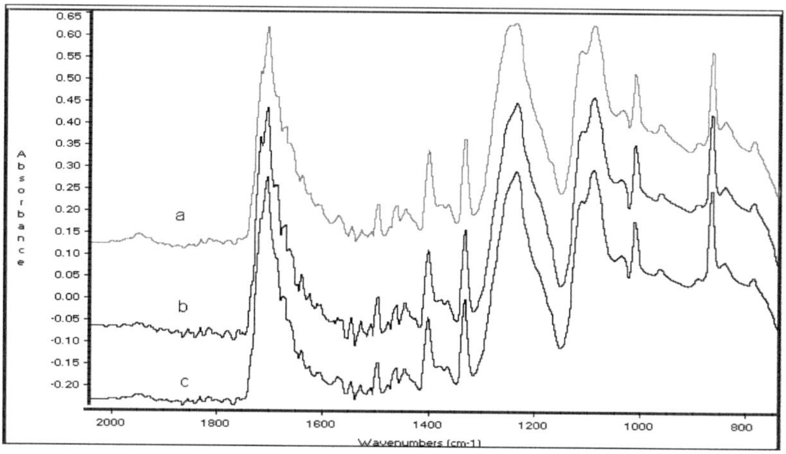

FIGURE 3.6 ATR-FTIR spectra of PET foil: (a) treated by DCSBD plasma in O_2, (b) treated by DCSBD plasma in N_2, and (c) untreated.

3.7.5 XPS

The XPS survey scans C1s, O1s peaks of the untreated PET, and scans C1s, O1s, and N1s peaks of the modified PET are shown in Figures 3.7–3.10 in Table 3.2.

The main peaks of untreated PET C1s and O1s are centered at 284.70 and 531.90 eV, respectively (Fig. 3.7). The C1s peak of untreated PET after deconvolution is composed of three peaks: peak A belonging to the C-C and C-H bonds, peak B belonging to the C-O bond, and peak C belonging to the O=C-O bond.[9]

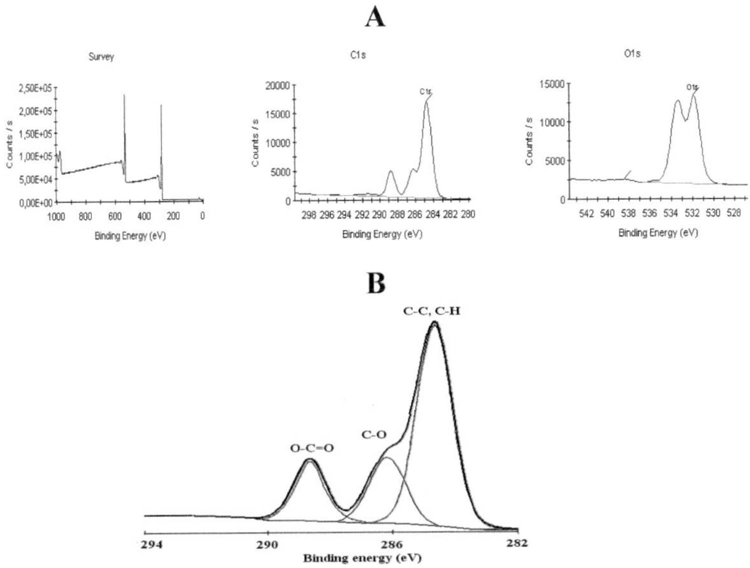

FIGURE 3.7 XPS measurement of the untreated PET: A—survey scan, C1s, O1s; B—deconvolution of C1s peak.

For the PET treated by the DCSBD plasma in N_2 (Fig. 8a and b) the survey scan is dominated by sharp C1s peak centered at 284.70 eV, and O1s peak at 531.80 eV, which significantly increases due to the surface oxidation of PET by the DCSBD plasma. The oxygen content increased from 24.41 to 30.12 atm. % after the DCSBD plasma treatment in N_2 for 5s, but the carbon concentration decreased from 75.59 to 68.23 atm. %. Nitrogen is also present on the PET surface (N1s at 399.70 eV) of the PET modified by the DCSBD plasma in N_2 (5 s), and its content was 1.65 atm. %. For a longer time (10 s) of PET modification by the DCSBD plasma in nitrogen its content increased to 3.99 atm. %. In case of the treatment of the PET by the DCSBD plasma in nitrogen an effective attachment of the nitrogen-containing group has occurred. The nitrogen content in the PET as determined by XPS analysis increased with time of DCSBD plasma modification ($c_{DCSBD, N2, 5s} < c_{DCSBD, N2, 10s}$).

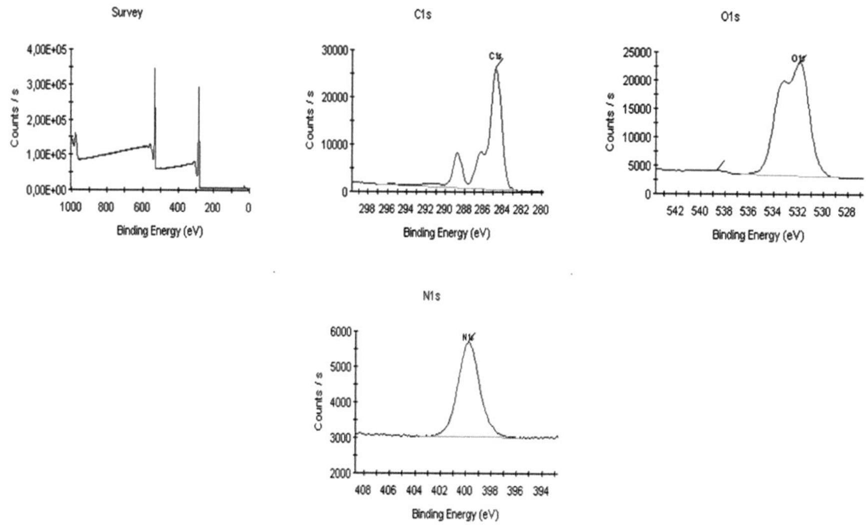

FIGURE 3.8 XPS of PET modified by DCSBD plasma in nitrogen, 10s: survey scan, C1s, O1s, N1s peaks.

The XPS results of the PET modified by the DCSBD plasma in oxygen are shown in Figure 3.9. The C1s and O1s peaks of the PET treated by the DCSBD plasma in oxygen show almost no effect of pretreatment on the surface composition of the polymer, but the treated sample reaches higher proportion of oxidized carbon type.

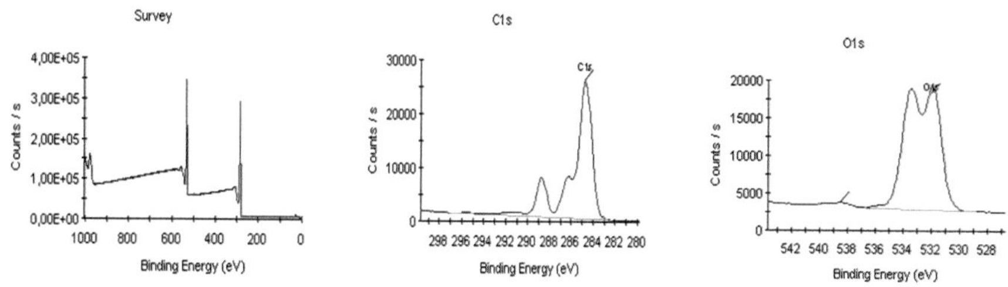

FIGURE 3.9 XPS of PET modified by DCSBD plasma in oxygen, 10s: survey scan, C1s, O1s peaks.

The XPS results of PET treated by RF plasma in air are illustrated in Figure 3.10. The oxygen content (O1s peak at 531.90 eV, 28.47 atm. %) increased compared to unmodified PET. The nitrogen concentration (N1s peak) after the modification of PET by DCSBD plasma was 0.34 atm. %, and this value was substantially lower than that after the modification by DCSBD plasma in nitrogen.

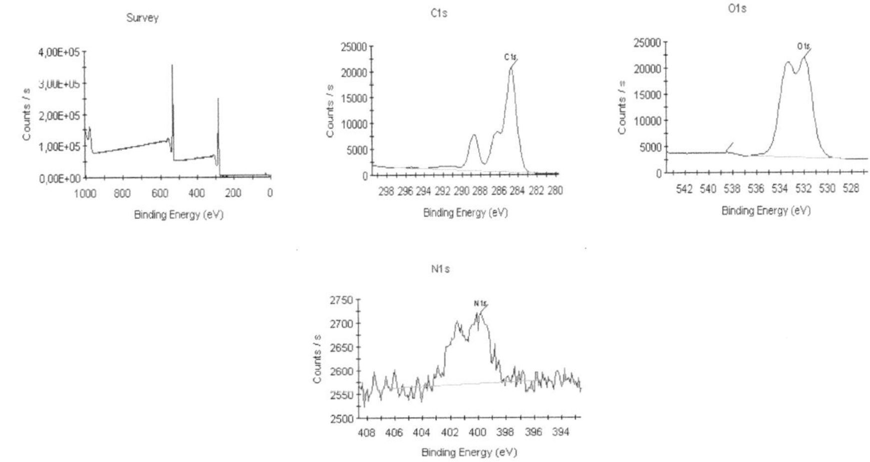

FIGURE 3.10 XPS of PET modified by RF plasma in air, 120s: survey scan, C1s, O1s, N1s peaks.

Table 3.3 shows the XPS results of the surface chemical composition of untreated PET and PET modified by DCSBD and RF plasma, as well as the O/C, N/C, and (N+O)/C ratios of the investigated PET samples.

TABLE 3.3 XPS Measurements of PET Modified by DCSBD and RF Plasma.

Parameter	Untreated PET	DCSBD, N_2, 5s	DCSBD, N_2, 10s	DCSBD, O_2, 5s	DCSBD, O_2, 10s	RF, air, 120s
C1s	75.59	68.23	70.06	76.51	76.67	71.19
O1s	24.41	30.12	25.95	23.49	23.33	28.47
N1s	0	1.65	3.99	0	0	0.34
O/C	0.323	0.35	0.370	0.416	0.441	0.399
N/C	0	0.036	0.057	0	0	0.006
(N+O)/C	0.323	0.388	0.427	0.416	0.441	0.405

The O/C ratio of pristine PET (0.32) modified by DCSBD plasma increased from 0.35 to 0.37 (modification in nitrogen, 5 and 10 s), or from 0.42 to 0.44 (modification in oxygen), and to 0.400 (modification by RF discharge plasma in air, 120 s). Higher values of the O/C ratio were found for PET modified by DCSBD plasma in oxygen (0.42 or 0.44), and by RF plasma in air (0.40) compared to the modification by DCSBD plasma in nitrogen (0.35 or 0.37). The increase in the N/C ratio was found in case of DCSBD plasma modification of PET in nitrogen (0.04, 5s or 0.06, 10 s), or after the modification of PET by RF plasma in air (0.006, 120 s). The increase in the (N+O)/C ratio (0.32 for pristine polymer) was mainly found for the PET sample modified by DCSBD plasma in nitrogen (0.43, 10 s), and for the same modification by plasma of PET in oxygen (0.44, 10 s), and/or by RF plasma in air (0.41, 120 s).

3.8 SURFACE ENERGY

The surface energy of PET foils modified by DCSBD plasma in O_2 and N_2 at atmospheric pressure, and by RF plasma in air at reduced pressure vs. the activation time is shown in Figure 3.11.

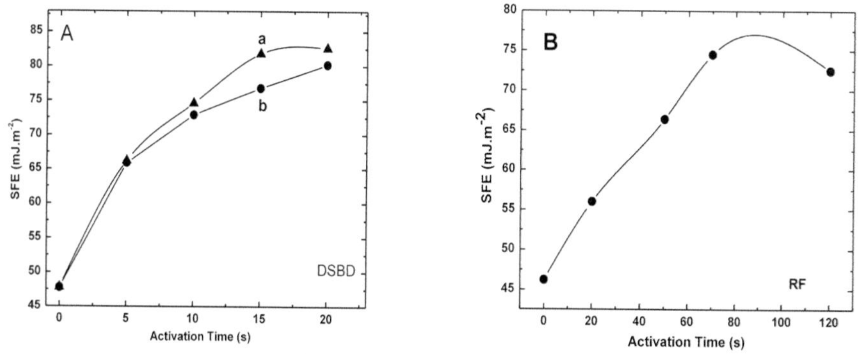

FIGURE 3.11 Surface energy of PET foil modified by DCSBD plasma (A), in O_2 (a) and in N_2 (b) or RF plasma (B) in air versus activation time.

The surface energy of PET during modification by DCSBD plasma in O_2 and N_2 (Fig. 3.11A) significantly increased in comparison with the untreated polymer. Figure 3.11A, plot a, shows that the surface energy of PET modified by DCSBD plasma in O_2 increases from the initial value of 47.8 mJ.m^{-2} for untreated PET to 82.6 J.m^{-2} for 20 s of plasma modification. Figure 3.11A, plot b, shows the surface energy of PET modified by DCSBD plasma in N_2. These values are lower than those of PET modified by the same method in O_2. The PET modified by the RF plasma in air (Fig. 3.11B) reached the same level of pretreatment for a significantly longer time of plasma activation. In this case the surface energy was 74.2 mJ.m^{-2} after modification for 60 s by plasma.

The surface energy results of PET modified by RF plasma, as well as its polar component (PC) and dispersive component (DC), are summarized in Figure 3.12 as a function of activation time. The total surface energy of PET treated by RF plasma increases with time of activation from 47.8 mJ.m^{-2} (the unmodified polymer) to 72.4 mJ.m^{-2} (modification by RF plasma, 120s).

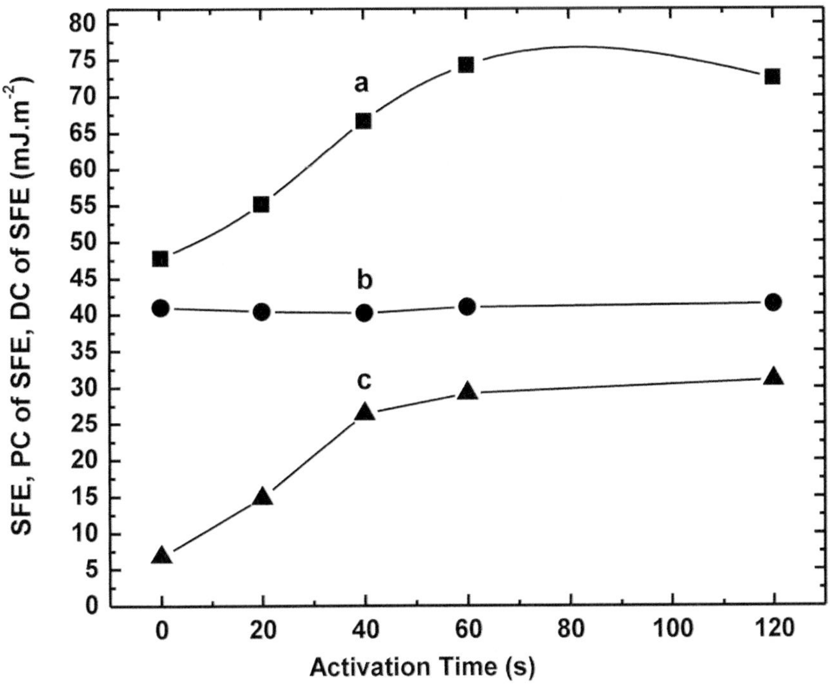

FIGURE 3.12 Surface energy (a), its dispersive (b) and polar (c) component of PET modified by RF plasma in air versus time of activation.

The polar component of the surface energy increased, after modification by RF plasma in argon, from 6.8 mJ.m^{-2} (the unmodified sample) to 31 mJ.m^{-2} after 120 s of treatment time. However, the dispersive component did not change significantly (from 41.0 mJ.m^{-2}, unmodified PET to 41.4 mJ.m^{-2}, RF, 120 s).

3.9 AGING OF MODIFIED POLYESTER

The process of the hydrophobic recovery takes place on the surface of polymers, modified by low-temperature plasma, during aging and it leads to a decrease in the surface energy and its polar component.[34] The hydrophobic recovery can be explained as the result of the:

(i) inter-reactions of chemical groups on the plasma-treated surface of a polymer having many functionalities available for specific interactions;
(ii) further oxidation and degradation with exposure to air, when unstable hydroperoxides and peroxides decompose forming more stable ketone and aldehyde groups;
(iii) migration of a low-molecular weight oxidized material, for example, polar fragments, into a polymeric film to reach a thermodynamically more stable state with a lower value of the surface free energy;
(iv) low-molecular weight compounds (processing additives, e.g., organic slip, antistatic, and antiblocking agents) in a polymer tend to exude to the surface of the polymer, due to its incompatibility with the polymer, during the hydrophobic recovery creating a low-molecular weight layer having a lower value of the surface energy on the surface. In the case of plasma-treated PET the aging effect is related to the movement of C=O groups in ester residues toward the topmost layer, and to the movement of C–O groups away from the topmost layer. Such movement of the C=O groups could occur within at least 3 nm from the surface.[34]

The influence of aging time on the surface energy and its polar component of PET foil, modified by DCSBD plasma in oxygen for 40 s, is shown in Figure 3.13. The surface energy of PET modified by RF plasma in oxygen decreased after 30 days of aging from 66.2 to 60.7 mJ.m^{-2}, and its polar component decreased from 26.4 to 20.9 mJ.m^{-2}. This process is called hydrophobic recovery, and it leads to a reduced value of the surface energy of the modified polymer.

If we compare hydrophobic recovery of PET modified by RF plasma with hydrophobic recovery of polyolefin, we can confirm higher stability of PET modification by DCSBD plasma, after 30 days of aging the surface energy was diminished by 10%, and its polar component was smaller by 20% in comparison with the freshly modified polymer value.

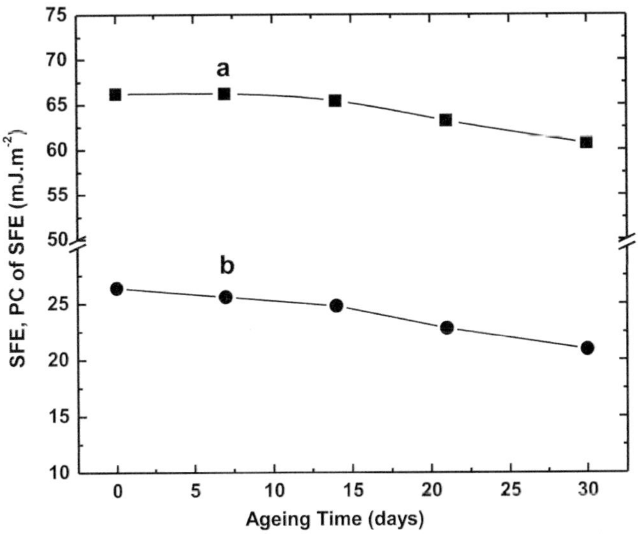

FIGURE 3.13 Surface energy (a), and its polar component (b) of PET modified by RF plasma in O_2, 40 s vs. aging time.

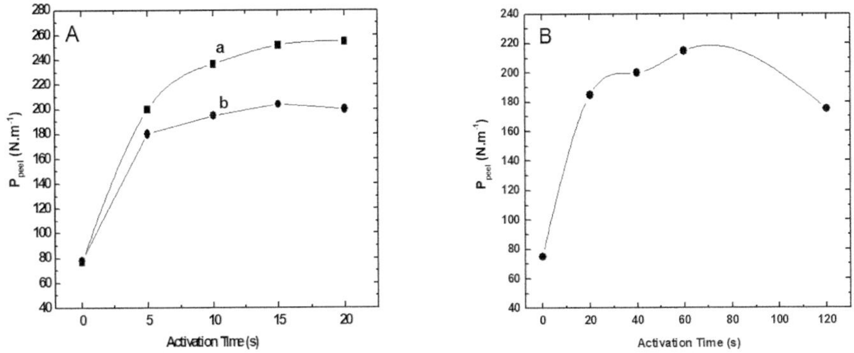

FIGURE 3.14 Peel strength of adhesive joint of PET modified by DCSBD plasma (A), a—O_2, b—N_2, and RF plasma in air (B) to polyacrylate vs. time of activation.

The peel strength of the adhesive joints of PET foils, modified by DCSBD plasma (Fig. 3.14A) in oxygen (14A, plot a), in nitrogen (Fig. 3.14A, plot b), and by RF plasma in air (Fig. 3.14B) to polyacrylate vs. time of activation, is shown in Figure 14A and B. The peel strengths of PET to polyacrylate after modification by DCSBD plasma significantly increased, and these increases were higher for the samples modified in oxygen. The peel strength of the adhesive joint increased from 77 $N.m^{-1}$ (unmodified PET) to 180 $N.m^{-1}$ (DCSBD, 10 s, N_2), and 237 $N.m^{-1}$ (DCSBD, 10 s, O_2). According to Figure 14B the peel strength of the adhesive joint of PET modified by RF plasma in air for 40 s increased from 75 to 200 $N.m^{-1}$, and after 60 s it reached a maximum value of 215 $N.m^{-1}$, after which it decreased to 175 $N.m^{-1}$ after 120 s. These results clearly indicate that plasma irradiation considerably improves the adhesive properties of PET for both kinds of plasmas.

Figures 3.15 and 3.16 show the relation between the peel strength of the adhesive joint of PET modified by DCSBD plasma to polyacrylate and the surface energy of polymer.

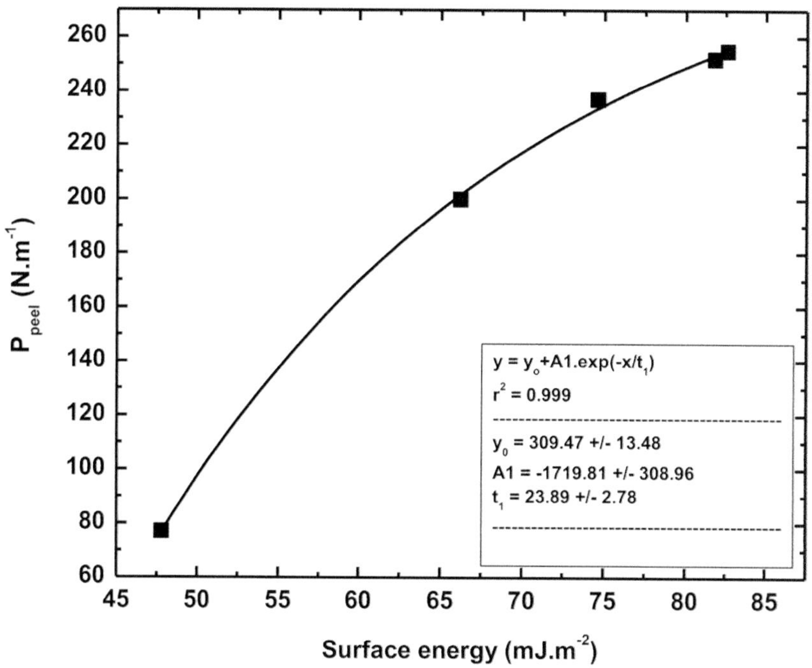

FIGURE 3.15 Relation between the peel strength of adhesive joint of PET modified by DCSBD plasma in O_2 to polyacrylate vs. surface energy of the polymer.

A first-order exponential decay was found for modification in oxygen (Fig. 3.15) and in nitrogen (Fig. 3.16). For both the processing gases the relation between peel strength of the adhesive joint and surface energy is given by the following relation: P_{peel} = 309.5—1719.8 exp(- SFE/23.9), where SFE represents the surface free energy.

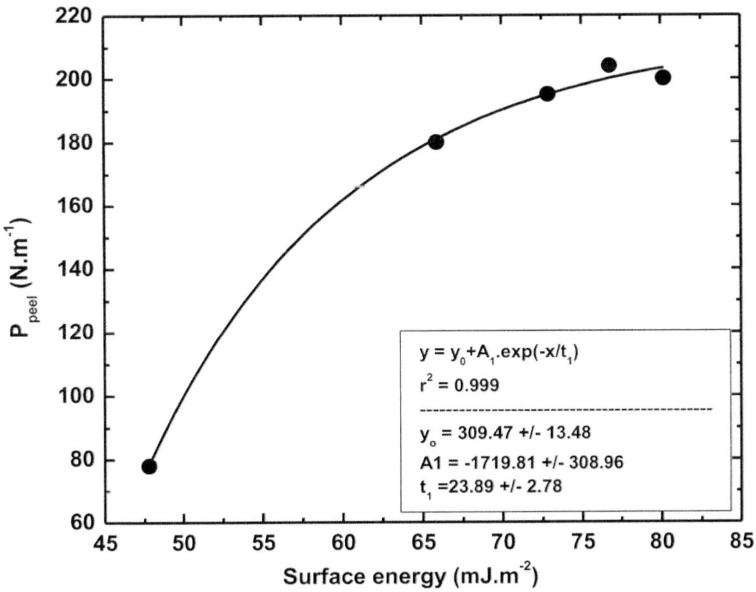

FIGURE 3.16 Relation between the peel strength of adhesive joint of PET modified by DCSBD plasma in N_2 to polyacrylate versus surface free energy of the polymer.

3.10 CONCLUSIONS

The roughness of the PET film modified by DCSBD plasma in N_2 and O_2, and by RF plasma in air increased. AFM results confirmed the creation of some heterogeneities due to local melting on the surface of PET. Free radicals created during modification by DCSBD plasma were identified by ESR measurements, which showed two main singlet signals, which resulted from radical breakdown of the polymer. Aging of PET modified by DCSBD plasma in N_2 and O_2 led to 5 times decrease in the free radicals concentration. DCSBD plasma modified PET showed broadening of the C=O stretch at 1710 cm^{-1} due to the creation of oxygen-containing sites. XPS showed an increase in oxygen and nitrogen content in the PET surface layers after modification by DCSBD plasma in N_2 or O_2, and by RF plasma in air. The surface energy of PET modified by DCSBD and RF plasma significantly increased, and this increase was higher for oxygen compared to nitrogen. The peel strength of PET modified by DCSBD plasma or RF plasma to polyacrylate significantly increased, and this value was higher in oxygen rather than in the case of nitrogen. A correlation between the peel strength of the adhesive joint of PET, modified by DCSBD plasma in oxygen or nitrogen to polyacrylate, and surface energy of polymer by the following relation: $P_{peel} = 309.5—1719.8\exp(\text{-SFE}/23.9)$ has been found.

ACKNOWLEDGMENTS

The authors are grateful for financial support to the Slovak Research and Development Agency projects No. APVV-18-0378, APVV-19-0269, APVV-20-159, and VEGA 2/0019/19.

KEYWORDS

- **polymer aging**
- **cold plasma**
- **peel strength of adhesive joint**
- **surface free energy**

REFERENCES

1. Hippler, R.; Kersten, H.; Schmidt, M.; Schoenbach, K. H. *Low-Temperature Plasmas: Fundamentals, Technologies, and Techniques*; Wiley: Weinheim, Germany, 2008.
2. Ostrikov, K.; Cvelbar, U.; Murphy, A. B. Plasma Nanoscience: Setting Directions, Tackling Grand Challenges. *J. Phys. D: Appl. Phys.* **2011**, *44*, 174001.
3. Ostrikov, K. Colloquium: Reactive Plasmas as a Versatile Nanofabrication Tool. *Rev. Modern Phys.* **2005**, *77* (4), 489–511.
4. Černáková, L.; Kováčik, D.; Záhoranová, A.; Černák, M.; Mazúr, M. Surface Modification of Polypropylene Non-Woven Fabrics by Atmospheric-Pressure Plasma Activation Followed by ACRYLIC ACID Grafting. *Plasma Chem. Plasma Process.* **2005**, *25*, 427–437.
5. Odrášková, M.; Ráheľ, J.; Záhoranová, A.; Tiňo, R.; Černák, M. Plasma Activation of Wood Surface by Diffuse Coplanar Surface Barrier Discharge Plasma Chemistry. *Plasma Chem. Plasma Process.* **2008**, *28*, 203–211.
6. Novák, I.; Pollák, V.; Chodák, I. Study of Surface Properties of Polyolefins Modified by Corona Discharge Plasma. *Plasma Processes. Polym.* **2006**, *3*, 355–364.
7. O'Hare, L. A.; Smith, J. A.; Leadley, S. R.; Parbhoo, B.; Goodwin, A. J.; Watts, J. F. Surface Physico-Chemistry of Corona-Discharge-Treated Poly(Ethylene Terephthalate) Film. *Surf. Interface Anal.* **2002**, *33*, 617–625.
8. Deshmukh, R. J.; Bhat, N. V. The Mechanism of Adhesion and Printability of Plasma Processed PET Films. *Mater. Res. Innov.* **2003**, *7*, 283–290.

9. Riccardi, C.; Barni, R.; Selli, E.; Mazzone, G.; Massafra, M. R.; Marcandalli B.; Poletti G. Surface Modification of Poly(Ethylene Terephthalate) Fibers Induced by Radio Frequency Air Plasma Treatment. *Appl. Surf. Sci.* **2003**, *211*, 386–397.
10. Barni, R.; Riccardi, C.; Selli, E.; Massafra, M.R.; Marcandalli, B.; Orsini, F.; Poletti, G.; Meda, L. Wettability and Dyeability Modulation of Poly(Ethylene Terephthalate) Fibers Through Cold SF_6 Plasma Treatment. *Plasma Process. Polym.* **2005**, *2*, 64–72.
11. Ademovic, Z.; Wei, J.; Jensen, B.W.; Hou, X.; Kingshott, P. Surface Modification of PET Films Using Pulsed AC Plasma Polymerisation Aimed at Preventing Protein Adsorption Plasma Process. *Plasma Process. Polym.* **2005**, *2*, 53–63.
12. Noeske, M.; Degenhardt, J.; Strudthoff, S.; Lommatzsch, U. Plasma jet treatment of Five Polymers at Atmospheric Pressure: Surface Modifications and the Relevance for Adhesion. *Int. J. Adhes. Adhes.* **2004**, *24*, 171–177.
13. Ráheľ, J.; Černák, M.; Hudec, I.; Štefečka, M.; Kando, M.; Chodák, I. Surface Modification of Polyester Monofilaments by Atmospheric-Pressure Nitrogen Plasma. *Plasmas Polym.* **2000**, *5*, 119–127.
14. Novák, I.; Chodák, I. Adhesion of Poly(Propylene) Modified by Corona Discharge. *Angew. Makromol. Chem.* **1998**, *260*, 47–51.
15. Šimor, M.; Ráhel, J.; Černák, M.; Imahori, Y.; Štefečka, M.; Kando, M. Atmospheric-Pressure Plasma Treatment of Polyester Nonwoven Fabrics for Electroless Metal Plating. *Surf. Coat. Technol.* **2003**, *172*, 1–6.
16. Janča, J.; Sťahel, P.; Krčma, F.; Lapčík, L. Plasma Surface Treatment of Textile Fibres for Improvement of Car Tires. *Czech. J. Phys.* **2000**, *50*, 449–452.
17. Tsuchyia, Y.; Akutu, K.; Iwata, A. Surface Modification of Polymeric Materials by Atmospheric Plasma Treatment. *Prog. Polym. Sci.* **1998**, *34*, 100–107.
18. Liu, C.; Cui, N.; Brown, N. M. D.; Meenan, B. J. Effects of DBD Plasma Operating Parameters on the Polymer Surface Modification. *Surf. Coat. Technol.* **2004**, *185*, 311–320.
19. Bhat, N. V., Upadhyay, D. J. Adhesion Enhancement and Characterization of Plasma Polymerized 1,2-Dichloroethane on Polypropylene Surface. *Plasma Chem. Plasma Process.* **2003**, *23*, 389–411.
20. Manenq, F.; Carlotti, S.; Mas, A. Some Plasma Treatment of PET Fibres and Adhesion Testing to Rubber. *Angew. Makromol. Chem.* **1999**, *271*, 11–17.
21. Cheng, T. S., Lin, H. T., Chuang, M. J. Surface Fluorination of Polyethylene Terephthalate Films with RF Plasma. *Mater. Lett.* **2004**, *58*, 650–653.
22. Toufik, M.; Mas, A.; Shkinev, V.; Nechev, A.; Elharfi, A.; Schue, F. Improvement of Performances of PET Track Membranes by Plasma Treatment. *Eur. Polym. J.* **2002**, *38*, 203–209.
23. Wang, J.; Pan, C. J.; Kwok, S. C. H.; Yang, P.; Chen, J. Y.; Wan, G. J.; Huang, N.; Chu, P. K. Characteristics and Anticoagulation Behavior of Polyethylene Terephthalate Modified by C_2H_2 Plasma Immersion Ion Implantation-Deposition. *J. Vac. Sci. Technol. A.* **2004**, *22*, 170–175.
24. Laurens, P.; Petit, S.; Arefi-Khonsari, R. Study of PET Surfaces After Laser or Plasma Treatment: Surface Modifications and Adhesion Properties Towards Al Deposition Plasma. *Plasmas Polym.* **2003**, *8*, 281–295.
25. Alamazan-Almazan, M. C.; Paredes, J. I.; Perez-Mendoza, M.; Domingo-Garcia, M.; Lopez-Garzon, F. J.; Martinez-Alonso, A.; Tascon, J. M. D. Surface Characterisation

of Plasma-Modified Poly(Ethylene Terephthalate) *J. Colloid Interface Sci.* **2006**, *293*, 353–363.

26. Almazán-Almazán, M. C.; Paredes, J. I.; Pérez-Mendoza, M.; Domingo-García, M.; Lopez-Garzon F.J.; Martinez-Alonso A.; Tascon J.M.D. Effects of Oxygen and Carbon Dioxide Plasmas on the Surface of Poly(Ethylene Terephthalate). *J. Colloid Interface Sci.* **2005**, *287*, 57–66.
27. Stern, T.; Cohn, D. Surface Tailoring of Biomedical Polymers: An FTIR Study. *J. Appl. Polym. Sci.* **2001**, *81*, 2203–2209.
28. Kumagai, H.; Hiroki, D.; Fujii, N.; Kobayashi, T. Poly (Ethylene Terephthalate) Decomposition Process in Oxygen Plasma; Emission Spectroscopic and Surface Analysis for Oxygen–Plasma Reaction. *J. Vac. Sci. Technol.* A. **2004**, *22*, 1–8.
29. Sakugo, N.; Mizutani, D.; Ohmura, Z.; Endo, H.; Yoneda, R.; Ikenaga, N.; Takikawa, H. Surface Modification of PET Film by Plasma-Based Ion Implantation. *Nucl. Instrum. Methods Phys. Res. Sect. B.* **2003**, *206*, 687–690.
30. Paynter, R. W. XPS Studies of the Ageing of Plasma-Treated Polymer Surfaces. *Surf. Interface Anal.* **2000**, *29*, 56–64.
31. Papakonstantinou, D.; Amanatides, E.; Mataras, D.; Ioannidis, V.; Nikopoulos, P. Improved Surface Energy Analysis for Plasma Treated PET Films. *Plasma Processes Polym.* **2007**, *4*, S1057–S1062.
32. Vesel, A.; Mozetic, M.; Zalar, A. XPS Study of Oxygen Plasma Activated PET. *Vacuum.* **2008**, *82*, 248–251.
33. Vesel, A.; Mozetic, M. Modification of PET Surface by Nitrogen Plasma Treatment. *J. Phys. Conf. Ser.* 2008, *100*, 1–4.
34. Inagaki, N.; Narushim, K.; Tuchida, N.; Mizayaki, K. Surface Characterization of Plasma-Modified Poly(Ethylene Terephthalate) Film Surfaces. *J. Polym. Sci., Part B: Polym. Phys.* **2004**, *42*, 3727–3740.
35. Carlotti, S.; Mas, A. Improvement of Adhesion of PET Fibers to Rubber by Argon-Oxygen Plasma Treatment. *J. Appl. Polym. Sci.* **1998**, *69*, 2321–2330.
36. Shin, Y.; Yoo, D. I. Surface Characterization of PET Nonwoven Fabric Treated by He/O2 Atmospheric Pressure Plasma. *J. Appl. Polym. Sci.* **2008**, *108*, 785–790.
37. Novák I.; Sysel, P.; Zemek, J.; Špírková, M.; Velič, D.; Aranyosiová, M.; Florián, Š.; Pollák, V.; Kleinová, A.; Lednický, F.; Janigová, I. Surface and Adhesion Properties of Poly(Imide-Siloxane) Copolymers. *Eur. Polym. J.* **2009**, *45*, 57–69.
38. Novák, I.; Elyashevich, G. K.; Chodák, I.; Olifirenko, A. S.; Števiar, M.; Špírková, M.; Saprykina, N.; Vlasova, E.; Kleinová, A. Polymer Matrix of Polyethylene Porous Films Functionalized by Electrical Discharge Plasma. *Eur. Polym. J.* **2008**, *44*, 2702–2707.
39. Novák, I.; Števiar, M.; Chodák, I.; Krupa, I.; Nedelčev, T.; Špírková, M.; Chehimi, M. M.; Mosnáček, J.; Kleinová, A. Study of Adhesion and Surface Properties of Low-Density Poly(Ethylene) Pre-Treated by Cold Discharge Plasma. *Polym. Adv. Technol.* **2007**, *18*, 97–105.

PART II
Applied Polymer Materials

CHAPTER 4

Effect of Bulk and Surface Properties of Polymeric Scaffolds on Its Applications in Tissue Engineering

DEEPAK PODDAR

Netaji Subhas University of Technology, Dwarka, New Delhi, India

ABSTRACT

In tissue engineering, a lack of attention on the interfacial milieu of the scaffolds frequently inhibits cellular interactions, resulting in poor cell attachment, survival, and intervention failure. Surface and bulk properties such as desired porosity and pore size for encouraging vascular invasion, suitable surface chemistry for mineral deposition along with augmented ability to support cell attachment, appropriate surface morphology and chemistry provide an active site and hydrophilicity for active cell attachment may necessitate amalgamation of fabricated scaffolds. These surface features promote cell signaling, proliferation, and differentiation, resulting in a favorable biological response. This chapter discusses the requisites of materials to be scaffolds and briefly elaborates the role of surface and bulk properties on the attachment of cell for proliferation and growth for the scaffolds in tissue engineering application.

4.1 INTRODUCTION

4.1.1 TISSUE ENGINEERING

There are several common clinical conditions due to which bone loses its functions such as bone disease, inflammation, trauma, and tumor that are

Tailored Functional Materials for Clean and Sustainable Development. Tamara Tatrishvili, PhD, Neha Kanwar Rawat, PhD, Ann Rose Abraham, PhD & A. K. Haghi, PhD (Eds.)

major concerns in the human health sector.[1] Globally, each year 1.5 million cases of bone injuries are registered which needs instant and long-lasting solution.[2] In the current clinical approaches, the bone diseases are commonly treated with bone grafting-techniques few of which are autograft, allograft, and xenograft.[3,4] A bone tissue engineering strategy for the development of scaffolds has been shown in Figure 1.1. These techniques possess higher failure rate due to immune rejection, donor crisis, and disease transfer.[5,6] Furthermore, bone grafting results in multiple surgeries, pain, and increased cost for patients with the risk of disease transmission and immune rejection which makes the process more difficult to execute.[7]

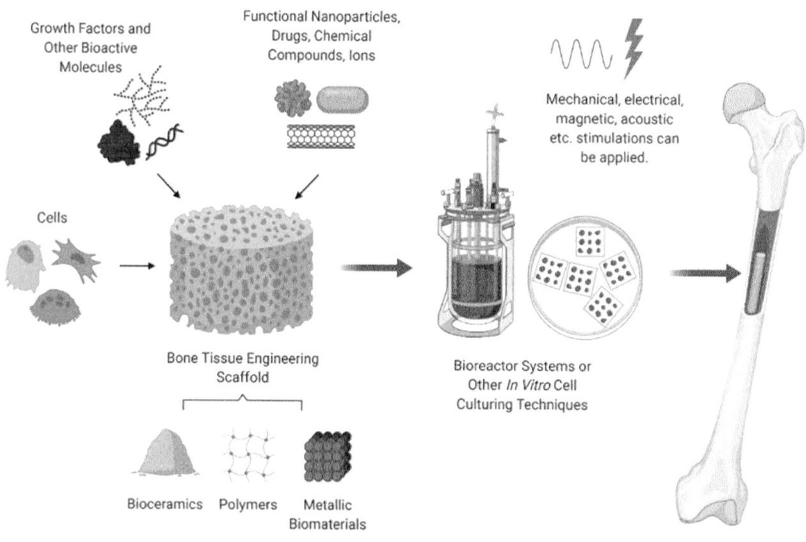

FIGURE 4.1 Bone tissue engineering strategy.

Source: Reprinted from Ref [96]. https://creativecommons.org/licenses/by/4.0/

This has encouraged an increased research interest in the scientific community for the alternative solutions to replace the conventional approaches such as transplantation or artificial prosthesis. Recent studies show the promising use of bone tissue-engineering as a potential alternative for the grafting technique.[8] The main principle of tissue engineering is to repair, replace the injured tissue, or to regenerate the desired tissue for the treatment of needed bone defect. Tissue engineering addresses the expansion and growth of cells (living) over the three-dimensional substrate having an interconnected porous nature in the presence of bioactive factors for the regeneration of the tissues.[9] This method provides the flexibility of direct implantation of the scaffolds at the damaged site, or the organ can be regenerated using an in vitro setup and later it can be transplanted. This provides an immense advantage over the other techniques and also overcomes the problem of organ donor for transplantation therapy.[10] The main goal of the technology is to solve the existing problem and provide the appropriate methods for the current medical treatment to overcome the limitations such as difficulty in regeneration of extensively damaged tissues, shortage of donor, immune rejection, and cost and pain associated with surgery.[11]

4.1.2 THREE-DIMENSIONAL SCAFFOLDS

Scaffolds are three-dimensional templates that mimic the ECM for cell attachment and growth by providing the mechanical support to the damaged site.[12] Scaffolds is considered a synthetic extracellular matrix that plays a vital role in tissue regeneration, by providing the appropriate microenvironment of the natural tissue to the cell to undergo attachment, proliferation, migration, and differentiation in the three-dimensional matrix.[13] The scaffolds serve the temporary platform and function as an artificial ECM for the tissues, and with time, they slowly degrade, and new regenerated tissues take their place. Therefore, the scaffolds should possess the characteristic features of the ECM to provide as appropriate biological and physiochemical environment for the cells. The characteristic properties for the scaffolds can be defined as follows.[14–17]

- **Architecture**
  a. To obtain high cell seeding efficiency, the scaffolds should have an open porous design.
  b. The scaffolds should have a high porosity, optimum pore size, and adequate pore interconnectivity to allow for efficient cell migration and cell–cell communication. Scaffolds with high porosity and excellent pore interconnectivity can provide an even cell distribution in the cell/scaffold construct. This will make the development of a homogenous tissue considerably simpler.
- **Biocompatibility**
  a. The scaffolds should be biocompatible, implying that after implantation, materials should not trigger a significant immune

reaction against the biological system. Prior to designing the scaffolds, it is necessary to assess the material's biocompatibility. There should be no chemical or biological impurities in the material that might trigger an immune reaction and result in the rejection of the product. The degrading property of the material should be carefully controlled for an optimal scaffold fabrication, for example, the rate of scaffold disintegration should be approximately equal to the pace of tissue regeneration. This is required to regenerate tissues that are suited to the exact damaged dimensions.

b. The adherence of cells is the primary step for cell growth over the layers. The scaffolds should therefore support adherence to the cell.

- **Mechanical strength**

  a. The outstanding mechanical strength. The scaffolds' physical characteristics, such as compressive Young's modulus and tensile strength, should be high enough to prevent the scaffolds from collapsing during the cell growth and implantation procedure. The mechanical strength should be sufficient to withstand the cell-mediated contractile forces' compression.

- **Degradability**

  a. Biodegradable scaffolds should be used. The breakdown of materials during tissue regeneration is referred to as biodegradation. Degradation might be hydrolytic, enzymatic, or a combination of both. The scaffolds should degrade into harmless products or compounds that can be conveniently guided to the body's normal metabolic pathways, such as glycolysis, TCA cycle, and so on. This step is necessary for the degradation products to be removed successfully. This can reduce the adverse effect that these breakdown products from scaffolds are anticipated to have.

- **Biological efficiency**

  a. Bioactive components are essential for cell growth, differentiation, and migration by providing appropriate biochemical signals. External supply of these bioactive substances in cell culture medium may not be adequate for delicate development and differentiation of cells inside scaffolds due to limited in vivo stability and uncertain biological effect. As a result, it is preferable to incorporate these essential cues into the scaffolds throughout the construction process, either by direct mixing with scaffolding materials or through the introduction of bioactive factors into the body via controlled drug delivery systems. This can facilitate a spatio-temporal release of these cues from the scaffolds. This may make it easier for these signals to be released from the scaffolds in a spatiotemporal manner. This type of integration can result in high-bioactivity scaffolds. The development of new biomaterial scaffolds for tissue engineering and regenerative medicine is driven by the design and production of bioactive scaffolds.

4.2 MATERIAL FOR SCAFFOLDS PREPARATIONS

The material section for the fabrication of the scaffolds plays a very vital role in the performance of the scaffolds. In general, material should be free from any harmful chemical or biological moieties, which may induce inflammation after implantation.[18] It should be nontoxic and bioresorbable in nature, should promote cell adhesion, should able to attain the cell–interface interaction, and most importantly material should have sufficient mechanical strength to provide the support to the host tissues.[19] There are two sets of materials, natural and synthetic, which are most often used for the fabrication of the scaffolds. The scaffolds made by the combination of the above two sets of material are known as hybrid scaffolds.[20] Because of their controlled breakdown, simplicity of production, and high bioactive characteristics, polymers have been widely explored for scaffold construction. Polymeric materials are produced in vast numbers in the chemical industry, either from biologic sources (natural polymers) or by known synthetic chemical processes (synthetic polymers). The characteristics of the properties such as tuneable surface chemistry, degradation are very useful in the field of tissue engineering.

Polymers are extensively studied for fabrication of scaffolds due to their controlled degradation, ease of preparation, and their high bioactive properties. Polymeric materials occur in abundant quantities either from biologic origin (natural polymers) or from synthesized (synthetic polymers) on large scale in chemical industries through established synthetic chemical reactions. The properties of the polymers (surface chemistry, degradation, and so on) can be precisely tuned based on their application (type of tissue to be regenerated).

4.2.1 NATURAL POLYMER

As name suggests, these polymers or their monomers are derived from the natural resources such as plants, vegetables, and animals. Naturally occurring polymers like collagen, chitosan, alginate, and cellulose-based materials are mostly preferred for biomaterial devices, due to their hydrophilic nature, biodegradability, biocompatibility, and property to promote wettability to enhance cell adhesion and osteoconductivity. The molecular structure of the polymers contains different functional groups such as -NH_2-, -OH-, COOH-, and many more, which cause friendly environment for the enhanced cell attachment, spreading, and proliferation on scaffolds. Due to the presence of such functional groups, the surface wettability of the scaffolds is also increased, making the process even simpler. Furthermore, the natural polymers showed rapid degradations and the obtained products do not offer any apparent immune reactions. However, in most cases, natural polymers lack in shape retention ability, are low in mechanical strength due to which it limits their solo use in TERM applications.

4.2.2 SYNTHETIC POLYMER

On the other hand, synthetic polymers like PCL, poly (lactic acid), and poly (urethane) are few of FDA-approved polymers to be used in tissue engineering application and are currently used as effective implants and reference[21] also shows potential application in tissue engineering as they have comparable and tuneable mechanical properties as bones, reproducibility, and controlled biodegradability.[22] Therefore, the scaffolds made from the synthetic polymer are well matched for the material used in hard and stiff tissue regeneration such as cartilages and bones.[23] However, due to low osteoconductivity that leads to a dearth in wettability, poor cell attachment, proliferation, and differentiation cause a constraint on the use of these polymers in tissue engineering.[24] Another huge constraint of these materials is that as they degrade, they lead to the production of toxic and harmful content for the cells, for example, PLGA releases glycolic acid and lactic acid, leads to tricarboxylic acid cycle, and thus causes disbalance in metabolism of cell.[25] To overcome such problems, combination of both synthetic and natural polymers by means of bioactive coating agent on the inner layer of the scaffold may give rise to a more suitable scaffold, which can provide biocompatibility and strength at the same time.

4.2.3 BIO-CERAMIC MATERIALS

Bio-ceramics are inorganic materials composed of the naturally occurring calcium and phosphorus molecules. These materials are very common and utterly used in tissue regeneration as they resemble the chemical and physical structure of the bone.[26, 27] The properties such as hardness, wear resistance, etc. make these materials a unique choice for the hard tissue engineering. Few examples of the bio-ceramics are calcium and phosphate molecules such as calcium HA, BCP, TCP, and wollastonite.[28–30] These materials' osteo-inductive (ability to induce new bone formation via bone cell growth) and osteoconductive (ability to enhance bone cell adhesion and differentiation) characteristics have also been demonstrated in bone tissue engineering.[31–33] These inorganic minerals possess the high quality and can be used alone or in combination with suitable polymeric materials to form hybrid materials.[34,35] In the field of dental the usability of ceramic is pretty common,[36] as maxillofacial and orthopaedic surgery are used to coat and fill the surgical surfaces of implant of host bone.[37] However, this material possesses high brittleness that makes the remolding process quite difficult because it limits its clinical applications special in BTEA. Also, bio-ceramics possess high mechanical strength and support cell growth.[38] Therefore, the combination of polymers (biopolymers) and ceramics may provide the appropriate performance for the enhanced cell adhesion and supportive properties with higher mechanical strength for BTEA.[39]

4.3 INFLUENCE OF SURFACE AND BULK PROPERTIES ON SCAFFOLDS

The adherence of cells and their development on scaffolds are highly dependent on the properties of the scaffolds. There are two types of qualities that must be considered while fabricating optimal scaffolds: bulk and surface properties. Both sets of features play a role in distinct areas of the scaffold, and an ideal scaffold should have a good balance of both sets of properties.

4.3.1 BULK PROPERTIES

Bulk properties are focused on the physical features of the scaffolds, such as mechanical strength, porosity, pore interconnectivity, biocompatibility, biodegradability, and the chemical structure of the molecules employed in scaffold fabrication. Additional, bulk properties affect the other key parameters' strength, offer a healthy environment for cell growth, and also engage in scaffold cell attachment behavior, although the interface of the scaffolds is thought to be the decisive element for cell contact and growth.

4.3.1.1 SCAFFOLD COMPOSITION

When the cell is seeded into the scaffolds, it largely interacts with the surrounding environment to adhere and proliferate. The scaffolds' material composition does have a substantial impact on the biological response of the cells, which is directly impacted by the scaffolds' physicochemical qualities such as strength, hydrophilicity, surface charge, and degradation rate.[40,41] In the recent past, several materials have been employed in the fabrication of scaffolds, mainly falling into two categories according to origin such as natural or synthetic polymers. Both kinds of materials have several advantages over one another, but each has its own set of limitations.[42] Natural polymers possess adequate biocompatibility, cell recognition, and cell attachment but are constrained in mechanical strength, rapid breakdown, difficult to handling, and tedious purification processes.[43,44] Synthetic materials, on the other hand, exhibit completely different behavior compared to natural polymers though they have adequate mechanical strength. However, it possesses low osteoconductivity that causes a lack of wettability in scaffolds, as well as poor cell adhesion, proliferation, and differentiation, limiting the usage of synthetic polymers.[45] Now, researchers adopted an approach to chemically modify the synthetic polymers with natural parts to enhance cell recognition and attachment behavior, as well as control suitable mechanical strength and controlled breakdown of the scaffold.[46] According to Vyas et al., the inclusion of biomolecules or bioactive substances in the scaffolds has a significant impact on cell adhesion and attachment to the scaffolds. The study showed the PCL biocompatibility was improved by increasing the SMP content in the PCL/SMP scaffolds to 20%, resulting in increased biocompatibility, cell adhesion, and calcium mineralization without reducing or changing the physical properties of the scaffolds.[47]

4.3.1.2 SCAFFOLD ARCHITECTURE

The architecture of scaffolds is concerned with pore size, porosity, shape, and pore interconnectivity. The optimal scaffold, according to researchers, should have a porous structure with extensively linked pores for homogeneous cell dispersion, cellular development, nutrient flow, and transportation.[48] The biocompatible, interconnected, extremely porous network serves as a template for tissues to initiate, develop, and restructure themselves into.[49] A suitable pore size is also highly important in cell penetration, as tiny pore sizes might impede cellular migration, and if the pore size is too large, cell adhesion suffers greatly owing to a loss in contact surface area.[50] Cell uses a bridge mechanism to migrate via the pores network, which is supported by nearby cells; however due to large pores, this process can be hampered.[51] Several scholars have reported on the influence of porosity and pore size on cell proliferation and tissue ingrowth, as indicated in Table 4.1. Pore interconnectivity is regarded as a vital feature for ensuring the mass transfer of oxygen and nutrients supplied by the cell (200 m in size) from the blood supply.[52–55] Sanchez et al. demonstrated the effect of porosity and pore size on the mechanical and biological behavior of scaffolds. The porous titanium scaffolds were fabricated using a powder metallurgy technique in two distinct porosities (55% and 70%) with varying pore sizes, and the influence on elastic modulus and cell attachment behavior of the scaffolds was investigated. Because of the scaffolds' compact and dense network, the scaffolds with smaller pore sizes displayed higher mechanical properties. Higher cell attachment and growth were obtained in scaffolds with pore sizes of 300 μm, since cells were unable to penetrate the surface with smaller size, and cells with larger pore sizes were spread across the scaffolds, limiting cell development.[56] Samourides et al. reported the fabrication of varied pore size PGSU scaffolds using the freeze drying process. The concentration of the polymer in the solution mixture was used to regulate the pore size and porosity. The scaffolds were made in three distinct pore diameters and porosity levels. The results showed that increasing the porosity nature of scaffolds reduced mechanical strength but had no effect on the scaffolds' structure and morphology. All the scaffolds showed the cell affinity, but the scaffolds with largest pore size and porosity {(28.2 ± 5.3 μm), (96.4 ± 0.3%)} stimulate the large amount of collagen production. All of the scaffolds demonstrated cell affinity, but the scaffolds with the biggest pore size and porosity (28.2 ± 5.3 μm), (96.4 ± 0.3%) stimulated far more collagen formation. However, in smaller pore size, cells were unable to penetrate to a significant depth in the scaffolds, resulting in cell attachment mainly on the surface of the sample, reducing cell growth and proliferation.[57] Another research explores the influence of pore pore size distribution on cell growth ab metal scaffold cell growth. Chen et al. used an alloy to construct scaffolds with varying pore size and porosity. The scaffolds with the smallest pore size and porosity (500 μm, 60%) were deemed to be the best among the study group because they had improved cell adhesion, proliferation, and osteogenic differentiation. This might be due to the fact that the author used a material with larger pores, and pores larger than 300 μm produced difficulties in cell migration and transportation.[58]

TABLE 4.1 Favorable Pore Sizes for Different Types of Cells for Tissue Regeneration.[59]

Pore size	Cell/tissue	Reference
5 μm	Neovascularization	[60]
5–15 μm	Fibroblast ingrowth	[61]
20 μm	Ingrowth of hepatocyte	[62]
20–125 μm	Regeneration of adult mammalian skin	[63]
200–350 μm	Osteoconduction	[64]

4.3.1.3 *MECHANICAL STRENGTH*

The mechanical characteristics of the scaffolds ought to be appropriate for effective implantation. To facilitate fast tissue regeneration, the scaffolds should have enough biostability (strength, elasticity, and toughness) and a low relative degradation rate in comparison to the host tissue.[65] The stiffness of human tissues ranges from 4.25 kPa for soft tissues (brain) to 2.6 MPa for stiffer tissues (cartilage) to 15–20 GPa for hard tissues (bones).[66,67] The ability of scaffolds to retain mechanical strength after implantation for tissue regeneration is a significant and differentiating characteristic. A previous study has also linked the mechanical strength of the scaffolds to cell attachment and proliferation behavior, as stem cells display the greatest neurogenesis on softer scaffolds (1 kPa) and the greatest osteogenesis on stiffer scaffolds (24 kPa).[68] Mechanical parameters such as elastic modulus are important in stem cell development on the scaffold's surface. MSC differentiate into osteoblasts on harder surfaces, nerve cells on softer surfaces, and muscle cells on medium hard surfaces.[69] Every cell has distinct lineages that survive in different environments, and so has diverse surface compatibility of varying stiffness.

Ma et al. proposed a study in which the stiffness of the scaffolds affected the behavior of neural stem cells. Two types of scaffolds with different elastic moduli of 30 and 64 KPa were built, and cell attachment behavior was found to be related to scaffold stiffness. When compared to the softer scaffolds, the stiffer scaffolds enhanced cell attachment, proliferation and increased integrin and vinculin gene expression by 1.5 and 2.3 times, respectively. Cell adhesion is determined by the adhesive force at the cell–matrix interface, and on stiffer surfaces, focal adhesion spreads more than on softer surfaces, resulting in greater attachment and cell proliferation.[70]

4.3.2 SURFACE PROPERTIES

The surface characteristics of the scaffolds are clearly defined as critical factors, as the surface is the principal region of cell interaction with the scaffolds.[71] The surface environment regulates cell interactivity, adhesion, proliferation, and differentiation following attachment to the surface.[72] As a result, the presence of a favorable biological and chemical environment on the surface of the scaffolds is critical since it may directly alter cell activity. Surface properties deal with surface qualities such as surface charge, energy, tropology, polarity, and chemistry, and by regulating them through various techniques, the cell attachment may be easily tailored to meet the demands, as illustrated in Figure 4.2.[73]

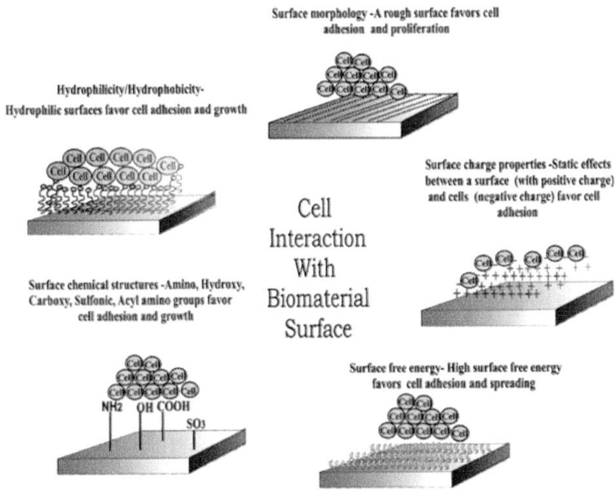

FIGURE 4.2 Symmetric illustration of different types of cell interaction dependent on surface properties of scaffolds.

4.3.2.1 SURFACE CHEMISTRY

The presence of various functional chemical groups on the surface of the scaffolds demonstrates a significant impact on cell adhesion at the scaffold interface. The variation in surface chemistry of the scaffold contact definitely influences endothelial cell adhesion and proliferation.[74] The inclusion of various functional groups such as -NH_2-, -OH-, COOH-, and so on improves wettability and cell adhesion on the surface of the scaffolds, as seen in Figure 4.3.[75] The introduction of natural polymers, ceramics, and other growth factors on the surface of scaffolds can introduce such type of functional groups on the surface, which can modify the alignment of binding protein and further considerably promote the scaffolds cell development. It is widely proven in the literature that natural polymers such as collagen, chitosan, or gelatine have a higher affinity for cell attachment and proliferation than synthetic polymers such as PCL, PLA, or PU. The presence of numerous functional groups such as -NH2-, -OH-, and COOH- aids in cell attachment and adhesion to the substrate.

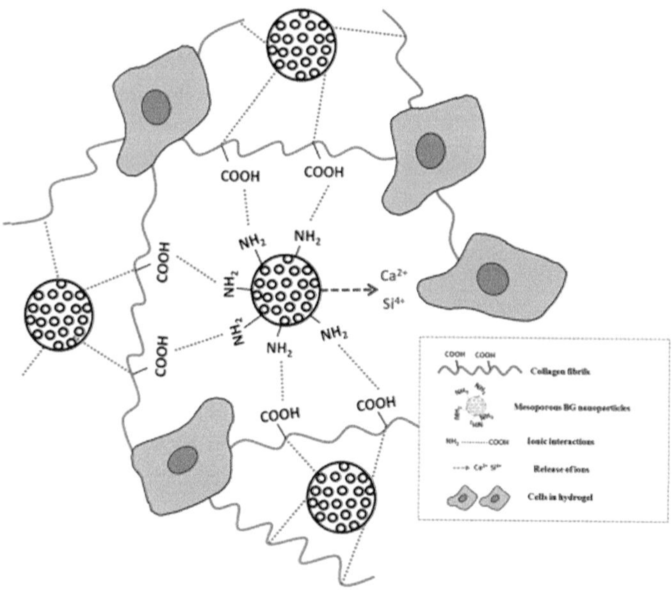

FIGURE 4.3 Schematic illustration, presence of different functional group participates in the interaction with cell, which is effective for bone tissue engineering, as well as providing bone bioactivity, such as bone-cell-favoring ionic releases and apatite-forming ability, which helps in cellular mineralization.

Cui et al. investigated the HA nucleation and growth process on fibrous scaffolds functionalized with various chemical groups. Biomineralization is critical in the BTEA; however, there is a scarcity of data on the influence of HA development on distinct functional groups. PDLLA electrospun fibers were functionalized (carbonyl, hydroxyl, amino group, and their combination) to investigate HA nucleation and growth. The electrospinning processes were used to create the various functionalized fibers mats. According to the findings, fibers with a high carboxyl group content, a combination of hydroxyl and carboxyl group (3/7), and a combination of amino, hydroxyl, and carboxyl group (2/3/5) favored nucleation and HA growth over the interface of scaffolds, resulting in increased concentration and reduced HA crystal size. According to the results of the investigation, the charge density of the engaged chemical groups and the degree of interaction among the polar/ionic groups have a major influence on the mineralization process, as shown in Figure 4.4. The carbonyl group formed an ionic connection with the calcium ions (–$COOCa^{2+}OOC$-), resulting in the development of HA on the surface. A similar interaction was identified in the combination of both hydroxyl and carboxyl groups to calcium ions such as [–$COOCa^{2+}$-O– and –COOCa+ or $(–COO)_2$ Ca], as the combination of hydroxyl and carboxyl groups holds negative charge which bound to the positive charge of the calcium ion to form HA. The electrostatic interaction of the hydroxyl group is weaker than that of the carboxyl group, yet it is adequate for calcium ion adsorption. However, when the amino group was introduced into the picture, the amino group created the inner ionic link with the carboxyl group, further restricting the interaction of HA molecules and reducing mineralization.[76]

Judith et al. reported the influence of distinct functional groups on the interaction of human mesenchymal stem cell differentiation in similar research. In vitro tests were used to evaluate cell adhesion and differentiation after the surface of the glass cover slip was modified with different functional groups such as methyl, amino, silane, hydroxyl, and carboxyl. According to the study's findings, a substrate treated with a methyl group maintained viable cell adhesion under all test circumstances. The amine and silane-coated substrate stimulated and sustained cell osteogenesis, whereas the hydroxyl and carboxyl group-coated substrate supported and maintained MSC cell chondrogenic differentiation. In addition, this hydroxyl-coated substrate had the narrowest water contact angle (63.06° ± 2.10°) of any of the existing groups.[77]

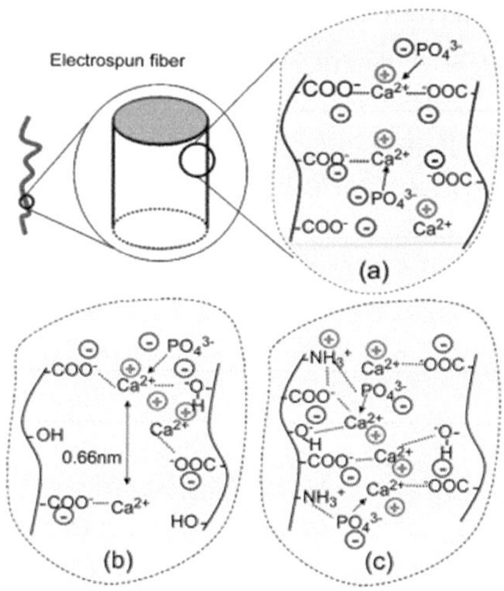

FIGURE 4.4 Schematic illustration of the mechanism of apatite formation on electrospun fibers (a) C-PLA, (b) HC-PLA, and (c) AHC-PLA.

Source: Reprinted with permission from Ref. [46]. Copyright © 2010 Elsevier Ltd.

4.3.2.2 SURFACE TOPOGRAPHY

Surface topography is concerned with the texture and roughness of the scaffolds' surface. It has been discovered that the roughness of the surface influences the type of cell response, such as epithelial and periodontal fibroblast cells being attached to smoother surfaces, whereas osteoblast cells prefer rougher surfaces for adherence, where texturing deals with the surface patterning of the scaffolds. Pattering may be classified into two types: isotropic (no specified directional orientation) and anisotropic (systematic and directional defined orientation of grooves on the surface of scaffolds). As shown in Table 4.2, surface texture and roughness influence cell attachment and influence cell behavior and attachment ability to the surface by considering the fact that cells form focal attachment to rougher surfaces and allow the cell to grow and proliferate in between the surface peak spaces.[78] Furthermore, comparing cell behavior on diverse surface topographies is challenging, owing to a paucity of scientific data and the variable behavior of distinct cells.[79] The microroughness of the surface, on the other hand, encourages osteoblast development for differentiation by promoting mineralization.[80] The roughness of the surface increases the overall exposed area for cell attachment and also improves the spatial organization of the adsorbed mediating sections.[81] Chen et al. investigated the influence of surface nano-roughness on electrospun scaffold cell attachment behavior. The surface roughness was customized by varying the humidity of the electrospinning chamber from 20% to 70%. Changes in humidity cause changes in the rate of evaporation of the solvent, which has a direct impact on the surface shape. Variations in % humidity cause the generation of an increasing number of pores, increasing porosity and pore size while decreasing the fiber water contact angle. Furthermore, the highly porous surface upregulated osteogenic genes such as OPN, BMP2, and RUNX2, whereas the lower porous surface increased the expression of osteogenic genes such as BSP, COL1A1, and OCN.[82] Carnes et al. conducted a similar work in which they scratched the surface of fibrin microthreads to incorporate aligned, submicron grooves for the aligned development of myoblast. The anisotropic groves were formed by immersing micro threads in a solution of 2-(N-morpholino) ethane-sulfonic acid at pH 5.0, causing etching over the surface and increasing total surface area, roughness, and mechanical strength. Additionally, surface roughness increases myotube development, which may be because of myoblast cells preferring aligning on the rough and etched surface. Shuai et al. improved the PLLA scaffold's weak interfacial bonding by modifying the surface roughness of the scaffolds with MgO nanoparticles. The scaffolds that were constructed were examined for their suitability for bone tissue engineering. The incorporation of MgO nanoparticles into the scaffolds via the bulk mixing procedure significantly decreases the water contact angle and improves the scaffold's biocompatibility.

TABLE 4.2 General Trend in the Response of Osteoblasts and Chondrocytes to Surface Roughness.

	MG63	**RC**	**GC**
Cell number	↓ R	↓ R	↓ R
[^{3}H]-Thymidine incorporation	↓ R	↓ R	↑ R
Alkaline phosphate-specific activity	↓ R	↓ R	↑ R
[^{3}H]-Uridine incorporation	↑ R	↑ R	↑ R
Collagenase digestible protein	↓S	↓ S	↓ R
Non-collagenase digestible protein	NC	NC	↓ R
% Collagen production	↓	↓ S	↓R
[^{35}S]-Sulfate incorporation	↓All	↓ All	↓ All

Notes: ↑, increase; ↓, decrease.

Abbreviations: R, roughness; S, smooth; NC, no change; All, charge was surface independent; RC, resting zone chondrocyte; MG63, human osteoblast like cell; GC, growth zone chondrocyte.

4.3.2.3 SURFACE ENERGY

The surface energy of a biomaterial is defined as the amount of overall polarity or charge density on the interface of biomolecules. Surface energy can be positive, neutral, or negative, and it can be obtained by several mechanisms such as ion adsorption, electrostatics, or surface chemistry alteration through other techniques.[83] In general, the greater the number of cells, the more likely they are to connect to a positive charge material as opposed to a neutral or negative charge substance.[84] Furthermore, surface charge has a role in the encapsulation of growth factors in the form of polymers or nano particles, as well as medicines, on the surface of scaffolds.[85] A little modification in the surface charge of the scaffolds may significantly improve cell adhesion, and hence it plays a critical role in cell attachment.[86] The inclusion of nano-particles can have a significant impact on the surface energy of the scaffolds.

The surface charge of 3D-printed scaffolds was fine-tuned in a study by Lin et al. by changing the feed ratio of alginate into the polylysine scaffolds, as illustrated in Figure 4.5. The addition of negatively charged alginate containing carboxyl groups modified the positive charge of polylysine due to the presence of amino groups in its backbone. The combination of these two chemicals causes electrostatic interaction, resulting in better and more stable printing and increased mechanical stability of the scaffolds. The author investigated three types of films, one by direct blending of alginate and polylysine (wt% of alg/pl; polylysine (4:1.46)), and the other two by coating polylysine on the film in different weight ratios (1:1 (alg/pl-PL 1:1) and 2:1 (alg/pl-PL 2:1) of prepared film: addition polylysine). The net charge of the synthesized film was measured using a zeta potential, and it was discovered that adding polylisine to the prefabricated film results in a net positively charged surface, but the blended film revealed a negative charge.The net charge of the synthesized film was measured using a zeta potential, and it was discovered that adding polylisine to the prefabricated film results in a net positively charged surface, but the blended film revealed a negative charge.[87]

A separate strategy was also used to alter the surface charge of the scaffolds by altering the voltage polarities throughout the scaffold creation process utilizing the electrospinning method. This approach changes the surface charge of the scaffolds, as well as their shape and cell attachment behavior. Metwally et al. conducted a two-part investigation to assess the physical and chemical effects of varying charges on PCL scaffolds, as well as biomineralization. The opposite charges scaffolds were created by swiping the needle's 14 KV voltage polarities on the ground, as seen in Figure 4.6. The results indicated a little decline in fiber diameter and contact angle (water, PBS, DMEM), as well as a very minor rise in roughness and porosity in negatively charged PCL fibers. Interestingly, in the positively charged PCL scaffolds, cells were not fully attached to the surface and preserved their spheroid form, but in the negatively charged PCL scaffolds, a greater number of cells were adherent in a more spreaded pattern, resulting in a flattened cell morphology.[88] Furthermore, negative surface potential aided calcium mineralization because positively charged calcium ions firmly and quickly attached to the negatively charged surface, resulting in intermolecular immobilization of ions, which caused nucleation and fast mineralization of HA.[89]

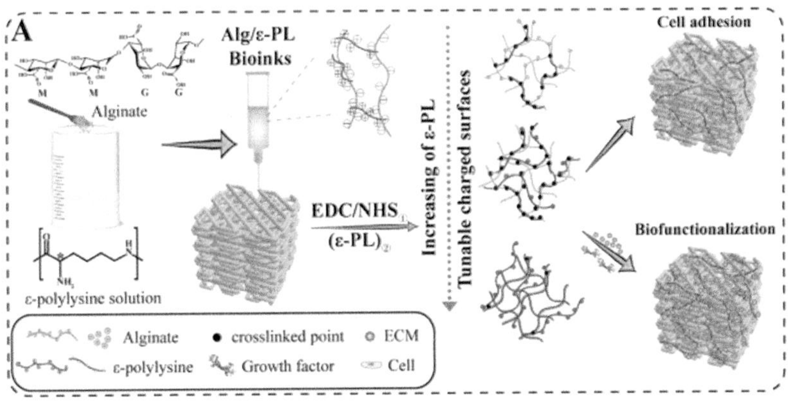

FIGURE 4.5 Schematic preparation of Alg/ε-PL bioinks and 3D scaffold printing with tunable surface charges with a view to improved cell adhesion

Source: Reprinted from Ref. [59]. Copyright © 2001 John Wiley & Sons

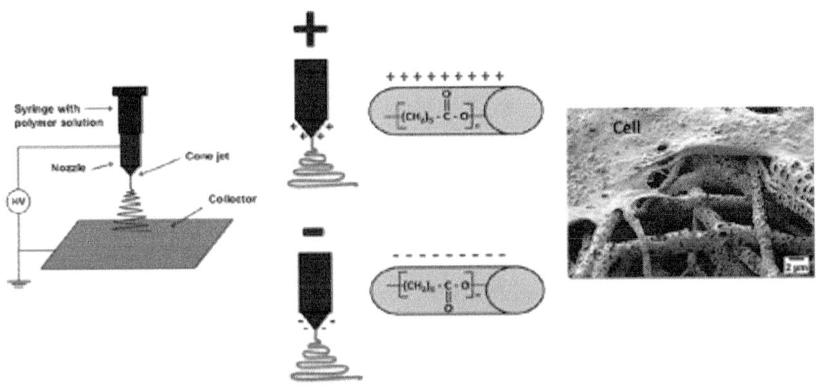

FIGURE 4.6 Schematic illustration of different charges induced on the surface of PCL scaffolds using electrospinning technique.

Source: Source: Reprinted from Ref. [62]. https://creativecommons.org/licenses/by/4.0/

4.3.2.4 SURFACE POLARITY/WETTABILITY

Although the polarity of any material is directly reliant on its surface energy, the polarity of the surface is highly important and may also be a defining element for cell attachment; thus, it must be described separately.[90] A surface with a total net charge that is either negative or positive may act hydrophilically in nature, whereas a surface with a neutral charge may behave more hydrophobically. Nevertheless, there are a few other parameters that characterize the surface's polarity, such as the existence of functional groups and surface topography. Wettability is also described as the adhesion force between a liquid and a solid substance, which causes the liquid to spread across the solid surface. The greater the dispersion of liquids, the greater the surface area.[91] The researchers described the cell as primarily attaching to a hydrophilic surface, whereas proteins require a hydrophobic surface for attachment.[92] Surface wettability has a large impact on cell adhesion and proliferation and may be altered via surface modification strategies. Zizar et al. demonstrated the influence of surface wettability on PLA scaffolds. When the PLA scaffolds were treated with varying weight ratios of HNT, the findings revealed a significant change in the surface wettability of the scaffolds. The water contact angle of the scaffolds treated with 5% HNT changed from 107.13° to 58.63° owing to capillary action of HNT as the hydrophilic groups present on the surface of the nanomaterial enable the creation of hydrogen bonds..[93] In a similar work, Jaidev et al. coated the surface of PLA with modified in-situ mineralization to investigate the effect of surface coating on surface wettability and its implications on scaffold cell adhesion and proliferation.

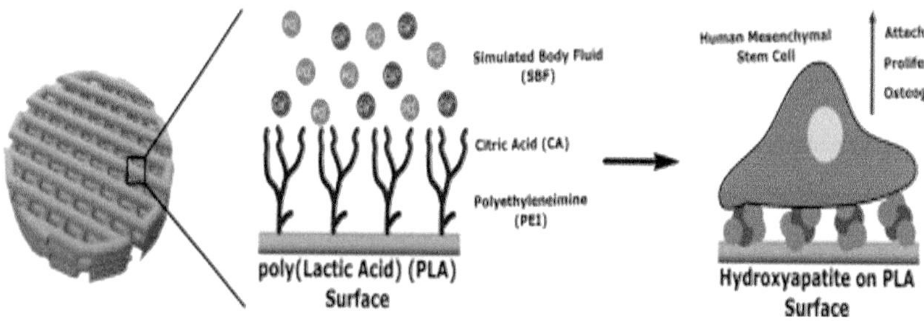

FIGURE 4.7 The attached citric acid over the PLA surface leads to the encroachment in cell adhesion and proliferation of scaffolds.

The as-formed PLA-CA was submerged in 1.5 SBF with an ionic concentration comparable to human blood for 7 days at 37°C, resulting in the introduction of HA on the PLA surface. The inclusion of carboxylic acid in the critic acid promotes mineralization and HA attachment to it. The addition of a -OH- group and citric acid lowers the water contact angle of the scaffolds; however, the addition of HA across the surface increases the contact angle from 67 ± 1° to 78 ± 2°, which was ascribed to increased surface roughness, while citric acid reduces the contact angle to extremely low values. In addition, the addition of HA enhanced surface wettability, imparted nanoscale roughness, and increased the mechanical strength of the scaffolds. When compared to as-printed scaffolds, the coated scaffolds had greater cell adhesion, proliferation, and osteogenic differentiation due to their improved wettability and roughness.[94] Similarly, the polarity of polymers was modified by the addition of polysaccharide. Martinus et al. created a chitosan/pectin membrane and discovered that the inclusion of pectin influences the interaction between the surface and cell by introducing polar groups (carbonyl). The inclusion of pectin increased wettability, swelling capacity, and cell attachment and spreading on the scaffolds.[95]

4.4 CONCLUSIONS

This chapter briefly discusses the basic of tissue engineering and the role of types of materials used in fabrication of polymeric scaffolds for tissue engineering applications. Additionally, surface (tropology, charge, roughness) and bulk properties (structure, mechanical strength) play a vital role in the interaction of cell to the scaffolds and define the effectiveness of scaffolds. Furthermore, in this chapter, the role of individual properties has been briefly discussed along with related examples. It had been found that the interface of the scaffolds is the far most important site and responsible for the attachment and growth of cells. However, each individual type of cell showed different preference for the attachment over scaffolds, but the interface with suitable surface chemistry, rough tropology with positive charge, and hydrophilic surface lead to effective cell attachment, growth, and proliferation of cell.

KEYWORDS

- **tissue engineering**
- **surface properties**
- **bulk properties**
- **surface properties**
- **tropology**
- **scaffolds**

REFERENCES

1. Amani, H.; Habibey, R.; Shokri, F.; Hajmiresmail, S. J.; Akhavan, O.; Mashaghi, A.; Pazoki-Toroudi, H. Selenium Nanoparticles for Targeted Stroke Therapy through Modulation of Inflammatory and Metabolic Signaling. *Sci. Re.* **2019,** *9* (1), 1–15. https://doi.org/10.1038/s41598-019-42633-9.
2. Langer, R.; Vacanti, J. P. Tissue Engineering. *Science (80-.).* **1993,** *260* (5110), 920–926. https://doi.org/10.1126/SCIENCE.8493529.
3. Jayakumar, R.; Prabaharan, M.; Shalumon, K. T.; Chennazhi, K. P.; Nair, S. V. Biomedical Applications of Polymer/Silver Composite Nanofibers. *Adv. Polym. Sci.* **2011,** *246* (1), 263–282. https://doi.org/10.1007/12_2011_123.
4. Chung, S.; King, M. W. Design Concepts and Strategies for Tissue Engineering Scaffolds. *Biotechnol. Appl. Biochem.* **2011,** *58* (6), 423–438. https://doi.org/10.1002/BAB.60.
5. Richbourg, N. R.; Peppas, N. A.; Sikavitsas, V. I. Tuning the Biomimetic Behavior of Scaffolds for Regenerative Medicine through Surface Modifications. *J. Tissue Eng. Regen. Med.* **2019,** *13* (8), 1275–1293. https://doi.org/10.1002/TERM.2859.
6. Kumar, P.; Dehiya, B. S.; Sindhu, A. Bioceramics for Hard Tissue Engineering Applications: A Review. *Int. J. Appl. Eng. Res.* **2018,** *13* (5), 2744–2752.
7. Van Vlierberghe, S.; Dubruel, P.; Schacht, E. Biopolymer-Based Hydrogels as Scaffolds for Tissue Engineering Applications: A Review. *Biomacromolecules* **2011,** *12* (5), 1387–1408. https://doi.org/10.1021/BM200083N.
8. Biazar, E.; Kamalvand, M.; Avani, F. Recent Advances in Surface Modification of Biopolymeric Nanofibrous Scaffolds. *Int. J. Polym. Mater. Polym. Biomater.* **2020.** https://doi.org/10.1080/00914037.2020.1857383.
9. Eltom, A.; Zhong, G.; Muhammad, A. Scaffold Techniques and Designs in Tissue Engineering Functions and Purposes: A Review. *Adv. Mater. Sci. Eng.* **2019,** *2019.* https://doi.org/10.1155/2019/3429527.
10. Sachlos, E.; Czernuszka, J. T.; Gogolewski, S.; Dalby, M. Making Tissue Engineering Scaffolds Work. Review: The Application of Solid Freeform Fabrication Technology to the Production of Tissue Engineering Scaffolds. *Eur. Cell. Mater.* **2003,** *5*, 29–40. https://doi.org/10.22203/ECM.V005A03.

11. Maeda, H.; Kasuga, T.; Nogami, M.; Ueda, M. Preparation of Bonelike Apatite Composite for Tissue Engineering Scaffold. *Sci. Technol. Adv. Mater.* **2005,** *6* (1), 48–53. https://doi.org/10.1016/J.STAM.2004.07.003/XML.
12. Nair, A.; Thevenot, P.; Dey, J.; Shen, J.; Sun, M. W.; Yang, J.; Tang, L. Novel Polymeric Scaffolds Using Protein Microbubbles as Porogen and Growth Factor Carriers. *Tissue Eng.—Part C Methods* **2010,** *16* (1), 23–32. https://doi.org/10.1089/ten.tec.2009.0094.
13. Castner, D. G.; Ratner, B. D. Biomedical Surface Science: Foundations to Frontiers. *Surf. Sci.* **2002,** *500* (1–3), 28–60. https://doi.org/10.1016/S0039-6028(01)01587-4.
14. Ponsonnet, L.; Reybier, K.; Jaffrezic, N.; Comte, V.; Lagneau, C.; Lissac, M.; Martelet, C. Relationship between Surface Properties (Roughness, Wettability) of Titanium and Titanium Alloys and Cell Behaviour. *Mater. Sci. Eng. C* **2003,** *23* (4), 551–560. https://doi.org/10.1016/S0928-4931(03)00033-X.
15. Yao, X.; Peng, R.; Ding, J.; Yao, X.; Peng, R.; Ding, J. D. Cell–Material Interactions Revealed Via Material Techniques of Surface Patterning. *Adv. Mater.* **2013,** *25* (37), 5257–5286. https://doi.org/10.1002/ADMA.201301762.
16. Bauer, S.; Schmuki, P.; von der Mark, K.; Park, J. Engineering Biocompatible Implant Surfaces: Part I: Materials and Surfaces. *Prog. Mater. Sci.* **2013,** *58* (3), 261–326. https://doi.org/10.1016/J.PMATSCI.2012.09.001.
17. Windt, T. S. de; Hendriks, J. A. A.; Xing Zhao, L. A. V.; Creemers, L. B.; Dhert, W. J. A.; Randolph, M. A.; Saris, D. B. F. Concise Review: Unraveling Stem Cell Cocultures in Regenerative Medicine: Which Cell Interactions Steer Cartilage Regeneration and How? *Stem cell Transl. Med.* **2014,** *3*, 723–733. https://doi.org/https://doi.org/10.5966/sctm.2013-0207.
18. Ayala, R.; Zhang, C.; Yang, D.; Hwang, Y.; Aung, A.; Shroff, S. S.; Arce, F. T.; Lal, R.; Arya, G.; Varghese, S. Engineering the Cell–Material Interface for Controlling Stem Cell Adhesion, Migration, and Differentiation. *Biomaterials* **2011,** *32* (15), 3700–3711. https://doi.org/10.1016/J.BIOMATERIALS.2011.02.004.
19. Martínez, A.; Blanco, M. D.; Davidenko, N.; Cameron, R. E. Tailoring Chitosan/Collagen Scaffolds for Tissue Engineering: Effect of Composition and Different Crosslinking Agents on Scaffold Properties. *Carbohydr. Polym.* **2015,** *132*, 606–619. https://doi.org/10.1016/J.CARBPOL.2015.06.084.
20. Sridhar, R.; Lakshminarayanan, R.; Madhaiyan, K.; Barathi, V. A.; Limh, K. H. C.; Ramakrishna, S. Electrosprayed Nanoparticles and Electrospun Nanofibers Based on Natural Materials: Applications in Tissue Regeneration, Drug Delivery and Pharmaceuticals. *Chem. Soc. Rev.* **2015,** *44* (3), 790–814. https://doi.org/10.1039/C4CS00226A.
21. Hiew, V. V.; Simat, S. F. B.; Teoh, P. L. The Advancement of Biomaterials in Regulating Stem Cell Fate. *Stem Cell Rev. Rep.* **2017,** *14* (1), 43–57. https://doi.org/10.1007/S12015-017-9764-Y.
22. Xue, J. M.; Tan, C. H.; Lukito, D. Biodegradable Polymer–Silica Xerogel Composite Microspheres for Controlled Release of Gentamicin. *J. Biomed. Mater. Res. Part B Appl. Biomater.* **2006,** *78B* (2), 417–422. https://doi.org/10.1002/JBM.B.30503.
23. Dawson, E.; Mapili, G.; Erickson, K.; Taqvi, S.; Roy, K. Biomaterials for Stem Cell Differentiation. *Adv. Drug Deliv. Rev.* **2008,** *60* (2), 215–228. https://doi.org/10.1016/J.ADDR.2007.08.037.
24. Vyas, C.; Zhang, J.; Øvrebø, Ø.; Huang, B.; Roberts, I.; Setty, M.; Allardyce, B.; Haugen, H.; Rajkhowa, R.; Bartolo, P. 3D Printing of Silk Microparticle Reinforced

Polycaprolactone Scaffolds for Tissue Engineering Applications. *Mater. Sci. Eng. C* **2021,** *118*. https://doi.org/10.1016/J.MSEC.2020.111433.
25. Léon Y León, C. A. New Perspectives in Mercury Porosimetry. *Adv. Colloid Interface Sci.* **1998,** *76–77*, 341–372. https://doi.org/10.1016/S0001-8686(98)00052-9.
26. Coombes, A. G. A.; Rizzi, S. C.; Williamson, M.; Barralet, J. E.; Downes, S.; Wallace, W. A. Precipitation Casting of Polycaprolactone for Applications in Tissue Engineering and Drug Delivery. *Biomaterials* **2004,** *25* (2), 315–325. https://doi.org/10.1016/S0142-9612(03)00535-0.
27. Kim, T. G.; Shin, H.; Lim, D. W. Biomimetic Scaffolds for Tissue Engineering. *Adv. Funct. Mater.* **2012,** *22* (12), 2446–2468. https://doi.org/10.1002/ADFM.201103083.
28. Salem, A. K.; Stevens, R.; Pearson, R. G.; Davies, M. C.; Tendler, S. J. B.; Roberts, C. J.; Williams, P. M.; Shakesheff, K. M. Interactions of 3T3 Fibroblasts and Endothelial Cells with Defined Pore Features. *J. Biomed. Mater. Res.* **2002,** *61* (2), 212–217. https://doi.org/10.1002/JBM.10195.
29. Dhandayuthapani, B.; Yoshida, Y.; Maekawa, T.; Kumar, D. S. Polymeric Scaffolds in Tissue Engineering Application: A Review. *Int. J. Polym. Sci.* **2011,** *2011*. https://doi.org/10.1155/2011/290602.
30. Yang, S.; Leong, K. F.; Du, Z.; Chua, C. K. The Design of Scaffolds for Use in Tissue Engineering. Part I. Traditional Factors. *Tissue Eng.*. Mary Ann Liebert, Inc. July 9, 2001, pp 679–689. https://doi.org/10.1089/107632701753337645.
31. Salgado, A. J.; Coutinho, O. P.; Reis, R. L. Bone Tissue Engineering: State of the Art and Future Trends. *Macromol. Biosci.* **2004,** *4* (8), 743–765. https://doi.org/10.1002/MABI.200400026.
32. Khodaei, M.; Fathi, M.; Meratian, M.; Savabi, O. The Effect of Porosity on the Mechanical Properties of Porous Titanium Scaffolds: Comparative Study on Experimental and Analytical Values. *Mater. Res. Express* **2018,** *5* (5), 055401. https://doi.org/10.1088/2053-1591/AABFA2.
33. Fang, Q.; Zhang, Y.; Chen, X.; Li, H.; Cheng, L.; Zhu, W.; Zhang, Z.; Tang, M.; Liu, W.; Wang, H. et al. Three-Dimensional Graphene Enhances Neural Stem Cell Proliferation Through Metabolic Regulation. *Front. Bioeng. Biotechnol.* **2020,** *0*, 436. https://doi.org/10.3389/FBIOE.2019.00436.
34. Chen, Z.; Yan, X.; Yin, S.; Liu, L.; Liu, X.; Zhao, G.; Ma, W.; Qi, W.; Ren, Z.; Liao, H. et al. Influence of the Pore Size and Porosity of Selective Laser Melted Ti6Al4V ELI Porous Scaffold on Cell Proliferation, Osteogenesis and Bone Ingrowth. *Mater. Sci. Eng. C. Mater. Biol. Appl.* **2019,** *106*, 110289–110289. https://doi.org/10.1016/J.MSEC.2019.110289.
35. Nair, L. S.; Laurencin, C. T. Biodegradable Polymers as Biomaterials. *Prog. Polym. Sci.* **2007,** *32* (8–9), 762–798. https://doi.org/10.1016/J.PROGPOLYMSCI.2007.05.017.
36. Stolz, M.; Raiteri, R.; Daniels, A. U.; VanLandingham, M. R.; Baschong, W.; Aebi, U. Dynamic Elastic Modulus of Porcine Articular Cartilage Determined at Two Different Levels of Tissue Organization by Indentation-Type Atomic Force Microscopy. *Biophys. J.* **2004,** *86* (5), 3269–3283. https://doi.org/10.1016/S0006-3495(04)74375-1.
37. Miller, K.; Chinzei, K.; Orssengo, G.; Bednarz, P. Mechanical Properties of Brain Tissue In-Vivo: Experiment and Computer Simulation. *J. Biomech.* **2000,** *33* (11), 1369–1376. https://doi.org/10.1016/S0021-9290(00)00120-2.
38. Kim, T. H.; An, D. B.; Oh, S. H.; Kang, M. K.; Song, H. H.; Lee, J. H. Creating Stiffness Gradient Polyvinyl Alcohol Hydrogel Using a Simple Gradual Freezing-Thawing

Method to Investigate Stem Cell Differentiation Behaviors. *Biomaterials* **2015,** *40,* 51–60. https://doi.org/10.1016/J.BIOMATERIALS.2014.11.017.
39. Even-Ram, S.; Artym, V.; Yamada, K. M. Matrix Control of Stem Cell Fate. *Cell* **2006,** *126* (4), 645–647. https://doi.org/10.1016/J.CELL.2006.08.008.
40. Ma, Q.; Yang, L.; Jiang, Z.; Song, Q.; Xiao, M.; Zhang, D.; Ma, X.; Wen, T.; Cheng, G. Three-Dimensional Stiff Graphene Scaffold on Neural Stem Cells Behavior. *ACS Appl. Mater. Interfaces* **2016,** *8* (50), 34227–34233. https://doi.org/10.1021/ACSAMI.6B12305.
41. Awad, H. A.; Wickham, M. Q.; Leddy, H. A.; Gimble, J. M.; Guilak, F. Chondrogenic Differentiation of Adipose-Derived Adult Stem Cells in Agarose, Alginate, and Gelatin Scaffolds. *Biomaterials* **2004,** *25* (16), 3211–3222. https://doi.org/10.1016/J.BIOMATERIALS.2003.10.045.
42. Liu, X.; Feng, Q.; Bachhuka, A.; Vasilev, K. Surface Modification by Allylamine Plasma Polymerization Promotes Osteogenic Differentiation of Human Adipose-Derived Stem Cells. *ACS Appl. Mater. Interfaces* **2014,** *6* (12), 9733–9741. https://doi.org/10.1021/AM502170S.
43. El-Fiqi, A.; Lee, J. H.; Lee, E. J.; Kim, H. W. Collagen Hydrogels Incorporated with Surface-Aminated Mesoporous Nanobioactive Glass: Improvement of Physicochemical Stability and Mechanical Properties Is Effective for Hard Tissue Engineering. *Acta Biomater.* **2013,** *9* (12), 9508–9521. https://doi.org/10.1016/J.ACTBIO.2013.07.036.
44. Margel, S.; Vogler, E. A.; Firment, L.; Watt, T.; Haynie, S.; Sogah, D. Y. Peptide, Protein, and Cellular Interactions with Self-Assembled Monolayer Model Surfaces. *J. Biomed. Mater. Res.* **1993,** *27* (12), 1463–1476. https://doi.org/10.1002/JBM.820271202.
45. Yu, T. T.; Cui, F. Z.; Meng, Q. Y.; Wang, J.; Wu, D. C.; Zhang, J.; Kou, X. X.; Yang, R. L.; Liu, Y.; Zhang, Y. S. et al. Influence of Surface Chemistry on Adhesion and Osteo/Odontogenic Differentiation of Dental Pulp Stem Cells. *ACS Biomater. Sci. Eng.* **2017,** *3* (6), 1119–1128. https://doi.org/10.1021/ACSBIOMATERIALS.7B00274/SUPPL_FILE/AB7B00274_SI_001.PDF.
46. Cui, W.; Li, X.; Xie, C.; Zhuang, H.; Zhou, S.; Weng, J. Hydroxyapatite Nucleation and Growth Mechanism on Electrospun Fibers Functionalized with Different Chemical Groups and Their Combinations. *Biomaterials* **2010,** *31* (17), 4620–4629. https://doi.org/10.1016/J.BIOMATERIALS.2010.02.050.
47. Park, E. J.; Gevrek, T. N.; Sanyal, R.; Sanyal, A. Indispensable Platforms for Bioimmobilization: Maleimide-Based Thiol Reactive Hydrogels. *Bioconjug. Chem.* **2014,** *25* (11), 2004–2011. https://doi.org/10.1021/BC500375R/SUPPL_FILE/BC500375R_SI_001.PDF.
48. Groessner-Schreiber, B.; Tuan, R. S. Enhanced Extracellular Matrix Production and Mineralization by Osteoblasts Cultured on Titanium Surfaces in Vitro. *J. Cell Sci.* **1992,** *101* (1), 209–217. https://doi.org/10.1242/JCS.101.1.209.
49. Bettinger, C. J.; Langer, R.; Borenstein, J. T. Engineering Substrate Topography at the Micro- and Nanoscale to Control Cell Function. *Angew. Chemie Int. Ed.* **2009,** *48* (30), 5406–5415. https://doi.org/10.1002/ANIE.200805179.
50. Lossdörfer, S.; Schwartz, Z.; Wang, L.; Lohmann, C. H.; Turner, J. D.; Wieland, M.; Cochran, D. L.; Boyan, B. D. Microrough Implant Surface Topographies Increase Osteogenesis by Reducing Osteoclast Formation and Activity. *J. Biomed. Mater. Res. A* **2004,** *70A* (3), 361–369. https://doi.org/10.1002/JBM.A.30025.

51. Vagaská, B.; Bačáková, L.; Filová, E.; Balík, K.; Bačáková, L. Osteogenic Cells on Bio-Inspired Materials for Bone Tissue Engineering. *Physiol. Res.* **2010,** *59*, 309–322.
52. Chen, H.; Huang, X.; Zhang, M.; Damanik, F.; Baker, M. B.; Leferink, A.; Yuan, H.; Truckenmüller, R.; van Blitterswijk, C.; Moroni, L. Tailoring Surface Nanoroughness of Electrospun Scaffolds for Skeletal Tissue Engineering. *Acta Biomater.* **2017,** *59*, 82–93. https://doi.org/10.1016/J.ACTBIO.2017.07.003.
53. Carnes, M. E.; Pins, G. D. Etching Anisotropic Surface Topography onto Fibrin Microthread Scaffolds for Guiding Myoblast Alignment. *J. Biomed. Mater. Res. Part B Appl. Biomater.* **2020,** *108* (5), 2308–2319. https://doi.org/10.1002/JBM.B.34566.
54. Shuai, C.; Zan, J.; Yang, Y.; Peng, S.; Yang, W.; Qi, F.; Shen, L.; Tian, Z. Surface Modification Enhances Interfacial Bonding in PLLA/MgO Bone Scaffold. *Mater. Sci. Eng. C* **2020,** *108*, 110486. https://doi.org/10.1016/J.MSEC.2019.110486.
55. Jing, D.; Bhushan, B. Quantification of Surface Charge Density and Its Effect on Boundary Slip. *Langmuir* **2013,** *29* (23), 6953–6963. https://doi.org/10.1021/LA401168W.
56. Verma, A.; Stellacci, F. Effect of Surface Properties on Nanoparticle-Cell Interactions. In *Small*; John Wiley & Sons, Ltd, January 4 2010; pp 12–21. https://doi.org/10.1002/smll.200901158.
57. Gentile, P.; Bellucci, D.; Sola, A.; Mattu, C.; Cannillo, V.; Ciardelli, G. Composite Scaffolds for Controlled Drug Release: Role of the Polyurethane Nanoparticles on the Physical Properties and Cell Behaviour. *J. Mech. Behav. Biomed. Mater.* **2015,** *44*, 53–60. https://doi.org/10.1016/J.JMBBM.2014.12.017.
58. Tidwell, C. D.; Ertel, S. I.; Ratner, B. D.; Tarasevich, B. J.; Atre, S.; Allara, D. L. Endothelial Cell Growth and Protein Adsorption on Terminally Functionalized, Self-Assembled Monolayers of Alkanethiolates on Gold. *Langmuir* **1997,** *13* (13), 3404–3413. https://doi.org/10.1021/LA9604341.
59. Lin, Z.; Wu, M.; He, H.; Liang, Q.; Hu, C.; Zeng, Z.; Cheng, D.; Wang, G.; Chen, D.; Pan, H. et al. 3D Printing of Mechanically Stable Calcium-Free Alginate-Based Scaffolds with Tunable Surface Charge to Enable Cell Adhesion and Facile Biofunctionalization. *Adv. Funct. Mater.* **2019,** *29* (9), 1808439. https://doi.org/10.1002/ADFM.201808439.
60. Jaidev, L. R.; Chatterjee, K. Surface Functionalization of 3D Printed Polymer Scaffolds to Augment Stem Cell Response. *Mater. Des.* **2019,** *161*, 44–54. https://doi.org/10.1016/J.MATDES.2018.11.018.
61. Metwally, S.; Karbowniczek, J. E.; Szewczyk, P. K.; Marzec, M. M.; Gruszczyński, A.; Bernasik, A.; Stachewicz, U. Single-Step Approach to Tailor Surface Chemistry and Potential on Electrospun PCL Fibers for Tissue Engineering Application. *Adv. Mater. Interfaces* **2019,** *6* (2), 1801211. https://doi.org/10.1002/ADMI.201801211.
62. Metwally, S.; Ferraris, S.; Spriano, S.; Krysiak, Z. J.; Kaniuk, Ł.; Marzec, M. M.; Kim, S. K.; Szewczyk, P. K.; Gruszczyński, A.; Wytrwal-Sarna, M. et al. Surface Potential and Roughness Controlled Cell Adhesion and Collagen Formation in Electrospun PCL Fibers for Bone Regeneration. *Mater. Des.* **2020,** *194*, 108915. https://doi.org/10.1016/J.MATDES.2020.108915.
63. Boyan, B. D.; Hummert, T. W.; Dean, D. D.; Schwartz, Z. Role of Material Surfaces in Regulating Bone and Cartilage Cell Response. *Biomaterials* **1996,** *17* (2), 137–146. https://doi.org/10.1016/0142-9612(96)85758-9.
64. Lai, Y.; Pan, F.; Xu, C.; Fuchs, H.; Chi, L.; Lai, Y. K.; Pan, F.; Fuchs, H.; Chi, L. F.; Xu, C. In Situ Surface-Modification-Induced Superhydrophobic Patterns with

Reversible Wettability and Adhesion. *Adv. Mater.* **2013,** *25* (12), 1682–1686. https://doi.org/10.1002/ADMA.201203797.

65. Mohd Nizar, M.; Hamzah, M. S. A.; Abd Razak, S. I.; Mat Nayan, N. H. Thermal Stability and Surface Wettability Studies of Polylactic Acid/Halloysite Nanotube Nanocomposite Scaffold for Tissue Engineering Studies. In *IOP Conference Series: Materials Science and Engineering*, Vol. 318; IOP Publishing, 2018; p 012006. https://doi.org/10.1088/1757-899X/318/1/012006.
66. Martins, J. G.; Camargo, S. E. A.; Bishop, T. T.; Popat, K. C.; Kipper, M. J.; Martins, A. F. Pectin-Chitosan Membrane Scaffold Imparts Controlled Stem Cell Adhesion and Proliferation. *Carbohydr. Polym.* **2018,** *197*, 47–56. https://doi.org/10.1016/J.CARBPOL.2018.05.062.
67. Richbourg, N. R.; Peppas, N. A.; Sikavitsas, V. I. Tuning the Biomimetic Behavior of Scaffolds for Regenerative Medicine through Surface Modifications. *J. Tissue Eng. Regen. Med.* **2019,** *13* (8), 1275–1293. https://doi.org/10.1002/TERM.2859.
68. Chung, S.; King, M. W. Design Concepts and Strategies for Tissue Engineering Scaffolds. *Biotechnol. Appl. Biochem.* **2011,** *58* (6), 423–438. https://doi.org/10.1002/BAB.60.
69. Van Vlierberghe, S.; Dubruel, P.; Schacht, E. Biopolymer-Based Hydrogels As Scaffolds for Tissue Engineering Applications: A Review. *Biomacromolecules* **2011,** *12* (5), 1387–1408. https://doi.org/10.1021/BM200083N.
70. Biazar, E.; Kamalvand, M.; Avani, F. Recent Advances in Surface Modification of Biopolymeric Nanofibrous Scaffolds. *Int. J. Polym. Mater. Polym. Biomater*. **2020**. https://doi.org/10.1080/00914037.2020.1857383.
71. Maeda, H.; Kasuga, T.; Nogami, M.; Ueda, M. Preparation of Bonelike Apatite Composite for Tissue Engineering Scaffold. *Sci. Technol. Adv. Mater.* **2005,** *6* (1), 48–53. https://doi.org/10.1016/J.STAM.2004.07.003/XML.
72. Nair, A.; Thevenot, P.; Dey, J.; Shen, J.; Sun, M. W.; Yang, J.; Tang, L. Novel Polymeric Scaffolds Using Protein Microbubbles as Porogen and Growth Factor Carriers. *Tissue Eng.—Part C Methods* **2010,** *16* (1), 23–32. https://doi.org/10.1089/ten.tec.2009.0094.
73. Yao, X.; Peng, R.; Ding, J.; Yao, X.; Peng, R.; Ding, J. D. Cell–Material Interactions Revealed Via Material Techniques of Surface Patterning. *Adv. Mater.* **2013,** *25* (37), 5257–5286. https://doi.org/10.1002/ADMA.201301762.
74. Vasita, R.; Katti, D. S.; Shanmugam, K. Improved Biomaterials for Tissue Engineering Applications: Surface Modification of Polymers. *Int. J. Nanomed.* **2008,** *8* (4), 341. https://doi.org/10.2174/156802608783790893.
75. Park, E. J.; Gevrek, T. N.; Sanyal, R.; Sanyal, A. Indispensable Platforms for Bioimmobilization: Maleimide-Based Thiol Reactive Hydrogels. *Bioconjug. Chem.* **2014,** *25* (11), 2004–2011. https://doi.org/10.1021/BC500375R/SUPPL_FILE/BC500375R_SI_001.PDF.
76. Amani, H.; Arzaghi, H.; Bayandori, M.; Dezfuli, A. S.; Pazoki-Toroudi, H.; Shafiee, A.; Moradi, L. Controlling Cell Behavior through the Design of Biomaterial Surfaces: A Focus on Surface Modification Techniques. *Adv. Mater. Interfaces* **2019,** *6* (13), 1900572. https://doi.org/10.1002/ADMI.201900572.
77. Shi, C.; Yuan, W.; Khan, M.; Li, Q.; Feng, Y.; Yao, F.; Zhang, W. Hydrophilic PCU Scaffolds Prepared by Grafting PEGMA and Immobilizing Gelatin to Enhance Cell Adhesion and Proliferation. *Mater. Sci. Eng. C* **2015,** *50*, 201–209. https://doi.org/10.1016/J.MSEC.2015.02.015.

78. Zhang, H.; Lin, C. Y.; Hollister, S. J. The Interaction between Bone Marrow Stromal Cells and RGD-Modified Three-Dimensional Porous Polycaprolactone Scaffolds. *Biomaterials* **2009,** *30* (25), 4063–4069. https://doi.org/10.1016/J.BIOMATERIALS.2009.04.015.
79. Jiao, Y. P.; Cui, F. Z. Surface Modification of Polyester Biomaterials for Tissue Engineering. *Biomed. Mater.* **2007,** *2* (4), R24. https://doi.org/10.1088/1748-6041/2/4/R02.
80. Keselowsky, B. G.; Collard, D. M.; García, A. J. Surface Chemistry Modulates Fibronectin Conformation and Directs Integrin Binding and Specificity to Control Cell Adhesion. *J. Biomed. Mater. Res. Part A* **2003,** *66A* (2), 247–259. https://doi.org/10.1002/JBM.A.10537.
81. Li, Q.; Xing, D.; Ma, L.; Gao, C. Synthesis of E7 Peptide-Modified Biodegradable Polyester with the Improving Affinity to Mesenchymal Stem Cells. *Mater. Sci. Eng. C* **2017,** *73*, 562–568. https://doi.org/10.1016/J.MSEC.2016.12.088.
82. King, W. J.; Krebsbach, P. H. Growth Factor Delivery: How Surface Interactions Modulate Release in Vitro and in Vivo. *Adv. Drug Deliv. Rev.* **2012,** *64* (12), 1239–1256. https://doi.org/10.1016/J.ADDR.2012.03.004.
83. Zhou, J.; Guo, X.; Zheng, Q.; Wu, Y.; Cui, F.; Wu, B. Improving Osteogenesis of Three-Dimensional Porous Scaffold Based on Mineralized Recombinant Human-like Collagen via Mussel-Inspired Polydopamine and Effective Immobilization of BMP-2-Derived Peptide. *Colloids Surfaces B Biointerfaces* **2017,** *152*, 124–132. https://doi.org/10.1016/J.COLSURFB.2016.12.041.
84. Hong, M. H.; Hong, H. J.; Pang, H.; Lee, H. J.; Yi, S.; Koh, W. G. Controlled Release of Growth Factors from Multilayered Fibrous Scaffold for Functional Recoveries in Crushed Sciatic Nerve. *ACS Biomater. Sci. Eng.* **2018,** *4* (2), 576–586. https://doi.org/10.1021/ACSBIOMATERIALS.7B00801/SUPPL_FILE/AB7B00801_SI_001.ZIP.
85. Chen, G.; Ushida, T.; Tateishi, T. A Hybrid Network of Synthetic Polymer Mesh and Collagen Sponge. *Chem. Commun.* **2000,** *0* (16), 1505–1506. https://doi.org/10.1039/B003508O.
86. Chen, Q. Z.; Thompson, I. D.; Boccaccini, A. R. 45S5 Bioglass®-Derived Glass–Ceramic Scaffolds for Bone Tissue Engineering. *Biomaterials* **2006,** *27* (11), 2414–2425. https://doi.org/10.1016/J.BIOMATERIALS.2005.11.025.
87. Mohamad Yunos, D.; Bretcanu, O.; Boccaccini, A. R. Polymer-Bioceramic Composites for Tissue Engineering Scaffolds. *J. Mater. Sci.* **2008,** *43* (13), 4433–4442. https://doi.org/10.1007/S10853-008-2552-Y/FIGURES/9.
88. Motealleh, A.; Eqtesadi, S.; Perera, F. H.; Pajares, A.; Guiberteau, F.; Miranda, P. Understanding the Role of Dip-Coating Process Parameters in the Mechanical Performance of Polymer-Coated Bioglass Robocast Scaffolds. *J. Mech. Behav. Biomed. Mater.* **2016,** *64*, 253–261. https://doi.org/10.1016/J.JMBBM.2016.08.004.
89. Poddar, D.; Jain, P.; Rawat, S.; Mohanty, S. Influence of Varying Concentrations of Chitosan Coating on the Pore Wall of Polycaprolactone Based Porous Scaffolds for Tissue Engineering Application. *Carbohydr. Polym.* **2021,** *259*, 117501. https://doi.org/10.1016/j.carbpol.2020.117501.
90. Gao, A.; Wu, Q.; Wang, D.; Ha, Y.; Chen, Z.; Yang, P.; Gao, A.; Wu, Q.; Wang, D.; Ha, Y. et al. A Superhydrophobic Surface Templated by Protein Self-Assembly and Emerging Application toward Protein Crystallization. *Adv. Mater.* **2016,** *28* (3), 579–587. https://doi.org/10.1002/ADMA.201504769.

91. Sapp, M. C.; Fares, H. J.; Estrada, A. C.; Grande-Allen, K. J. Multilayer Three-Dimensional Filter Paper Constructs for the Culture and Analysis of Aortic Valvular Interstitial Cells. *Acta Biomater.* **2015,** *13*, 199–206. https://doi.org/10.1016/J.ACTBIO.2014.11.039.
92. Fu, J.; Li, X. B.; Wang, L. X.; Lv, X. H.; Lu, Z.; Wang, F.; Xia, Q.; Yu, L.; Li, C. M. One-Step Dip-Coating-Fabricated Core-Shell Silk Fibroin Rice Paper Fibrous Scaffolds for 3D Tumor Spheroid Formation. *ACS Appl. Bio Mater.* **2020,** *3* (11), 7462–7471. https://doi.org/10.1021/ACSABM.0C00679/SUPPL_FILE/MT0C00679_SI_001.PDF.
93. Chen, W.; Nichols, L.; Brinkley, F.; Bohna, K.; Tian, W.; Priddy, M. W.; Priddy, L. B. Alkali Treatment Facilitates Functional Nano-Hydroxyapatite Coating of 3D Printed Polylactic Acid Scaffolds. *Mater. Sci. Eng. C* **2021,** *120*, 111686. https://doi.org/10.1016/j.msec.2020.111686.
94. Liu, W. G.; Li, X. W.; Li, Y. S.; Cai, K. Y.; De Yao, K.; Yang, Z.; Li, X. Effects of Baicalin-Modified Poly(D,L-Lactic Acid) Surface on the Behavior of Osteoblasts. *J. Mater. Sci. Mater. Med.* **2003,** *14* (11), 961–965. https://doi.org/10.1023/A:1026398532009.
95. Yang, S.; Han, X.; Jia, Y.; Zhang, H.; Tang, T. Hydroxypropyltrimethyl Ammonium Chloride Chitosan Functionalized-PLGA Electrospun Fibrous Membranes as Antibacterial Wound Dressing: In Vitro and In Vivo Evaluation. *Polym.* **2017,** *9* (12), 697. https://doi.org/10.3390/POLYM9120697.
96. Collins, M. N., et al. "Scaffold fabrication technologies and structure/function properties in bone tissue engineering." Advanced functional materials 31.21 (2021): 2010609.

CHAPTER 5

Intricate Dissection of the Realm of Biopolymer Films in Regenerative Medicine and Other Pioneering Therapeutic Advances

DARIEO THANKACHAN[1], JACOB MATHEW[2], and
NEHA KANWAR RAWAT[3]

[1]*Department of Materials and Science Engineering, National Taiwan University of Science and Technology, Taiwan, Republic of China*

[2]*Department of Chemical Engineering, National Taiwan University of Science and Technology, Taiwan, Republic of China*

[3]*Materials Research Laboratory, Department of Chemistry, Jamia Millia Islamia, New Delhi, India*

ABSTRACT

Biopolymer films are employed in the medical industry today for a variety of significant applications, including cardiovascular implants, selective and nonselective drug delivery, and tissue engineering to restore the functionality of damaged tissues. This chapter discusses some of the key thoughts on these changes. A popular class of materials used today as a replacement for traditional materials to address complex and challenging problems in the health industry is biopolymers. Hydrogels, other biopolymer-based materials, and 3D bio printed scaffolds are essential for developing bio platforms for the generation of new cells and tissue components in tissue engineering. New biocompatible, green/natural

Tailored Functional Materials for Clean and Sustainable Development. Tamara Tatrishvili, PhD, Neha Kanwar Rawat, PhD, Ann Rose Abraham, PhD & A. K. Haghi, PhD (Eds.)

polymer materials that may also be biodegradable and non-immunogenic are being sought for in order to overcome the persistent problems that regenerative medicine faces. Also, the critical importance of surface nano- and microstructuring for improving biocompatibility and in vitro and in vivo performances is generally acknowledged. This chapter also covers how to optimize the structural design, bioactivity, and biodegradability of biopolymer films for use in the hemostatic domain. Smart polymers have great promise for the future of biomedicine and will provide an advantageous setting for research and development in tissue engineering and regenerative medicine.

5.1 INTRODUCTION

Films and membranes are 2D flat structures having lateral dimensions that are greater than their thickness. The principal use of these 2D membranes and films is in many sectors. Membranes are well known in the separation process since they are used in a variety of industrial processes such as purification and filtration. Despite the extensive use of synthetic polymers as membranes in this industry, biopolymer-based membranes have grown more important for applications such as the removal of heavy metals, dyes, and pesticides from industrial effluents, biocatalysis, and hazardous organic dechlorination.[1]

In food packaging industry, the term "film" is commonly used. Edible films and edible coatings are frequently used interchangeably in this field; however, they differ in terms of preparation. Edible films are created directly onto food surfaces after contact with the film-forming solution, whereas edible coatings are generated directly onto food surfaces after contact with the film-forming solution. These films and coatings are frequently used to give additional barrier protection to food products to increase their shelf life. They can, however, be tailored to carry antimicrobials, antioxidants, minerals, tastes, and colorants, thus increasing the food's overall quality. In food applications, homogeneous and dense architectures impact mechanical resistance, barrier properties, and active ingredient release control. Separation techniques, on the other hand, may have a low permeate flow through the thick membrane, lowering efficiency. As a result, pore size and pore size distribution impact membrane separation performance. Cell development and adhesion are affected by the porosity and connection of pores in biomaterials such as scaffolds for tissue engineering. Porosity contributes to the permeability to water vapor and exudates absorption in wound-dressing membranes, which modulates the healing process. Porosity impacts the healing process by contributing to water vapor permeability and exudates absorption in wound-dressing membranes. Membranes for separation processes can be classified as symmetrical (isotropic) or asymmetrical based on their structure (anisotropic). Symmetric membranes have a constant structure over their whole thickness. Asymmetric membranes are divided into two categories: phase separation membranes and thin-film composite membranes. Phase-separated membranes, sometimes called integrally skinned membranes, have a heterogeneous structure but a consistent chemical composition. These structures have a gradual change in pore size and porosity over the membrane's thickness, resulting in a denser "skin" layer on one surface that can be porous or nonporous. Thin composite membranes are made up of a porous substrate covered with a thin dense coating of a distinct polymer that is both structurally and chemically heterogeneous.[2]

5.2 BIOPOLYMER FILMS IN TISSUE ENGINEERING APPLICATION

For the last 20 years, researchers have been attempting to replicate the finest bio-based polymeric material for medical purposes, particularly in tissue engineering discipline. Bone, tendon, cartilage, skeletal muscle, skin, ligament, vascular tissues, and neural tissues are all scaffolds utilized in tissue healing. As a result of the employment of advanced technology, we can construct a systematic porous structure that can be utilized to regenerate organs or tissue. The biopolymer is used because of its intrinsic biodegradability, which allows for a series of cell and tissue remodeling on the surface without toxicity and acts as a good host for foreign bodies.[3] Taking advantage of the use of biopolymer, biopolymer-based scaffolds for bone, tendon, cartilage, skeletal muscle, skin, ligament, vascular tissues, and neural tissues have been developed.[4] As a result, we may divide biopolymers for tissue engineering into two categories: synthetic polymers and naturally occurring polymers. When compared with natural polymers, synthetic polymers are more hydrophobic. A few examples include poly (lactic-co-glycolic acid), polycaprolactone (PCL), and polyethylene glycol (PEG).[5] Natural polymers such as collagen, chitosan, gelatin, silk fibroin, alginate, cellulose, and starch have been produced as scaffolds in tissue regeneration applications. Despite the fact that these biopolymers display chemical heterogenicity and high dispersity,[6] their biocompatibility, low toxicity, and metabolic performance nevertheless place them higher in the tissue engineering hierarchy than synthetic polymers.[7]

However, to produce superior scaffolds for tissue engineering, the biopolymer must have particular qualities. Those criteria are as follows: first, the matrix must acquire an appropriate topology for superior cell adhesion. Furthermore, the scaffolds filled the various porosity areas with adequate pore size to promote nutrient and cell diffusion. Along with pore connectivity for better cell propagation, the matrix should have strong mechanical stability to aid cell development. Finally, the polymer must be highly proficient in protein distribution and growth factors to energize the cell variety.[8–10] Take through, herein we discussed about natural biopolymer scaffolds for the tissue engineering.

Moving on to natural polymers, chitin and chitosan are common compounds that are used in a variety of medicinal applications. Chitin and chitosan are both N-acetyl-glucosamine and N-glucosamine linear polymers. Chitin is a polymer in which the fraction of N-acetyl-glucosamine surpasses 50%. Chitosan, on the other hand, is a polymer in which N-glucosamine accounts for more than 50% of the total weight. The use of these biopolymers can be linked to their desired nontoxicity, biocompatibility, nonimmunogenicity, and relative affordability.[11] These materials have the benefit of being simple to make gel and scaffolds for a variety of applications in tissue engineering. Chitin has been employed in a variety of applications such as bone, cartilage, wound,[12] and depending on the use, various additives have been added to the matrix. However, the use of martial arts as a whole is becoming increasingly rare these days. Initially, the design was created using core material,[13,14] but further improvements were made by incorporating composite into the polymer matrix. In the matrix, for example, a compound like calcium might be employed as a hard tissue replacement material. Because the presence of calcium compounds enhances osteogenic potential, hydroxyapatite (HA) or calcium compounds have been added to the polymer matrix for bone or orthopedic therapeutic therapy.[15] In a series of studies, Wan et al. have investigated the interaction of chitin with HA and the studies depicted that it was favorable for the bone replacement.[16] Xing et al. recently created chitin-hydroxyapatite (HAP)-collagen composite scaffolds (CHCS) that used epichlorohydrin (ECH) as a crosslinking agent. According to the studies, the CHCS model offers a lot of potential in terms of bone regeneration.[17]

Chitosan's chemical structure has a significant number of primary amines and acetamido groups, resulting in highly strong inter- and intramolecular interactions, resulting in increased crystallinity and, as a result, more rigid mechanical properties.[18] It is well known as a fantastic biopolymer that, among other things, can be easily transformed into membranes, gels, sponges, scaffolds, and beads. As a result, it may be used in some specialized applications in its original or modified form. The hydrogel form of chitosan is most widely used in tissue engineering because it has excellent cell adhesion, cell survival, cell interaction, and neurite outgrowth.[19] Considering the advantages it offers, several modifications were carried out to improve the properties to withstand several conditions. Nezhad-Mokhtari et al. developed a unique injectable hydrogel design using collagen, aldehyde-modified nanocrystalline cellulose, and chitosan loaded with gold nanoparticles (Collagen/ADH-CNCs/CS-Au).[20] The proportion of CS-Au in the hydrogels has a significant influence on the microscopic morphology, equilibrium swelling, in vitro degradation, and mechanical properties, according to the study. Furthermore, the developed material exhibits excellent nontoxigenicity, making it a viable substrate for tissue engineering. Besides, Reddy et al.[21] modeled a novel scaffold for targeting chronic wound. The ideas of curcumin–chitosan loaded nanofibrous antibacterial scaffold was to enhance the wound healing for deep cut and to regulate the healing time to a definite interval. The experiments showed that the curcumin—chitosan NFs were effective therapeutically; other than that the matrix could contain the entire potency of curcumin to maximize efficiency.

Hyaluronic acid (HA), commonly known as hyaluronan, is an anionic glycosaminoglycan that contains repeating units of the disaccharides 1,3-N-acetyl-D-glucosamide and 1,4-D-glucuronic acid. It is derived from extracellular tissues in numerous parts of the human body, where it aids in lubrication. Hyaluronic acid is a biopolymer with a high molecular weight and a backbone chain of 5000–30,000 sugar molecules that are found in nature. HA has been shown to be effective for tissue engineering due to its tunable properties such as biodegradability, bioresorbability, biocompatibility, and ability to create hydrogels.[22] One of the first modifications investigated was the crosslinking of carboxylic acid groups with amino terminal groups. Barbucci et al. examined several di-amine crosslinking agents, such as 1,3-diaminepropane, 1,6-diaminohexane, PEG500 di-amine, and PEG800 di-amine, to make hydrogels. Studies suggest that synthesized material has a favorable physiological environment as well as excellent biocompatibility.[23] The interaction of hyaluronic acid carboxylic acid moieties with mono- and polyfunctional hydrazides resulted in biochemical probes and biopolymers, similar to previous research. The researchers discussed that the material had an excellent regulated drug release. As a result, it is regarded as an efficient scaffold for tissue engineering.[24] In relation to the aforementioned models, a few novel design concepts have recently been proposed, one of which was by Rusu et al. They created a nanostructured hydrogel as a biomaterial for wound healing here. The matrix was created by incorporating maleoyl-chitosan/poly (aspartic acid) (MAC5/PAS) nanogels into a polymer network based on thiolated hyaluronic acid (HASH). The results suggest that the micro porosity size of HASH hydrogel allows for the controlled release of medicines with short half-lives, such as antibodies.[25] In addition, 3D-bioprinting can create anatomically formed implants for tissue engineering. In addition to animal-based biomaterials like gelatin and collagen, functional biomaterial material bio ink must be produced as a result of the necessity. Thanh and colleagues developed an amine-hyaluronic acid (HA-NH2) molecule that was covalently crosslinked to the aldehyde-alginate material (Alg-CHO). According to the findings, HA-Alg hydrogel is a suitable composite for 3D printing. Furthermore, HA-Alg hydrogel increased cartilage tissue formation and may have aided other tissue syntheses due to its tailorable mechanical and degradable properties.[26] As a result, we may infer that hyaluronic acid is required in tissue engineering applications to reconstruct an effective matrix with desirable qualities.

When compared to other stiff hydrogels and real tissue, hydrogel films exhibit superior mechanical stability and other potential performances when used instead of hydrogel directly. Sara et al.[27] produced a prospective design with excellent stability compared with other forms of hydrogels by taking these characteristics into mind. The corresponding designs are composed of stiffer and stronger biopolymer chitosan hydrogel films generated by integrating nanoscale architecture with templated microstructure. These templated nanopillar chitosan hydrogel films outperform unstructured flat hydrogel in terms of moduli, with moduli increasing by 160%, up to 20 MPa, and work at break increasing by 450%, up to 8.5 MJ m^3. Furthermore, the strength at break improves by 350%, reaching 37 MPa, making it one of the strongest hydrogel films on the market. The manufacturing method is as follows: they created templated nanopillar hydrogel films using a popular scalable fabrication approach. The results indicated that the templated microstructure resulted in improved mechanical performance. Furthermore, this film demonstrated improved compatibility, making it a viable material for tissue engineering applications.

Similarly, a few biologically produced polymer polymers, such as polylactide (PLA), polyhydroxybutyrate (PHB), and polyhydroxyalkanoates (PHA), have some intriguing characteristics. However, these are too costly for usage in the targeted applications. As a result, a desirable film, which is less expensive and more durable for the applications, is unavoidable. Consider gelatin as a viable choice based on Prochon et al.'s research,[28] which can be converted into a thin, transparent film that is flexible and has stable and high mechanical qualities. The gelatin composite material is as follows. They used animal gelatin (pig skin) with the addition of casein, food gelatin, glycerin, and enzymes as biocatalysts of chemical modification and further extraction of gelatin from collagen to make such material. For further characterization, compositions yielding films with homogenous morphologies and acceptable mechanical characteristics were chosen. Tensile strength reaches 3.11 MPa, with a maximum elongation at the break of 97.96%. According to microbial scaring investigations, the composites fully dissolved under the activity of microorganisms after 30 days, proving their biodegradability. Taking all of these features into account, this material is a promising option for nanostructural growth as well as a foundation for novel biomimetic cement as bone replacement and biodegradable implantations for improved biocompatibility.

5.3 BIOPOLYMER FILMS FOR CARDIOVASCULAR APPLICATIONS

Cardiovascular disease is one of the world's major causes of death. As a result, numerous systematic ways were devised to address such a problem. One such significant advancement occurred in the realm of material science, where some of the afflicted sections were effectively supported using material that is bio sustainable to the body and is commonly referred to as biomaterials. The use of biomaterials is justified since they serve an inert role in the body, essentially acting as a compactable system and providing some function comparable to the one replaced.[29] The use of biopolymer film or coating as a recognized approach in cardiovascular treatment is owing mostly to its low thrombogenicity and nontoxic nature. As a result, depending on the application, numerous strategical methods, such as a pure polymer membrane or surface-modified biomaterials, have evolved. Materials used in cardiovascular implants to enhance physiological conditions include chitosan, hyaluronic acid, polyamide, and polyvinyl alcohol. The next section discusses the numerous biopolymers that have been employed in cardiovascular applications.

Because of their tailored properties and great biocompatibility, biopolymers are regarded as a good alternative. Various polymers were used in a number of applications ranging from vascular grafts to stents, prosthetic heart valves, catheters, and so on, depending on the application.

Considering polyamides (PA), which are used as transparent tubes for cardiovascular applications, hemodialysis membranes, and the manufacture of percutaneous transluminal coronary angioplasty (PTCA) catheters, the PTCA balloon catheters were used to reopen the stenotic coronary arteries. Several changes were made to this system to improve its efficiency. One such design is to increase the heat resistance of catheters to prevent ischemia-reperfusion harm in acute ischemic stroke. Thus, Jiang et al. optimized a cooling catheter by sealing air into closed-tube chambers.[30] Instead of using traditional methods, different techniques are being taken in the field of hemodialysis membranes, which give regulated and consistent thinness, emphasizing the permeability of the membrane.[31] Moving on to polyethylene-terephthalate (PET), also known as dacron, which is utilized as a fabric in heart valves to cover the annulus of the ring, its primary application was to prevent infection and enhance the hemocompatibility of heart valves.[32] PET was also a viable choice for vascular graft replacement due to its tiny pore size and biocompatibility. However, essential changes were made to increase the effectiveness of the vascular transplant. The connection of systematic antibody response to PET substrate against infection is one such technique. To improve the anti-infectious ratio, the prothesis was covered with bovine collagen and gelatin.[33]

Polyurethane (PU) is also commonly employed in cardiovascular applications due of its excellent physiochemical and mechanical characteristics. Furthermore, polyurethane is extremely biocompatible, allowing for unlimited use in blood-contacting devices. It possesses a high shear strength, flexibility, and transparency. Additionally, the surface of polyurethane is resistant to microorganisms, and the production of thrombosis by PU is essentially identical to that of the versatile cardiovascular biomaterial PTFE. Segmented polyurethanes (SPUs) have traditionally been used in a variety of cardiovascular applications such as valve constructions, pacemaker leads, and ventricular-aiding devices.[34] However, numerous synergistic designs were developed to improve the prothesis performance. The invention of flexible scaffolds for artificial vascular grafts was one such design. In this case, a mixture of titanium dioxide nanotubes (TNT) and PU was used. TNT was added to the PU matrix to increase its flexibility, biocompatibility, and porosity. The results suggest that endothelialization of open-ended

(OE) TNT-PU scaffolds was shown to be two times larger than that of the standard PU scaffolds. Furthermore, as compared with PU grafts, platelet adherence was relatively low.[35] Similarly, significant changes have occurred in the region of heart valves. For example, the design by Dehghani et al. in polyurethane–heparinized carbon nanotubes helped to improve the anti-clotting property in leaflets.[36] Similarly, despite its poor structural strength, polysiloxane-based polyurethane demonstrated unprecedented strength. Herein, Wang and colleagues added two urethane groups to the ends of the polysiloxane chains. As a result, the design offers a good microphase separation structure, unparalleled strength, reproducible processing, noncytotoxicity, shape memory qualities, and 3D printing capabilities.[37]

We now move on to the most current polymer composite film models, which have recently been employed to enhance the concepts for future exposure in the field of cardiovascular applications. Consider the Natalia Mihailescu groups'[38] concept for regenerative medical applications. In this study, compositional gradient coatings for surface functionalization were created using synchronized laser irradiation of two separate cryogenic targets, one made of sulfated halomonas levan and the other of quaternate low molecular weight chitosan. When compared with the pure material, the resulting output has better surface qualities. Furthermore, they provide thin coatings capable of providing a favorable contact for cells, resulting in better osseointegration while also demonstrating superior antibacterial qualities. When compared with other biopolymer films, Hongping Wan's model[39] demonstrated previous advances in the field of catheter use. The goal was to create a lubricious hydrophilic covering with reduced friction. However, when particles become loose and float freely in the bloodstream, these coatings might create iatrogenic consequences. Nevertheless, the design of a polyethylene glycol functionalized hyaluronic acid coating negated such problems and gave a well-lubricated surface. Furthermore, the great biocompatibility indicates that the coating was a safe and lubricant advantage to the cardiovascular catheter.

5.4 BIOPOLYMER FILMS FOR HEMOSTATIC APPLICATIONS

Hemostatic is a process to prevent the blood flow by enhancing the coagulation procedure if the person is suffering from severe cut or injury. Hemostatic procedures can be performed using a variety of techniques such as monopolar or bipolar electrocautery, argon–plasma coagulation, cutting of accessible tubular structures, suturing, administration of topical hemostatic chemicals, and wound tamponade.[40] As a result, hemostatic devices are critical in medical treatment because they assist in the formation of an efficient cure in critical situations such as surgical procedures and urgent trauma, particularly injuries sustained in battlefields and other difficult situations.[41] The operating principle of hemostasis may be separated into two parts: primary and secondary. In the first stage, a sudden block formed on the damaged vessel wall as a result of vasocontraction, platelet adhesion, and aggregation. However, the fibrin layer was not formed at this point. As a result, primary hemostasis is regarded as a temporary treatment for the damaged portion. Unless subsequent hemostasis supports (secondary) the platelet plug by forming a durable fibrin clot, the bleeding may reoccur. Finally, the fibrin clot was broken up and normal blood flow was restored thanks to the operations of the fibrinolytic system.[42] The currently available hemostatic materials include collagen (Col), gelatin (GE), alginate (AG), chitosan (CS), oxidized cellulose, and porous zeolite. All of them have strong hemostasis properties; however, they all have downsides. Collagen is a poor hemostatic agent because it relies on platelet activation to stop bleeding and has poor tissue adhesion.[43] Similarly, chitosan has an active hemostatic propensity by initiating coagulation without any active intrinsic mechanism, as in the conventional hemostatic technique.[44,45]. We may undertake some studies on various hemostatic materials that are now available in the following part, with the goal of providing a reference for future study and development of unique hemostatic materials.

5.4.1 *CHITOSAN-BASED FILM HEMOSTATIC MATERIALS*

The most significant characteristics of an excellent wound-dressing material are that it is nontoxic and nonallergenic to the biological system, that it is nonadherent to the skin, that it absorbs excess wound exudates, and that it has a strong antibacterial activity.[46] Chitosan was thought to be a better hemostatic agent because of its biocompatibility, antibacterial characteristics, biodegradability, wound healing, and nontoxicity.[47] To achieve an efficient hemostatic effect, CS is combined with functional materials, which improves the hemostatic capabilities of CS-based materials. CS has the ability to be processed into a variety of forms, including hydrogels, films, sponges, and other hemostatic materials. In recent years, hemostatic materials based on chitosan-based composite materials have been developed. Gelatin is an animal protein made from collagen that has been denaturated. And gelatin is the structural protein for tendon, skin like many connective tissues. Due to these specialties' gelatin used with chitosan in many films based wound-dressing applications. In this composite film material chitosan promotes wound healing and tissue development. Moreover, the physicochemical properties of CS improved by the addition of gelatin. Also, the antibacterial and hemostatic properties of CS-based wound-dressing materials could be improved by the loading of antibiotic to this composition. The material as created has a high degree of swelling, antimicrobial action, no evident cytotoxicity, and reduces bleeding.[48] Chitosan-sodium alginate-yunnabaiyo composite films have shown synergistic activity as a wound-dressing material with high hemostatic performance in surgical and traumatic bleeding situations.[49] Another set of researchers created wound-healing mesoporous bio-glass-chitosan porous composite films.[50] Using an electrospinning method, Hong et al. created very thin mesoporous bio-glass fibers with a diameter of a few nanometers.[51] High porosity and great biocompatibility define the wound-dressing films as produced. It also has excellent biodegradability and water adsorption characteristics.[52] Sara et al. created wound-dressing materials made of gallium-containing mesoporous bioactive glass-chitosan composite films. These newly created materials have a strong hemostatic effectiveness as well as a high antibacterial activity. It also has good water adsorption characteristics, making it an effective wound-dressing material.[53]

Hemostatic mechanism of chitosan is different from the intrinsic coagulation method and it involves the following steps

RBC aggregation, platelet stimulation, contact system activation, and spatial network formation.

5.4.1.1 RBC AGGREGATION

The most prevalent kind of hematocyte in the blood, the red blood cell (RBC), increases blood viscosity and facilitates platelet transport to the vascular wall, both of which are necessary for physiological hemostasis. Negatively charged proteins and glycolipids can be found on the surface of the RBC cell membrane. In a healthy condition, electrostatic repulsion inhibits RBC aggregation and adherence. The positive charge of $-NH3+$ on the CS chain interacts electrostatically with the anions on the surface of RBC, causing significant aggregation of RBC near the wound site and the formation of blood clots that halt bleeding fast.[54,55] As a result, the degree of protonation of amino groups on chitosan chains has a considerable impact on red blood cell adhesion. Chitosan's capacity to begin coagulation was shown to be connected to the percentage of deacetylation and to be more reliant on the quantity of protonated amine groups in studies.[56,57] Furthermore, as the molecular weight of chitosan increased, so did its interaction with red blood cells, which might be explained by increased entanglement caused by a particular intermolecular hydrogen bonding force or electrostatic repulsion between polyelectrolyte molecules.[58,59]

5.4.1.2 PLATELET STIMULATION

Platelets do not attach to endothelial cells under normal physiological circumstances. Platelet adhesion and aggregation stimulation, on the other hand, are critical in the hemostasis process in wounds. Platelet adhesion and aggregation are caused by biopolymers, which is a complicated process including a number of factors such as surface chain mobility, surface chemical composition, hydrogen bonding characteristics, charge density, hydrophobicity/hydrophilicity, and so on. In investigations, CS has been demonstrated to enhance platelet activation as well as platelet adhesion and aggregation.[60] Sagnella et al. discovered that neutralizing chitosan's positive charge reduced adherent platelets and aggregates while having no effect on coagulation activation, suggesting that platelet adhesion and aggregation need a high positive charge density in chitosan.[61,62]

5.4.1.3 CONTACT SYSTEM ACTIVATION

The contact activation pathway depicts how the interaction of a biomaterial surface with blood affects blood coagulation by changing protein activity after adsorption. Coagulation factors XII (FXII), coagulation factors XI (FXI), high-molecular-weight-kininogen (HMWK), and pre-kallikrein are all part of this pathway. When blood comes into touch with functional biomaterials, two biological processes occur: platelet adhesion and contact system activation, both of which have a synergistic influence on biomaterial-induced blood coagulation.[63,64] Fischer et al. discovered that in vivo, CS had a biochemical sequence that included virtually all plasma proteins as well as some of the most critical blood coagulation components, resulting in substantially stronger blood clots.[65,66]

5.4.1.4 SPATIAL NETWORK FORMATION

In terms of molecular structure, CS is a linear glycosaminoglycan that contributes in the production of robust blood clotting by aiding network formation, boosting blood component interaction with CS, and stimulating blood component contact with CS. Dowling et al. hypothesized a gelling hemostatic mechanism based on the self-assembly of hydrophobically modified chitosan (HM-CS) with blood cells. When HM-CS hydrophobes came into contact with blood cells, they used hydrophobic interactions to attach themselves to the hydrophobic interiors of blood cell membranes. As a result, between the CS chains and the blood cells, a 3D gel network developed, which might possibly halt blood flow.[67]

5.5 BIOPOLYMER-BASED MULTILAYERED FILMS FOR DRUG DELIVERY

Thin films are being studied by pharmaceutical professionals from all over the world as a breakthrough drug delivery approach. Thin films have shown to be a viable alternative to established dose forms. The thin films are intended to be simple to swallow, self-administrable, and dissolve fast, making them a flexible drug delivery platform. Oral, buccal, sublingual, ocular, and transdermal routes have all been employed to offer this delivery method for both systemic and local effects.[68]

5.5.1 CLINICAL BENEFITS

Thin film is preferred by patients due to its attractive look and convenience of administration. Oral-dissolving film is especially beneficial for juvenile, geriatric, and psychiatric patients since it is simple to use and eliminates the risk of choking or asphyxia. Ophthalmic films have been shown to boost medication absorption from the anterior part of the eye by extending the time a medicine is maintained in the eye. Furthermore, patients who are immobile or unable to comply may benefit from the polymeric thin films since they are simple to apply and rarely vomited up. When a quick start is necessary, such as in motion sickness, rapid allergic reactions or coughing, bronchitis, or asthma, a thin covering is preferable.

5.5.2 THE PROMISE OF LAYER BY LAYER DELIVERY

The introduction of an electrostatic assembly technique for generating thin films with controllable composition down to tens of nanometers is a game-changing finding for drug delivery in biomedical applications. This innovative assembly technique for the creation of drug carriers has a number of significant advantages.

The biointerface's characteristics are crucial for integrating (bio)materials into live organisms. From protein rejection to selective cell attachment to antibacterial properties, polymer coatings have proven to be effective in managing this interface.[69] As a result, polymer thin films are employed in a variety of applications, including tissue engineering and drug delivery. To change a variety of materials, spray coating, dip coating, self-assembled monolayers, surface grafted or tethered polymer layers, and polymer multilayer assembly can all be used. The layer-by-layer (LbL) assembly technique, for example, is one of such systems that has the inherent power to succeed due to its simplicity, flexibility, and reproducibility.[70]

Depending on the kind of polymer, cross-linking density, termination layer, and other factors, LbL films made entirely of polymers can be utilized to control cell adhesion and differentiation. While polymer coatings are certainly beneficial, simple polymer films are insufficient to solve a variety of problems. One technique for improving the function of LbL coatings is to embed active molecules, such as proteins or genes, inside substrate-mediated drug delivery (SMDD) coatings, albeit the types and quantities of substances that may be trapped in this way are limited. It is especially problematic for small-molecular payloads or cargo that rely on their 3D shape, such as enzymes.[71,72]

5.5.3 DRUG DELIVERY APPLICATIONS

On the inside of a polyelectrolyte multilayer (PEM) structure, cyclodextrin (CDs) has a cup-like shape on the surface and a hydrophobic interior. CDs and PE have been used in the field of medication delivery for a long time, with novel structural and functional alterations.[73] Jessel et al. provided one of the earliest instances, demonstrating that anti-inflammatory films could be made by including the medication piroxicam into PEMs via inclusion complexes based on monomeric CDs. These films were made using three layers of CD–drug complexes, as well as poly(L-lysine) (PLL) and poly(L-glutamic acid) (PGA).[74] The Hammond group built on previous work by using charged CD-polymers to stably trap drug molecules such as ciprofloxacin, flurbiprofen, and diclofenac, demonstrating that release kinetics are independent of the therapeutic agent incorporated and can be controlled by the choice of degradable polycations. They presented a thin permanent nondegradable microbicidal base film made of linear N,N-dodecyl,methyl poly(ethyleneimine) (DMLPEI) and poly(acrylic acid), as well as a hydrolytically degradable top multilayer film made of cationic poly(-amino ester) (PBAE) (PAA). PLL covalently bonded with CDs was employed as a polycationic vector for risedronate, a popular anticancer medicine used to prevent bone metastases, in another successful preclinical trial.[75]

5.5.4 FREESTANDING MULTILAYER FILMS

The technique of sequential layer adsorption can be repeated until the required number of layers is reached. Tens to hundreds of polyelectrolyte multilayers (PEMs) can be built, resulting in thin coatings as thin as a few nanometers and thick films as thick as a few micrometers. Most 2D multilayer films are permanently bonded to the underlying substrate and cannot be removed without causing damage to the film. Certain substrates, on the other hand, may be used to create films that can be peeled away from the underlying surface. A freestanding film is the one that is separated from its substrate while maintaining structural integrity.

5.5.5 POLYELECTROLYTE MULTILAYER FILMS AND STEM CELL BEHAVIOR

Herring first employed endothelial cell seeding in a vascular transplant.[76] Harvesting and cultivating functioning cells requires invasive surgery (biopsy) as well as a significant amount of time and knowledge, making the method costly and confined to specialist research teams. As a result, developing a new source of rapidly replicating, ready-to-use ECs that lack tissue-specific phenotypic expression might be a top goal. Because of their potential value in regenerative medicine, progenitor and stem cells with substantial differentiation capacities are now the topic of various investigations. Circulating blood, bone marrow, adipose tissue, and perinatal tissues have all been discovered to contain these cells. The importance of biochemical, mechanical, shape, micro/nanopatterning, and mechanical cues in stem cell activity and differentiation is becoming increasingly clear. According to Discher et al., the mechanical qualities of the substrate can differentiate bone marrow-mesenchymal stem cells into the stromal lineage. According to tissue elasticity, soft matrices are neurogenic, stiffer matrices are myogenic, and rigid matrices are osteogenic. When stem cells are cultivated on a substrate that contains the target tissue, they take on the physiological form and phenotype of the tissue. Unfortunately, to our knowledge, no elasticity domain suitable for vascular EC differentiation has been discovered. Menu's group has spent a lot of time researching stem cell activation in PEMs.[77] Berthelemy et al. investigated the effect of PAH/PSS films on vascular progenitor cell proliferation for the first time. They found that circulating endothelial progenitors differentiated faster on PSS/PAH multilayer films than on any other protein coating they had tried previously.[1]

5.6 CONCLUSIONS

In the field of biomaterials, the terms films and membranes are frequently interchanged, as biopolymers are prized for their biodegradability and biocompatibility. This is especially true for free-standing wound-dressing devices. Bone implants and cardiovascular devices can benefit from biopolymer-based coatings. Depending on their form, films and membranes can have dense (nonporous) or porous structures. The future of smart polymers in biomedicine is bright, and it will prove to be a fertile ground for innovation and advancement. Of course, this is dependent on ongoing research. Many issues confronting this field of research must be solved to bring these findings and advancements to fruition in the near future. The low-cost creation of improved diagnostic applications and smarter stimuli-responsive polymers is one concern. Obtaining further toxicity data for these smart polymers, improving drug delivery response times, doing more research into the biodegradability of these polymers, and extending our understanding of enzyme, magnetic, and ultrasound-responsive polymers are just a few of the others.

Equal contributors: Darieo Thankachan and Jacob Mathew contributed equally in this book chapter.

KEYWORDS

- **regenerative medicine**
- **hemostatic material**
- **drug delivery**
- **layer by layer delivery**
- **multilayer films**
- **tissue engineering**

REFERENCES

1. Berthelemy, N. et al. O_2 Level Controls Hematopoietic Circulating Progenitor Cells Differentiation into Endothelial or Smooth Muscle Cells. *PLOS ONE* **2009,** *4* (5), e5514.
2. El-Ghaffar, M. A. A.; Tieama, H. A. A *Review of Membranes Classifications, Configurations, Surface Modifications, Characteristics and Its Applications in Water Purification*, 2017.
3. Kohane, D. S.; Langer, R. Polymeric Biomaterials in Tissue Engineering. *Pediatric Res.* **2008,** *63* (5), 487–491.
4. Chinga-Carrasco, G. Novel Biocomposite Engineering and Bio-Applications. *Bioengineering* **2018,** *5* (4), 80.
5. Wei, Q. et al. Synthetic Polymers for Biomedical Applications. *Int. J. Biomater.* **2018,** *2018*, 7158621.
6. Lee, E. J.; Kasper, F. K.; Mikos, A. G. Biomaterials for Tissue Engineering. *Ann. Biomed. Eng.* **2014,** *42* (2), 323–337.
7. Stoppel, W. L. et al. Clinical Applications of Naturally Derived Biopolymer-Based Scaffolds for Regenerative Medicine. *Ann. Biomed. Eng.* **2015,** *43* (3), 657–680.
8. Stratton, S. et al. Bioactive Polymeric Scaffolds for Tissue Engineering. *Bioactive Mater.*, **2016,** *1* (2), 93–108.
9. Ratheesh, G. et al. 3D Fabrication of Polymeric Scaffolds for Regenerative Therapy. *ACS Biomater. Sci. Eng.* **2017,** *3* (7), 1175–1194.
10. Hsu, S.-h.; Hung, K.-C.; Chen, C.-W. Biodegradable Polymer Scaffolds. *J. Mater. Chem. B* **2016,** *4* (47), 7493–7505.
11. Zahir-Jouzdani, F. et al. Chitosan and Thiolated Chitosan: Novel Therapeutic Approach for Preventing Corneal Haze After Chemical Injuries. *Carbohydr. Polym.* **2018,** *179*, 42–49.
12. Jayakumar, R. et al. Chitin Scaffolds in Tissue Engineering. *Int. J. Mol. Sci.* **2011,** *12* (3), 1876–1887.
13. Hutmacher, D. W. Scaffolds in Tissue Engineering Bone and Cartilage. *Biomaterials* **2000,** *21* (24), 2529–2543.
14. Madihally, S. V.; Matthew, H. W. T. Porous Chitosan Scaffolds for Tissue Engineering. *Biomaterials* **1999,** *20* (12), 1133–1142.

15. Ito, M. In Vitro Properties of a Chitosan-Bonded Hydroxyapatite Bone-Filling Paste. *Biomaterials* **1991,** *12* (1), 41–45.
16. Wan, A. C.; Khor, E.; Hastings, G. W. Hydroxyapatite Modified Chitin as Potential Hard Tissue Substitute Material. *J. Biomed. Mater. Res.* **1997,** *38* (3), 235–241.
17. Xing, F. et al. Chitin-Hydroxyapatite-Collagen Composite Scaffolds for Bone Regeneration. *Int. J. Biol. Macromol.* **2021,** *184*, 170–180.
18. Nishimura, S. et al. Chemospecific Manipulations of a Rigid Polysaccharide: Syntheses of Novel Chitosan Derivatives with Excellent Solubility in Common Organic Solvents by Regioselective Chemical Modifications. *Macromolecules* **1991,** *24* (17), 4745–4748.
19. Zahir-Jouzdani, F. et al. Chitosan and Thiolated Chitosan: Novel Therapeutic Approach for Preventing Corneal Haze After Chemical Injuries. *Carbohydr. Polym* **2018,** *179*, 42–49.
20. Nezhad-Mokhtari, P.; Akrami-Hasan-Kohal, M.; Ghorbani, M. An Injectable Chitosan-Based Hydrogel Scaffold Containing Gold Nanoparticles for Tissue Engineering Applications. *Int. J. Biol. Macromol.* **2020,** *154*, 198–205.
21. Kumar Reddy, T. S. et al. Curcumin And Chitosan Loaded Nano Scaffold For Targeting Chronic Wounds Through Tissue Engineering In Regenerative Medicine. *NVEO-NATURAL VOLATILES & ESSENTIAL OILS J.| NVEO* **2022,** 1049–1066.
22. Necas, J. et al. Hyaluronic Acid (Hyaluronan): A Review. *Vet. Medicina* **2008,** *53* (8), 397–411.
23. Barbucci, R. et al. Synthesis, Chemical and Rheological Characterization of New Hyaluronic Acid-Based Hydrogels. *J. Biomater. Sci. Polym. Ed.* **2000,** *11* (4), 383–399.
24. Prestwich, G. D. et al. Controlled Chemical Modification of Hyaluronic Acid: Synthesis, Applications, and Biodegradation of Hydrazide Derivatives. *J. Controll. Release* **1998,** *53* (1), 93–103.
25. Rusu, A. G. et al. Nanostructured Hyaluronic Acid-Based Hydrogels Encapsulating Synthetic/ Natural Hybrid Nanogels as Promising Wound Dressings. *Biochem. Eng. J.* **2022,** *179*, 108341.
26. Thanh, T. N. et al. Hyaluronic Acid Crosslinked with Alginate Hydrogel: A Versatile and Biocompatible Bioink Platform for Tissue Engineering. *Eur. Polym. J.* **2022,** *166*, 111027.
27. Heedy, S. et al. Nanopillar Templating Augments the Stiffness and Strength in Biopolymer Films. *ACS Nano* **2022,** *16* (2), 3311–3322.
28. Prochon, M. et al. Biodegradable Composites with Functional Properties Containing Biopolymers. *Catalysts* **2022,** *12* (1), 77.
29. Black, J.; Hastings, G. *Handbook of Biomaterial Properties*; Springer Science & Business Media, 2013.
30. Jiang, M. et al. Design and Evaluation of an Air-Insulated Catheter for Intra-Arterial Selective Cooling Infusion from Numerical Simulation and In Vitro Experiment. *Med. Eng. Phys.* **2022,** *99*, 103736.
31. Thiam, B. G. et al. 3D Printed and Conventional Membranes—A Review. *Polymers* **2022,** *14* (5), 1023.
32. Gott, V. L., Alejo, D. E.; Cameron, D. E. Mechanical Heart Valves: 50 Years of Evolution. *Ann. Thorac. Surg.* **2003,** *76* (6), S2230–S2239.
33. Köhler, R. et al. Association of Systemic Antibody Response Against Polyethylene Terephthalate with Inflammatory Serum Cytokine Profile Following Implantation of

Differently Coated Vascular Prostheses in a Rat Animal Model. *J. Biomed. Mater. Res. A* **2022,** *110* (1), 52–63.
34. Zdrahala, R. J.; Zdrahala, I. J. Biomedical Applications of Polyurethanes: A Review of Past Promises, Present Realities, and a Vibrant Future. *J. Biomater. App.* **1999,** *14* (1), 67–90.
35. Kianpour, G. et al. Synergy of Titanium Dioxide Nanotubes and Polyurethane Properties for Bypass Graft Application: Excellent Flexibility and Biocompatibility. *Mater. Design* **2022,** *215*, 110523.
36. Dehghani, F.; Khorasani, M. T.; Movahedi, M. Fabrication of Polyurethane—Heparinized Carbon Nanotubes Composite for Heart Valves Application. *Mater. Chem. Phys.* **2022,** *280*, 125819.
37. Wang, W. et al. Unprecedented Strength Polysiloxane-Based Polyurethane for 3D Printing and Shape Memory. *ACS Appl. Mater. Interfaces* **2022,** *14* (2), 3324–3333.
38. Mihailescu, N. et al. Gradient Multifunctional Biopolymer Thin Film Assemblies Synthesized by Combinatorial MAPLE. *Appl. Surface Sci.* **2019,** *466*, 628–636.
39. Wan, H. et al. A Polyethylene Glycol Functionalized Hyaluronic Acid Coating for Cardiovascular Catheter Lubrication. *Mater. Design* **2020,** *196*, 109080.
40. Butler, N.; Toogood, G. Principles of Liver Resection. *Surgery (Oxford)* **2017,** *35* (12), 692–699.
41. Seyednejad, H. et al. Topical Haemostatic Agents. *Br. J. Surg.* **2008,** *95* (10), 1197–225.
42. Hiller, E. Basic Principles of Hemostasis. In *Modern Hematology: Biology and Clinical Management*; Munker, R. et al., Ed.; Humana Press: Totowa, 2007; pp 327–345.
43. Kunio, N. R. et al. Chitosan based Advanced Hemostatic Dressing Is Associated with Decreased Blood Loss in a Swine Uncontrolled Hemorrhage Model. *Am. J. Surg.* **2013,** *205* (5), 505–510.
44. Benesch, J.; Tengvall, P. Blood Protein Adsorption onto Chitosan. *Biomaterials* **2002,** *23* (12), 2561–2568.
45. Thatte, H. S. et al. Mechanisms of Poly-N-Acetyl Glucosamine Polymer-Mediated Hemostasis: Platelet Interactions. *J. Trauma* **2004,** *57* (1 Suppl), S13–S21.
46. Liu, H. et al. A Functional Chitosan-Based Hydrogel as a Wound Dressing and Drug Delivery System in the Treatment of Wound Healing. *RSC Adv.* **2018,** *8* (14), 7533–7549.
47. Hardiansyah, A. et al. Electrospinning and Antibacterial Activity of Chitosan-Blended Poly (Lactic Acid) Nanofibers. *J. Polym. Res.* **2015,** *22* (4), 59.
48. Li, H. et al. Preparation, Characterization, Antibacterial Properties, and Hemostatic Evaluation of Ibuprofen-Loaded Chitosan/Gelatin Composite Films. *J. Appl. Polym. Sci.* **2017,** *134* (42), 45441.
49. Zhi-liang, Z. Experimental Study of Hemostasis Effect of Chitosan/sodium-Alginate-Yunnanbaiyao Composite Film. *Health Res.* **2011**.
50. Jia, T.-b. et al. Fabrication and Characterization of Chitosan/Mesoporous Bioactive Glasses Porous Films. *Chin. J. Tissue Eng. Res.* **2011,** *15* (42), 7877.
51. Hong, Y. et al. Preparation, Bioactivity, and Drug Release of Hierarchical Nanoporous Bioactive Glass Ultrathin Fibers. *Adv. Mater.* **2010,** *22* (6), 754–758.
52. Baino, F.; Fiorilli, S.; Vitale-Brovarone, C. Composite Biomaterials Based on Sol-Gel Mesoporous Silicate Glasses: A Review. *Bioengineering (Basel, Switzerland)*, **2017,** *4* (1), 15.

53. Pourshahrestani, S. et al. Potency and Cytotoxicity of a Novel Gallium-Containing Mesoporous Bioactive Glass/Chitosan Composite Scaffold as Hemostatic Agents. *ACS Appl. Mater. Interfaces* **2017,** *9* (37), 31381–31392.
54. Bennett, B. L.; Littlejohn, L. Review of New Topical Hemostatic Dressings for Combat Casualty Care. *Military Med.* **2014,** *179* (5), 497–514.
55. Huang, Y. et al. Hemostasis Mechanism and Applications of N-Alkylated Chitosan Sponge. *Polym. Adv. Technol.* **2017,** *28* (9), 1107–1114.
56. Edwards, J. V. et al. Positively and Negatively Charged Ionic Modifications to Cellulose Assessed as Cotton-Based Protease-Lowering and Hemostatic Wound Agents. *Cellulose* **2009,** *16* (5), 911–921.
57. Zhou, X. et al. An investigation of Chitosan and Its Derivatives on Red Blood Cell Agglutination. *RSC Adv.* **2017,** *7* (20), 12247–12254.
58. Hattori, H.; Ishihara, M. Changes in Blood Aggregation with Differences in Molecular Weight and Degree of Deacetylation of Chitosan. *Biomed. Mater.* **2015,** *10* (1), 015014.
59. Klokkevold, P. R. et al. The Effect of Chitosan (Poly-N-Acetyl Glucosamine) on Lingual Hemostasis in Heparinized Rabbits. *J. Oral Maxillofacial Surg.* **1999,** *57* (1), 49–52.
60. Hattori, H. et al. Hemostasis for Severe Hemorrhage with Photocrosslinkable Chitosan Hydrogel and Calcium Alginate. *Ann. Biomed. Eng.* **2010,** *38* (12), 3724–3732.
61. Sagnella, S.; Mai-Ngam, K. Chitosan Based Surfactant Polymers Designed to Improve Blood Compatibility on Biomaterials. *Colloids Surfaces B* **2005,** *42* (2), 147–155.
62. Chou, T.-C. et al. Chitosan enhances platelet adhesion and aggregation. *Biochem. Biophys. Res. Commun.* **2003,** *302* (3), 480–483.
63. Sperling, C. et al. Blood Coagulation on Biomaterials Requires the Combination of Distinct Activation Processes. *Biomaterials* **2009,** *30* (27), 4447–4456.
64. He, Q. et al. Positive Charge of Chitosan Retards Blood Coagulation on Chitosan Films. *J. Biomater. App.* **2011,** *27* (8), 1032–1045.
65. Fischer, T. H. et al. Comparison of Structural and Hemostatic Properties of the Poly-N-Acetyl Glucosamine Syvek Patch with Products Containing Chitosan. *Microsc. Res. Techn.* **2004,** *63* (3), 168–174.
66. Hu, Z. et al. Chitosan-Based Composite Materials for Prospective Hemostatic Applications. *Marine Drugs* **2018,** *16* (8), 273.
67. Dowling, M. B. et al. A Self-Assembling Hydrophobically Modified Chitosan Capable of Reversible Hemostatic Action. *Biomaterials* **2011,** *32* (13), 3351–3357.
68. Karki, S. et al. Thin Films as an Emerging Platform for Drug Delivery. *Asian J. Pharma. Sci.* **2016,** *11* (5), 559–574.
69. Stuart, M. A. C. et al. Emerging Applications of Stimuli-Responsive Polymer Materials. *Nat. Mater.* **2010,** *9* (2), 101–113.
70. Hammond, P. T. Building Biomedical Materials Layer-by-Layer. *Mater. Today* **2012,** *15* (5), 196–206.
71. Farris, E. et al. Micro- and Nanoparticulates for DNA Vaccine Delivery. *Exp. Biol. Med. (Maywood, N. J.)*, **2016,** *241* (9), 919–929.
72. Zhang, J.; Ma, P. X. Host-Guest Interactions Mediated Nano-Assemblies Using Cyclodextrin-Containing Hydrophilic Polymers and Their Biomedical Applications. *Nano Today* **2010,** *5* (4), 337–350.
73. Sun, T. et al. Self-Assembled Vesicles Prepared from Amphiphilic Cyclodextrins as Drug Carriers. *Langmuir* **2012,** *28* (23), 8625–8636.

74. Jessel, N. et al. Multiple and Time-Scheduled In Situ DNA Delivery Mediated by Beta-Cyclodextrin Embedded in a Polyelectrolyte Multilayer. *Proc Natl Acad Sci USA* **2006,** *103* (23), 8618–8621.
75. Smith, R. C. et al. Layer-by-Layer Platform Technology for Small-Molecule Delivery. *Angew. Chem. Int. Ed. Engl.* **2009,** *48* (47), 8974–8977.
76. Herring, M.; Gardner, A.; Glover, J. A Single-Staged Technique for Seeding Vascular Grafts with Autogenous Endothelium. *Surgery* **1978,** *84* (4), 498–504.
77. Engler, A. J. et al. Matrix Elasticity Directs Stem Cell Lineage Specification. *Cell* **2006,** *126* (4), 677–689.

CHAPTER 6

Recent Progress in Conducting Polymeric Films for Biomedical Application

KIRAN RANA[1], RINKY GHOSH[2], and NEHA KANWAR RAWAT[3]

[1]*Department of Plastics Engineering, Central Institute of Plastics, Engineering and Technology, Vatva, Gujarat, India*

[2]*Department of Materials Science and Engineering, Indian Institute of Technology Kanpur, Kanpur, India*

[3]*Department of Materials Engineering, Indian Institute of Science, Bangalore, India*

ABSTRACT

Conducting polymers which are also known as "synthetic metals" is one of the attractive branches of polymers. Over the past few years, as the biomedical research progresses, these materials have been widely utilized in biomedical applications like controlled drug release, neuronal therapies, tissue engineering, biosensors, and polymer-based artificial muscles due to its unique properties such as tunable electrical conductivity, environmental stability, easy processing, and biocompatibility.

6.1 INTRODUCTION

Biomedical materials such as polymers, ceramics, and metals possess excellent properties like biocompatibility, good mechanical strength, wear-resistant, and corrosion-resistant. But in certain cases where the electrically conductive properties are needed, it becomes the limitation of these

Tailored Functional Materials for Clean and Sustainable Development. Tamara Tatrishvili, PhD, Neha Kanwar Rawat, PhD, Ann Rose Abraham, PhD & A. K. Haghi, PhD (Eds.)

materials. Also, certain tissues are responsive to electrical stimuli. Electrical stimulation is very helpful in tissue engineering and regenerative medicine, as it affects cell migration, cell proliferation, and cell differentiation.[1]

In the recent years, conductive polymers have played an important role in the advancement of biomedical research. The conductive polymers are broadly categorized into four different types such as conjugated polymers, charge transfer polymers, ionic conducting polymers, and filled polymers by conductive fillers. Some of the commonly used conjugated polymers are polyacetylene, polypyrrole, polyaniline, poly (3,4-ethylene dioxythiophene), and polythiophene derivatives.[2] These conjugated polymers have an alternating double and single bonds in their backbone. This extended π-orbital system helps in delocalization of electrons along the chain length which makes them conductive. In the case of metals, the electrons in their valance band are less as compared to available energy bands which result in the free movement of electrons means the delocalization of electrons. Like metals, the conducting polymers also require delocalization of charge and partially filled bands.[3] But the electrical conductivity of these polymers is very low due to the presence of a lesser amount of charge carriers. For improving the electrical conductivity in the range of ~1–10^4 S/cm of these materials, a doping process is needed which could introduce the charge carriers in the polymeric chain. The doping of these conjugated polymers can be achieved by redox doping, nonredox doping, ion implantation, doping with heat treatment, and electrochemical doping. The redox doping method is done by partial oxidation and partial reduction with electron acceptors and electron donors, whereas the counter ions maintain charge neutrality of the polymer. Another type of redox doping is doping by no doping ions method such as photo doping[4] and charge injection doping.[5] The photo doping is done by the irradiation of photons on the conducting polymers which are have higher energy than the bandgap of the polymer. In charge injection doping, there is no generation of counter ions. Unlike redox doping, in nonredox doping the protonic acids are used and the number of electrons in the polymer by the process is the same. The chemical synthesis provides conductive polymers in powder forms. The other doping methods like the ion implantation process and doping with heat treatment are rarely used. Other than chemical doping, the electrochemical doping is a simple technique and easily reversible. It helps to control the doping level accurately. The DC power source is used which is applied in between the electrodes. This technique can produce free-standing films by using small amount of polymer, but the yield is less as compared with chemical synthesis. The casting process is also used for film formation by using suitable solvents. But this technique helps to produce thin films by peeling them from electrodes after polymerization.[6] In conducting polymers, electrical conductivity depends upon the type of dopant, concentration, and doping time. The dopants can be a small cation/anion or a large polymeric unit. The large polymeric dopants become the part of the main backbone chain of polymer and affect the conductivity as well as the physical properties of the polymer.[7] The electronic band levels of these electrically conductive polymers are termed as highest occupied molecular orbital that is valence band and lowest unoccupied molecular orbital that is conduction band. The conjugated polymers are intrinsically insulated polymers having a band gap of ≥1.5 eV.

6.1.1 CONDUCTING MECHANISM OF CONJUGATED POLYMERS

According to the bond structure at the ground state, the electrically conductive polymers are divided into the degenerated ground state and the nondegenerate ground state as illustrated in Figure 6.1. There is a huge difference between the energy of polymer structure for these ground states. The degenerated ground state comprises the identical structure on the interchange of the single and double bonds and having the same energy, for example, polyacetylene.

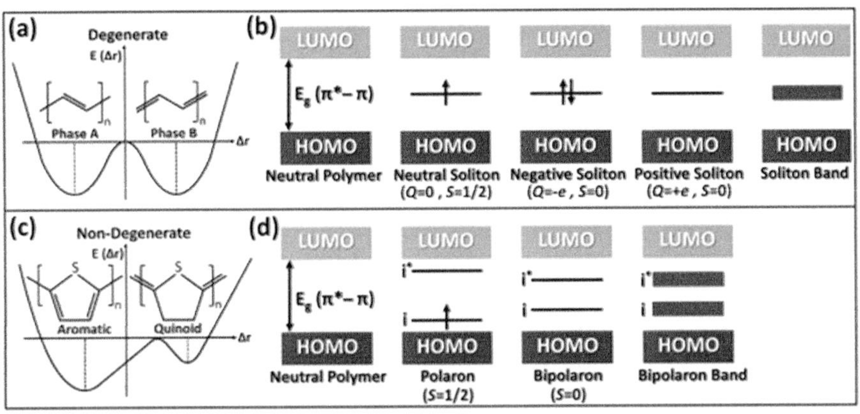

FIGURE 6.1 (a) Two degenerate states of polyacetylene: Phase A and Phase B (b) solitons lead to forming a band in between the HUMO and LUMO. (c) Nondegenerate state shows the different potential energy of aromatic and quinoid structures, (d) four different band structures at different doping concentrations.[9]

Source: Reprinted from R ef. [9]. https://creativecommons.org/licenses/by/4.0/

The polyacetylene has two isomeric configurations *cis*-polyacetylene and *trans*-polyacetylene. The *trans* configuration is more thermodynamically stable than the *cis* configuration. During the synthesis of polyacetylene at −78°C (low temperature) around 98% of *cis* polyacetylene configuration is formed, but as the temperature increases the *cis* isomer is changed to thermodynamically stable *trans* configuration.[8] The degenerate ground state is for *trans* isomer of polyacetylene because in *cis* isomer the hydrogen breaks the symmetry. Polyacetylene possesses sp^2 hybridization having the bond between the two carbon–carbons and one carbon–hydrogen sigma bonds along with one unpaired electron. In a neutral state, the linear polyacetylene would be a metal due to the half-filled π-band. But the experiments show that the polyacetylene behaves like a semiconductor with a 1.5 eV of the energy gap. This happens because of the Peierls distortion. With this instability, it gives the alternating bond pattern in *trans* polyacetylene. By this, the degenerate state shows two ground states that depend upon the dimerization formed by peierls distortion. Due to the bond alteration or peierls distortion, domain walls or solitons are formed in *trans*polyacetylene. These solitons exist in a neutral and charged state. They are formed when the molecular chain has an odd number of carbons. The electrically charged soliton is spinless and the neutral soliton is not spinless but has ½ spin. These species can be oxidized or reduced by the dopants to form negative and positive solitons. The solitons are easily delocalized in the polymeric chain. When two neutral solitons meet, they are vanished from the polymer backbone by forming a bond. In the case of charged soliton, they are quite stable and easily delocalized along the polymeric chain. During their interaction, a soliton band is formed between the HOMO (highest occupied molecular orbital) and LUMO (lowest unoccupied molecular orbital). As the doping level increases, the soliton population increases and the band becomes larger as shown in Figure 6.1.[10]

In nondegenerated structures, the bipolarons and polarons are the main charge carriers. The conjugated polymers such as polythiophene, polypyrrole, and polyaniline, and their derivatives are the aromatic structure-based polymers. These polymers show the nondegenerate ground state. In a nondegenerate state, the conjugated polymers have the energetically different structures on the interchange of alternating single and double bonds. The benzenoid and quinoid are the two different structures of nondegenerate state in which the energy of the quinoid structure is higher than the benzenoid structure. The quinoid structure is having a low ionization potential and also high electron affinity. The *p*-type and *n*-type doping of these conjugated polymers help to generate positive and negative polaron and bipolarons. In undoped states, a neutral polymer shows a large bandgap and behaves almost like an insulator. On oxidation, as the benzenoid structure is deformed into the quinoid structure on the removal of π electron, the formation of polaron takes place. Due to this, there is a rise of two localized bands in between the HOMO and LUMO. As a further increase in doping a second electron is removed from the polymeric chain and again there is a formation of polaron. As the population of polarons increases, the polarons may combine to each other and a bipolaron is formed. Again, with the increase in the dopant concentration, the overlap of bipolarons leads to the formation of two bipolaronic bands, and due to this, there is a decrease in the forbidden gap.[11]

This review discusses the applications of electrically conductive polymeric films in the biomedical field. These materials are very effective and are potentially used in biosensors, tissue engineering, drug delivery systems, biomedical implants, and other applications.[14,15]

6.2 APPLICATIONS OF CONDUCTING POLYMER FILMS IN BIOMEDICAL FIELD

6.2.1 POLYPYRROLE

Polypyrrole is an organic conjugated polymer and is widely used due to its properties such as environmentally stable, high electrical conductivity, easy synthesis processes, and availability of monomer. It can be synthesized by the electrochemical method or the chemical method. The chemical method for the synthesis of polypyrrole provides the yield in large amount as compared to the electrochemical method. But the electrochemical method for the preparation of polypyrrole provides the free-standing and very thin films with high purity. The resultant product from these methods is a conjugated conducting polypyrrole (Figure 6.2).

The properties of the resultant product are highly dependent upon the experimental conditions, polymerization method, type of dopant, monomer, and dopant concentration. Polypyrrole has potential applications in the field of biomedical due to its nontoxic nature as well as biocompatibility.[13] It can be synthesized at neutral pH and low oxidation potential which are also considered an advantage for its uses in the biomedical field. In biomedical applications, the prime requirement is to biologically modify the polypyrrole so as to increase the compatibility with tissues. Generally, physical adsorption and chemically modification are used to modify the polypyrrole. The physical method is quite simple but polypyrrole becomes inactive by this method. The chemical techniques are very helpful as compared to physical techniques and provide the stable functionalization of polypyrrole.[23]

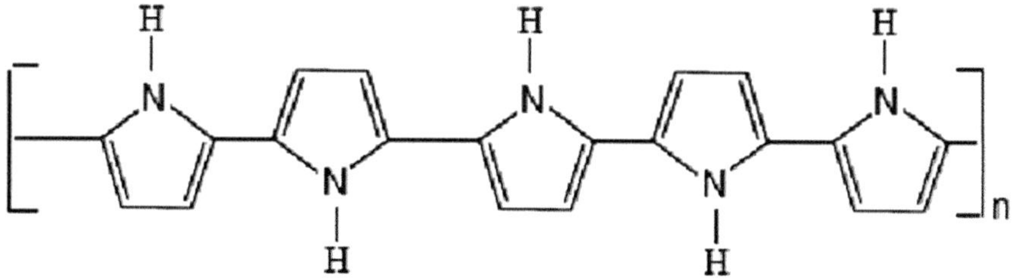

FIGURE 6.2 Structure of polypyrrole.[12]

Source: Reprinted from Ref. [12]. https://creativecommons.org/licenses/by/4.0/

6.2.1.1 *BIOSENSORS*

Polypyrrole is a widely used conducting polymer in the application of biosensors due to its unique properties which are highly recommended in biomedical applications such as good absorptive properties toward gases, DNA, proteins, chemical stability in the water as well as in air, easily synthesized by electrochemically and chemical methods. It is also redox-active, biocompatible (in vivo and in vitro), and corrosion-resistant.[16,17] Sensors are basically devices that are able to sense the changes in a physical quantity such as temperature, pressure, and humidity, and converts into a measurable and/or analyzable quantity while biosensors or biological sensors are analytical devices that consist of a transducer and a biological element (enzyme, antibody, microorganism, nucleic acid, etc). These biological elements interact with the analyte and detect the changes that are converted into electrical signals by the transducer. There are different types of biosensors based on biological elements and on the sensing devices. Some of the biosensors are enzyme-based biosensors, tissue-based biosensors, immunosensors, and DNA-based biosensors.[14,15] Despite these, another classification is based on the interaction between the biological element and the analyte. They are catalytic biosensor and affinity biosensors.

Aptamers are the small, stranded oligonucleotides which are able to bind a specifically targeted biomolecule with high affinity. They are not immunogenic and nontoxic.[18] Kwon et al. introduced a flexible FET aptasensor for the detection of vascular endothelial growth factor by using polypyrrole-converted nitrogen doped few-layer graphene (PPy-NDFLG). The PPy-NDFLG was fabricated by vapor deposition polymerization on the $CuCl_2$/Cu substrate. The chemical oxidative polymerization was started when the vapor-phased pyrrole monomers interacted with Cu cation. The thickness of the obtained polypyrrole film was about 20 nm. The PPy-NDFLG film was transferred into a transparent and flexible substrate.

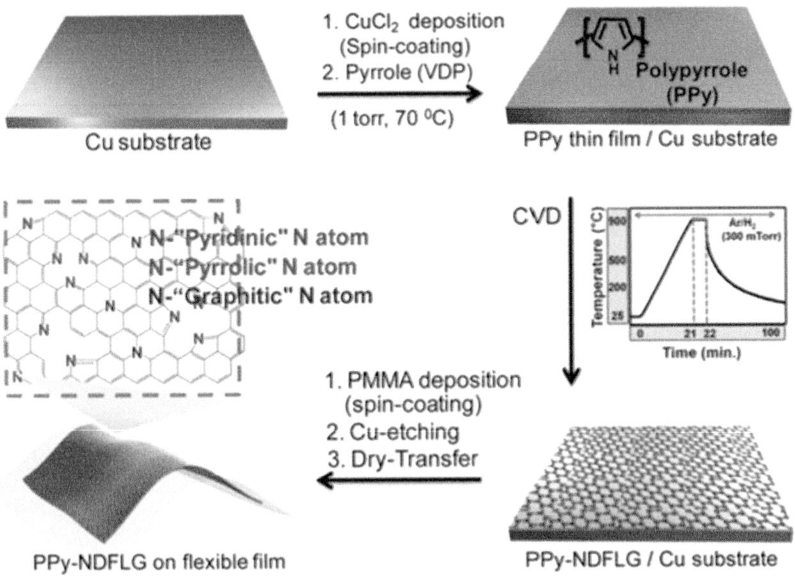

FIGURE 6.3 Synthesis of PPy-NDFLG flexible film.[19]

Source: Reprinted from R ef. [19] © 2012. American Chemical Society

For the fabrication of FET-type biosensors, the PPy-NDFLG was modified with glutaraldehydeconjugated 1,5-diaminonaphthalene (DAN) on its side plane. The aptamer introduced on to the DAN/PPy-NDFLG. The aptasensor showed the ultrasensitivity, high flexibility, and reusability. Also, the minimum detection level was ca. 100 fM.[19]

The affinity-based immunosensors are having a large potential in the bioanalytical field in near future. These biosensors are very sensitive and able to detect the concentration of biomolecules in nanomolar to femtomolar. They are capable of detecting the cancerous molecules, pathogens, bacteria, and viruses. The immunosensors work on the formation of antigen–antibody complexes and the response are converted into electrical signals by the transducer. The interaction between antigen–antibody is noncovalent in nature, but the affinity is very strong between them.[20] Mishra et al. proposed the immunosensor for C-reactive protein prepared by electrochemically synthesized ZnS nanocrystals-polypyrrole nano-composite film. It provided good biocompatibility and showed good binding efficiency toward the protein antibody molecules. This immunosensor was provided the linear impedance response to αCRP-Ag concentration from 10 ng.mL^{-1} to 10 μg.mL^{-1}.[22] Amouzadeh Tabrizi et al. developed the label-free and highly sensitive human immunoglobulin G immunosensor with overoxidized polypyrrole which was decorated with silver nanoparticles. The overoxidized polypyrrole film was porous and showed different permeability as compared to polypyrrole film. The electrochemical immunosensor was fabricated by the electrochemically synthesized OPPy-AgNP film on the surface of screen-printed electrode. The film was highly permeable to $[Fe(CN)_6]^{3-/4-}$. To immobilize the anti-HIgG on the surface of OPPy-AgNP/SPE, it was immersed in an anti-HIgG working solution. This immunosensor detected the HIg G in the range of 0.5–12.50 ng.mL^{-1} and MDL was 0.2 ng.mL^{-1}.[21]

6.2.1.2 TISSUE ENGINEERING

Polypyrrole is capable of supporting the cell adhesion and cell growth like glia cells dorsal root ganglia,[24, 25] mesenchymal stem cells,[26] and endothelial cells.[27,28] Wong et al. reported that electrically conducting polymers are suitable for controlling the cell function and its growth. For small- and large-scale cell cultures, the electrically conductive polymers including polypyrrole can be used as substrate as they are able to control the cells function. Also, the Fibronectin (FN) precoated polypyrrole helped to aortic endothelial cell attachment and its spreading in the serum-free medium.[29] Schmidt et al. reported the effectiveness of polypyrrole conductive polymer to increase the nerve cell differentiation. The autologous nerve grafting is a recent clinical approach for the peripheral nerve repairing but it has many disadvantages such as dimensions, limited tissues, and morbidity. To overcome the issues, the conducting polymers are successfully used which have biocompatibility and provide the neuronal guidance. The oxidized polypyrrole is an appropriate candidate for nerve cell culture and in vivo implantation. The PSS-doped polypyrrole improved and increased the neurite outgrowth of PC-12 cells. Also, the neurite length was increased by 90%.[30] Gomez et al. studied the modification of polypyrrole with the immobilization of nerve growth factor (NGF) on polypyrrole by using the PAA-azido linker. With this, the cell adhesion was enhanced and 50% of axon outgrowth was increased. It could be used in nerve regeneration applications.[31]

6.2.2 POLY (3,4-ETHYLENEDIOXYTHIOPHENE)

Poly (3,4-ethylenedioxythiophene (PEDOT) electrically conductive polymer is used in the biomedical field because of its chemical stability, high electrical conductivity, oxidative stability, optically transparent, low redox potential, and good electrochemical properties.[32] After oxidation, a long conjugated polymeric chain is formed with the positive charge. For the stabilization of a positive charge, an appropriate counterion is needed. The sulfonic groups are the best candidate for the stabilization of the positive charge of PEDOT. PEDOT: PSS is biocompatible in nature and has potential applications in the biomedical field.

Jeong et al. discussed the hydrothermal treatment for the electrical conductivity enhancement of PEDOT: PSS films used in biomedical applications. In the heat treatment of PEDOT: PSS film, the >80% relative humidity and the temperature >61°C were used. This results in a sharp increase in the electrical conductivity from 0.495 S/cm to 125.367 S/cm. The treated film also showed good mechanical flexibility.[33] Polypyrrole has been widely used in the biomedical industry due to its unique advantages. But it has some challenges such as, in some under biomedical environment, it becomes unstable and insulative, unstable under potential applications in long-term stability. To overcome these challenges the PEDOT polymer is widely accepted because of electrical properties, biocompatibility, and directly polymerized in living tissues.[34] Luo et al. studied the biocompatibility of PEDOT film by using the NIH3T3 fibroblast cell. The result showed the low intrinsic cytotoxicity and there was no inflammatory response at the implantation site.[34]

KEYWORDS

- **conducting polymers**
- **conjugated polymers**
- **biomedical**
- **electrical conductivity**
- **biosensors**
- **artificial muscles**

REFERENCES

1. Chen, C.; Bai, X.; Ding, Y.; et al. Electrical Stimulation as a Novel Tool for Regulating Cell Behavior in Tissue Engineering. *Biomater. Res.* **2019,** *23*, 25.
2. Kumar, R., Singh, S.; Yadav, B. C. Conducting Polymers: Synthesis, Properties and Applications. *Int. Adv. Res. J. Sci. Eng. Technol.* **2015,** *2* (11).
3. Gochnauer Daniel, L.; Gilani, T. H. Conduction Mechanism in Electrically Conducting Polymers. *Am. J. Undergrad. Res.* **2018,** *14* (4).
4. Aoai, T.; Nishio, R.; Hayashi, N.; Nomura, K. Photo Doping Process with PAG and Application for Thermoelectric Materials. *J. Photopolym. Sci. Technol.* **2016,** *29* (2), 335–341.
5. Hosseini, A. R.; Hoi Wong, M.; Shen, Y.; Malliaras, G. G. Charge Injection in Doped Organic Semiconductors. *J. Appl. Phys.* **2005,** *97*, 023705.
6. Monika, G. R. A Review on the Conducting Polymers. *Int. J. Adv. Sci. Res.* **2018,** *3* (3), 6–13.
7. Le, T.-H.; Kim, Y.; Yoon, H. Electrical and Electrochemical Properties of Conducting Polymers. *Polymers* **2017,** *9*, 150.
8. Rolland, M.; Aldissi, M.; Bernier, M.; Cadene, M.; Schue, F. Morphology of Iodine-doped Polyacetylene. *Nature* **1981,** *294* (5836), 60–61.
9. Heydari Gharahcheshmeh, M.; Gleason, K. K. Texture and Nanostructural Engineering of Conjugated Conducting and Semiconducting Polymers. *Mater. Today Adv.* **2020,** *8*, 1000862.
10. Semenof, G. W.; Sodano, P. Stretching the Electron As Far As It Will Go. *Electron. J. Theor. Phys.* **2006.**
11. Le, T.-H.; Kim, Y.; Yoon, H. Review on Electrical and Electrochemical Properties of Conducting Polymers. *Polymers* **2017,** *9*, 150.
12. Popli, S. A.; Patel, U. D. Electrochemical Decolourization of Reactive Black 5 in an Undivided Cell Using Ti and Graphite Anodes: Effect of Polypyrrole Coating on Anode. *J. Electrochem. Sci. Eng.* **2015,** *5* (2), 145–156.
13. Rahaman, M.; Aldalbahi, A.; Almoiqli, M.; Alzahly, S. Chemical and Electrochemical Synthesis of Polypyrrole Using Carrageenan as a Dopant: Polypyrrole/Multi-Walled Carbon Nanotube Nanocomposites. *Polymers* **2018,** *10*, 632.
14. Fatoyinbo, H. O.; Hughes M. P. Biosensors. In *Encyclopedia of Nanotechnology*; Bhushan, B. Ed.; Springer: Dordrecht, 2012.
15. Mehrotra, P. Biosensors and Their Applications - A Review. *J. Oral Biol. Craniofac. Res.* **2016,** *6* (2), 153–159.
16. Thomas, S.; Shanks, R.; Chandran, S. Polymers with Nano-encapsulated Functional Polymers. I. Kong. In *Design and Applications of Nanostructured Polymer Blends and Nanocomposite System*; William Andrew: UK, 2015; pp 126–147.
17. Ramanavicius, A.; Ramanaviciene, A.; Malinauskas, A. Electrochemical Sensors Based on Conducting Polymer – Polypyrrole (Review). *Electrochim. Acta* **2006,** *51*, 6025–6037.
18. Lakhin, A. V.; et al. Aptamers: Problems, Solutions and Prospects. *Acta Nat.* **2013,** *5* (4), 34–43.
19. Seok Kwon, O.; Joo Park, S.; Hong, J.-Y.; Han, A.-R.; Seop Lee, J.; Lee, J. S.; Hak Oh, J.; Jang, J. Flexible FET-Type VEGF Aptasensor Based on Nitrogen-Doped Graphene Converted from Conducting Polymer. *ACS Nano* **2012,** *6* (2), 1486–1493.

20. Moina, C.; Gabriel, Y. *Fundamental and Application of Immunosensors*; 2012.
21. Amouzadeh Tabrizi, M.; Shamsipur, M.; Mostafaie, A. A High Sensitive Label-free Immunosensor for the Determination of Human Serum IgG Using Overoxidized Polypyrrole Decorated with Goldnanoparticle Modified Electrode. *Mater. Sci. Eng. C.* **2016,** *59*, 965–969.
22. Mishra, S. K.; Pasricha, R.; Biradar, A. M.; Rajesh. ZnS nanocrystals/Polypyrrole Nanocomposite Film Based Immunosensor. *Appl. Phys. Lett.* **2012,** *100* (5).
23. Huang, Z.-B.; Yin, G.-F.; Liao, X.-M.; Gu, J.-W. Conducting Polypyrrole in Tissue Engineering Applications. *Front. Mater. Sci.* **2014,** *8* (1), 39–45.
24. Wang, X.; Gu, X.; Yuan, C.; et al. Evaluation of Biocompatibility of Polypyrrole In Vitro and In Vivo. *J. Biomed. Mater. Res. A* **2004,** *68A* (3), 411–422.
25. Cangellaris, O. V.; Gillette, M. U. Biomaterials for Enhancing Neuronal Repair. *Front. Mater.* **2018,** *5*, 21.
26. Castano, H.; O'Rear, E. A.; McFetridge, P. S.; Sikavitsas, V. I. Polypyrrole Thin Films Formed by Admicellar Polymerization Support the Osteogenic Differentiation of Mesenchymal Stem Cells. *Macromol. Biosci.* **2004,** *4* (8), 785–794.
27. Garner, B.; Hodgson, A. J.; Wallace, G. G.; Underwood, P. A. Human Endothelial Cell Attachment to and Growth on Polypyrrole-heparin is Vitronectin Dependent. *J. Mater. Sci. Mater. Med.* **1999,** *10*, 19–27.
28. Fahlgren, A.; Bratengeier, C.; Gelmi, A.; Semeins, C. M.; Klein-Nulend, J.; Jager, E. W. H.; Bakker, A. D. Biocompatibility of Polypyrrole with Human Primary Osteoblasts and the Effect of Dopants. *PLOS One* **2015.**
29. Wong, J. Y.; Langer, R.; Ingber, D. E. Electrically Conducting Polymers Can Noninvasively Control the Shape and Growth of Mammalian Cells. *Proc. Natl. Acad. Sci. USA* **1994,** *91* (8), 3201–3204.
30. Schmidt, C. E.; Shastri, V. R.; Vacanti, J. P.; et al. Stimulation of Neurite Outgrowth Using an Electrically Conducting Polymer. *Proc. Natl. Acad. Sci. USA* **1997,** *94* (17), 8948–8953.
31. Gomez, N.; Schmidt, C. E. Nerve Growth Factor-immobilized Polypyrrole: Bioactive Electrically Conducting Polymer for Enhanced Neurite Extension. *J. Biomed. Mater. Res. A* **2007,** *81A* (1), 135–149.
32. Sultana, N.; Chung Chang, H.; Jeferson, S.; Daniels, D. E. Application of Conductive Poly(3,4-ethylenedioxythiophene): Poly (styrenesulfonate) (PEDOT:PSS) Polymers in Potential Biomedical Engineering. *J. Pharm. Investig.* **2020.**
33. Jeong, W.; Gwon, G.; Ha, J.-H.; Kim, D.; Eom, K.-J.; Hyang Park, J.; Ju Kang, S.; Kwak, B.; Hong, J.-I.; Lee, S.; Choon Hyun, D.; Lee, S. Enhancing the Conductivity of PEDOT:PSS Films for Biomedical Applications via Hydrothermal Treatment. *Biosens. Bioelectron.* **2021,** 171.
34. Luo, S. C.; Mohamed Ali, E.; Tansil, N. C.; Yu, H. H.; Gao, S.; Kantchev, E. A.; Ying, J. Y. Poly(3,4-ethylenedioxythiophene) (PEDOT) Nanobiointerfaces: Thin, Ultrasmooth, and Functionalized PEDOT Films with In Vitro and In Vivo Biocompatibility. *Langmuir* **2008,** *24*, 8071–8077.

CHAPTER 7

Biopolymer Sustainable Films for Food Industries: Properties and Application Based on Chitosan

G. SANTHOSH[1] and AARTHI S. BHATT[2]

[1]*Department of Mechanical Engineering, NMAM Institute of Technology (Deemed to be University), Udupi, Karnataka, India*

[2]*Department of Chemistry, NMAM Institute of Technology (Deemed to be University), Udupi, Karnataka, India*

ABSTRACT

Use of biopolymers in food technology is creating a lot of attention in response to the challenges from the modernization of the society. Food packaging technology is a unique approach to improving the shelf life of perishables while keeping quality and safety in mind. Deacetylated chitosan derived from chitin is the second largest polysaccharide found in nature. Chitosan is a widely used environmentally friendly polymer in many scientific applications as it is nontoxic, biodegradable, and biocompatible. Further, chitosan can be a good inhibitor for the growth of bacteria and fungus and therefore is widely used in food packaging applications. Chitosan also demonstrates gas and aroma barrier behaviors in dry environment. Besides, chitosan can exhibit good antimicrobial behavior for food packaging applications as well. This chapter aims to present the properties and applications of chitosan polymer films used in food packaging applications.

Tailored Functional Materials for Clean and Sustainable Development. Tamara Tatrishvili, PhD, Neha Kanwar Rawat, PhD, Ann Rose Abraham, PhD & A. K. Haghi, PhD (Eds.)

7.1 INTRODUCTION

In the present-day scenario, food safety and environmental issues have raised some concerns among the scientific community. The challenges are longer shelf life and safe and healthy food without environmental concerns.[1] To address these issues, researchers have discovered bio-based polymer films for food-packing applications. Biopolymers are an excellent alternative which are widely investigated and are sustainable solutions to the plastic food packaging materials.[2,3] These days, one must be aware of the side effects of chemicals such as parabens, benzoic acids, and nitrites used in the food industry. Further, food spoilage because of microbial action is creating huge food loss; around 25% of the food is wasted before consumption.[4,5] Further, biopolymer films can be brilliant matrix for a variety of functional materials which can improve the quality and shelf life of the food.[6,7] Such biopolymers include cellulose, protein, lipids, chitosan, and starch. Chitosan, a derivative of chitin is predominantly obtained from the exoskeleton of shrimps and crabs.[8] Chitosan is obtained from the deacetylation process, partially from N-acetylated chitin and it is a linear amino polysaccharide with Dglucosamine and N-acetyl-D-glucosamine units.[9,10] Chitosan is an excellent biopolymer used in food, medicine, agriculture environmental, textile, and bioengineering fields due to its biocompatible, nontoxic; antimicrobial, and biodegradable properties.[11] Figure 7.1 shows a graphical representation of food being protected from microbial infection.

FIGURE 7.1 Protecting food from microbial infection.

7.2 SOURCES OF CHITOSAN

Chitosan is the most abundant polymer which is extracted from the exoskeleton of crustaceans, invertebrates, and fungi. Chitosan is composed of *β*-(1-4)-linked d-glucosamine and *N*-acetyl-d-glucosamine.[12] Chitosan molecules are of linear polysaccharide chains and have three reactive groups, "an amino group at C-2, and both primary at C-3 and secondary hydroxyl groups at C-6." Chitosan is nontoxic, biocompatible, and biodegradable polymer that provides easy access to modify the molecules.[13] The modified chitosan can be used in various industrial applications and biological processes like wastewater treatment and food packaging. The food packaging application of chitosan is shown in Figure 7.2.[14,15]

7.3 CHITOSAN EXTRACTION

7.3.1 CHEMICAL DEPROTEINIZATION

involves the distraction of various chemical bonds of chitin and proteins. The chemical deproteinization of chitosan undergoes heterogeneously with the help of chemicals that are also responsible for the depolymerization of biopolymers. In the chemical deproteinization process, a variety of chemicals are used to deproteinize the chitin which includes NaOH, Na_2CO_3, KOH, K_2CO_3, Na_2SO_3, Na_3PO_4, etc.[16]

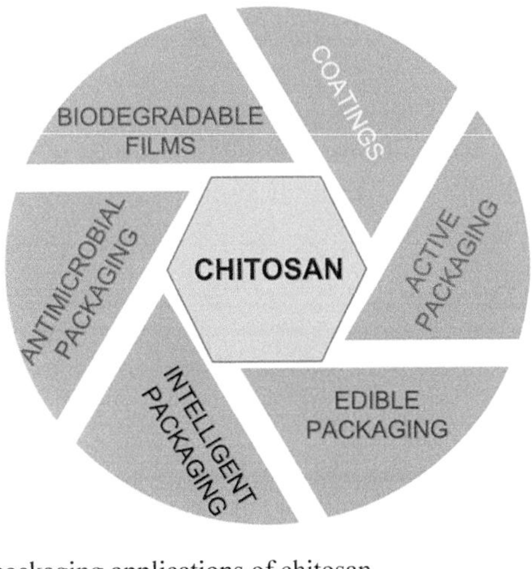

FIGURE 7.2 Food packaging applications of chitosan.

7.3.2 CHEMICAL DEMINERALIZATION

Demineralization is a process to remove minerals and calcium carbonate by acids such as HCl, HNO_3, and H_2SO_4, and others.[17] The following equation depicts the demineralization process. The demineralization is achieved by using dilute HCl, the acid absorbs the decomposition of calcium carbonate with the release of carbon dioxide as shown in the below equation.

$$2HCl + CaCO_3 \rightarrow CaCl_2 + H_2O \uparrow$$

7.3.3 BIOLOGICAL EXTRACTION OF CHITIN

The extraction of chitosan by means of the chemical route has some disadvantages which are as follows: (i) the chemical hydrolysis alters the physicochemical properties of chitin influencing its biological properties and leading to the decrease of deacetylation and molecular weight, (ii) increases the rate of the purification process, and (iii) stimulates the waste fluid flow with various chemicals. Hence, the use of green/eco-friendly extraction of chitosan has greater importance. The green extraction methods also favor the use of enzymes and microorganisms in extracting chitosan. The biological/green extraction techniques endorse high reproducibility with lesser solvent utilization.[18–20]

7.3.4 ENZYMATIC DEPROTEINIZATION

The enzymatic deproteinization process involves the proteases extracted from plants, animals, and microbes in the extraction of chitin. The proteases extracted from the biological sources remove proteins from the shells and reduce deacetylation and depolymerization during chitin extraction.[21] The deproteinization step is usually done before and after the demineralization process.

7.4 CHITOSAN PROPERTIES

Chitosan is a linear homopolymer that is made of β-(1,4)-linked N-acetyl-glucosamine units.[22–24] Chitosan is obtained by deacetylation of chitin,

chitin is a polysaccharide found in crustaceans, insect exoskeletons, and in a few cell walls.[25] Chitosan quality can be determined by the source of chitin and its degree of deacetylation.[26] Chitosan is an attractive polymer used in biomedical applications due to its excellent biological, nontoxic, biodegradable, and antimicrobial properties. One of the most notable properties of chitosan is that it does not cause severe inflammation or activate the immune system. The toxicity of chitosan can be reduced by the deacetylation process, the process is like that of succinyl-derived chitosan and chitosan nanoparticles.[27,28]

Deacetylated chitin is a natural polycationic polymer that has bactericidal and bacteriological properties. The positive charge of this polymer adheres to the bacterial surface which brings changes in the membrane wall preventing microbial growth.[29] The deacetylation process improves the antibacterial activity of the chitosan especially when chitosan having low pH. Reduction in molecular weight also increases antibacterial activities of chitosan.[30] The interaction of bacteria and chitosan depends on the hydrophilicity of the cell wall, which makes chitosan a better antitoxic element to mammalian cells.[31]

Antioxidants have proven to be most beneficial to health. chitosan and its derivatives as antioxidants have shown the ability to prevent free active oxygen radicals. The antioxidants are primarily used to prevent the membrane destruction, destruction of proteins, and DNA.[32] Especially, low-weight chitosan molecules have proven to be very effective in eliminating free radicals.[33]

7.5 CHITOSAN DERIVATIVES

As we all know that chitosan is a natural resource which is a deacetylated product of chitin; however, its insolubility in water and other organic solvents makes it difficult to use it in various applications. The better solution to this problem is its derivatives. Chitosan derivatives obtained from chemical modification may increase its solubility and other properties as well. Chitosan with hydroxyl and amino groups undergo many chemical reactions to form pendant groups by destroying the crystal structure of chitosan which ultimately increases the solubility of the chitosan. The chitosan derivatives formed by the chemical reactions such as hydroxylation, carboxylation, alkylation, acylation, and esterification will increase the physicochemical and biological behaviors which are essential for the biomedical applications.[34]

7.5.1 ACYLATED CHITOSAN

This is the most common and widely used to modify chitosan. The acylation process refers to the interaction of chitosan with organic acids and its derivatives such as anhydride and acyl chloride.[35] The water solubility of the chitosan can be increased by the acylation process, this process destroys inter and intramolecular hydrogen bonding which reduces the crystallinity of chitosan thereby enhancing the solubility.

7.5.2 ALKYLATED CHITOSAN

Alkylated chitosan having an alkyl group weakens the intermolecular hydrogen bonding leading to better solubility.[36] However, the alkyl group is a hydrophobic group, and the solubility of chitosan can be controlled by manipulating the length of an alkyl chain.[37,38] Alkylation of chitosan is caused by introducing the alkyl groups to the C_2–NH_2, C_6–OH, or C_3–OH groups of chitosan.[39,40]

7.5.3 HYDROPHILIC GROUP

Hydrophilic groups are the most compatible groups with water, hydrophilic groups include carboxylic acid group, the quaternary ammonium group, the sulfonic acid group, the phosphoric acid group, and many more.

7.5.3.1 CARBOXYLATED CHITOSAN

Carboxylated chitosan is produced by carboxylation reaction. In the reaction, glyoxylic acid or chloroalkanoic acid is reacted with either C_2–NH_2 or C_6–OH groups of chitosan, as a result –COOH groups are formed.[41] Carboxylated chitosan can dissolve in neutral and alkaline solutions with better thickening and heat-preventing properties.[42] Carboxylated chitosan is useful in agricultural, medical, health, and biochemical applications.[43]

7.5.3.2 QUATERNARY AMMONIUM CHITOSAN

The quaternary ammonia group is a positively charged hydrophilic group. The quaternary ammonia groups in chitosan increase the solubility and chargeability. The process of introducing quaternary groups into chitosan is referred as quaternization; the quaternization occurs with C_2–NH_2 groups of chitosan. The quaternary ammonia is formed via three different methods namely direct quaternary ammonium substitution, N-alkylation, and the epoxy derivative open loop method.[44–48]

7.5.3.3 ESTERIFIED CHITOSAN

The esterification process of chitosan takes place due to the presence of oxygen-containing inorganic acids, the commonly used acids in the esterification process are phosphoric acid, sulfuric acid, and chlorosulfonic acid.[49] One of the most widely used chitosan derivatives formed through esterification is sulfated chitosan. Sulfated chitosan is used as a substitute for heparin as an anticoagulant and antiviral drug. The sulfated chitosan exhibits the regulatory mechanism same as heparin.[50]

7.6 CHITOSAN FILMS

Chitosan is a linear homopolymer which offers huge benefits as an edible packaging material due to its good-film forming ability.[51–53] However, chitosan inability to extruding or molding behavior and the films cannot be heat sealed, this nature of chitosan limits their production at commercial scale.[54] However, chitosan can be mixed with other thermoplastics to enhance its thermal behavior, allowing it to process by extrusion.[55] The physical and functional properties of chitosan depend on various factors like molecular weight, degree of deacetylation, acid concentration, and plasticizers used. The acids and plasticizers used in the film preparation have a considerable effect on the mechanical property of the film.[56] Pure chitosan films are rigid and brittle in nature; hence plasticizers are used to reduce the rigidness of the polymer chains to improve the overall mechanical strength of the films.[57] However, this blending increases the water permeability of the films and sometimes reduces the gas and moisture barrier nature of the films. The use of plasticizers improves the free volume and molecular mobility in the polymer chain. The molecular mobility of the polymer chain makes it denser and more permeable.[58] Chitosan-based films have attracted huge attention as a food preservative due to its stronger antioxidant and antimicrobial nature.[59]

7.7 CHITOSAN FILM IN FOOD PACKAGING

The use of chitosan in food packaging application is either in the form of flexible films or as thin film coatings. The flexible films used in food packaging are either edible or inedible. However, the chitosan coatings are edible films coated directly on the perishables. Various additives have been used in various flexible food packaging films to impart the good properties, unlike coatings, which are constituted with minimum or no additives.

7.7.1 FLEXIBLE FILMS

Flexible food packaging films are prepared by using solvent casting technique; in this process, the chitosan is dissolved in acidified water.[60] The ultimate properties of the flexible films depend on degree of deacetylation, molecular weight, and category of acids used. Further, pH of the solvent will also decide the water vapor permeability of the film forming solution. From various research, it is found that the water vapor permeability value of the films obtained using acetic acid and propionic acid are better compared to that of other acids.[60]

The mechanical properties especially tensile strength of the flexible films are very crucial from the food packaging perspective. The tensile strength if the chitosan is found to be on par with high-density polyethylene, low-density polyethylene, and cellophane.[61] However, the use of chitosan as a food packaging film is restricted as it is sensitive to moisture and humid environment. To overcome these limitations, chitosan is blended with natural and synthetic polymers, few additives like fillers, plasticizers, and cross-linkers are also used.

7.7.1.1 CHITOSAN FILMS WITH NANOMATERIALS

Metal and metal oxide nanoparticles are one among various other materials used as fillers in food packaging applications. Metal ions such as silver, copper, and zinc are essential minerals abundantly available in nature; the use of these metal nanoparticles as fillers in polymers enhances their mechanical, thermal, and barrier properties. Further, these metal nanoparticles exhibit outstanding antimicrobial and antioxidant properties to enhance the life of the polymer film. Silver is one among such nanoparticles which is widely used in food packaging films due to its remarkable antimicrobial behavior.[62] The use of silver nanoparticles in chitosan films by Tripathi et al. shows improved antimicrobial behavior against different bacteria.[63] Vimala et al. used silver nanoparticles to enhance the mechanical properties with improved antimicrobial behavior of the chitosan films.[64] Priyadarshini et al. studied the effect of zinc oxide nanoparticles used in chitosan films, the study shows improved mechanical and antimicrobial activity of the film.[65] Another study reported by Rodrigues et al. also shows enhanced mechanical properties of the chitosan films after been incorporated with zinc nanoparticles.[66] These studies reveal that zinc oxide nanoparticles are safe and can be used to enhance the overall properties of the films. Multi-metal oxide nanoparticles are also considered safe and are being increasingly used in food packaging applications. Silver–zinc oxide nanoparticles used in chitosan were found to be very effective against various food pathogens.[67]

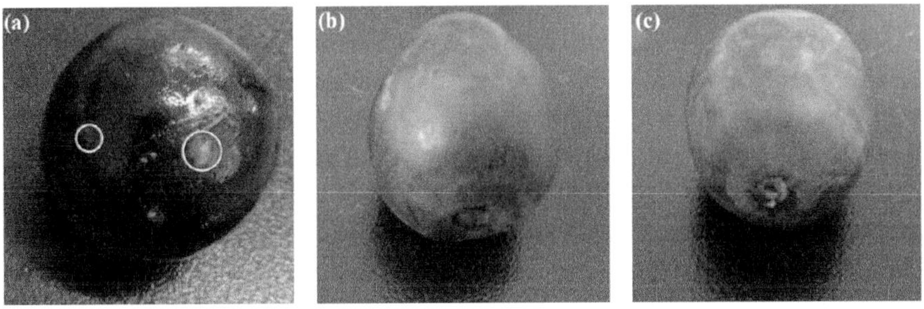

FIGURE 7.3 Grape coating: (a) plastic wrap; (b) pure chitosan film; (c) chitosan-TiO_2 film.

Titanium dioxide is another metal oxide that was studied by incorporating it in the chitosan film, Zhang et al., used titanium–chitosan films over grapes, grapes with titanium–chitosan film were kept for 6 days and found to be protected from microbial attack.[68] Figure 7.3 shows various coatings on grape, the grape coated with chitosan–titanium dioxide film is healthier compared to other two which is also protected from microbial attack. Cardenas et al. also prepared copper nanoparticle-incorporated chitosan films which exhibited excellent antimicrobial and mechanical properties.[69] Carbon allotropes like graphene, carbon nanotubes are also used as nanofillers in food packaging. Ahmed et al. fabricated chitosan/graphene nanocomposites films, the films exhibited better mechanical strength with increased barrier properties.[70] Multi-walled carbon nanotubes used as fillers in chitosan films, the composite films show a 40% increase in the mechanical properties.[71]

7.7.1.2 CHITOSAN FILMS WITH CLAY MINERAL

Polymer-clay composites have been considered promising candidates for food packaging applications due to amazing properties of clay minerals. The naturally available clay mineral's compact structure exhibits good barrier property as it exfoliates in the polymer matrix. The polymer–clay composites exhibit superior mechanical strength due to the intercalation of clay into the polymer matrix. The polymer–clay composites' good behavior makes them suitable in various applications. The use of clay mineral with chitosan for food packaging application is gaining popularity in recent decades even very small concentration of clay mineral can change the properties of the films. Casariego et al. prepared chitosan-nanoclay composite films with varying concentrations of nanoclay, the results suggest that the water vapor permeability of the film decreases with the increase in clay concentration.[72] Zhang et al. incorporated montmorillonite clay particles in chitosan–PVA matrix. It is found that the interfacial interaction between the filler and polymer matrix. The composite films showed better mechanical and barrier properties, it was found that the young's modulus and tensile strength of the composite film increases with the increase in filler content.[73] Many literatures have been found in the similar line, Rhim et al., Kusmono et al., and Swain et al. also studied the effect of clay mineral in chitosan-based films, it was observed that moisture-barrier properties, thermal stability, gas permeability, and mechanical properties of the composite films increase with the increase in filler content,[74–76] making them suitable for food packaging application.

7.8 APPLICATIONS OF CHITOSAN-BASED FILMS

7.8.1 SHELF-LIFE EXTENSION

The agricultural products such as vegetables and fruits have very short postharvest life as they exhibit continuous respiration and transpiration. The ripening process of fruits is a very complex process and is caused due to physiological and biological changes in the fruits. The source of these changes is chlorophyll, cell wall degradation, variation in sugar level, ethylene production, and the production of reactive oxygen species in the mitochondria are responsible for the ripening of perishables.[77–79] Therefore, it is very much essential to use natural nontoxic materials to protect and slow down the ripening process of the perishables postharvest. Martínez et al. reported that after the use of chitosan with Thymus capitatus oil on strawberry fruit increases its life for up to 15 days.[80]

TABLE 7.1 Chitosan-Based Nanocomposite Films Coated on Various Fruits and Vegetables.

Sl. No.	Chitosan-based composite films	Nanofiller	Fruits	References
1	Chitosan–montmorillonite	Montmorillonite	Tangerine fruits	81
2	Chitosan/TiO_2	TiO_2	Red grapes	82
3	Chitosan/nano ZnO	ZnO	Ladyfinger	83
4	Chitosan/gelatin/AgNPs	Ag	Red grapes	85

Coating of bio-polymer films has become more popular, inexpensive, and effective technique to preserve perishables. Chitosan-based films gaining more attention in food packaging applications due to their biodegradable nature and inertness to microbial activity.[84,85] Table 7.1 shows the chitosan-based film coatings with different nanofillers on various fruits to protect them from microbes.

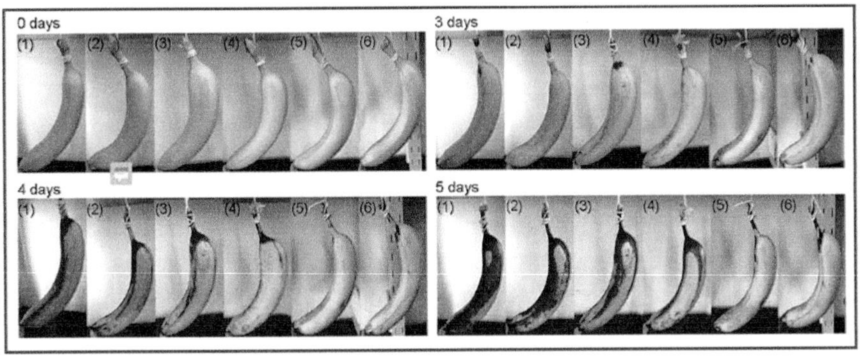

FIGURE 7.4 Ripening banana during storage: (1) uncoated, (2) coated with carboxymethyl only, (3) coated with chitosan/carboxymethyl (40/60), (4) coated with chitosan/carboxymethyl (70/30), (5) coated with chitosan/carboxymethyl (90/10), and (6) coated with chitosan only.

Source: Reprinted with permission from Ref. [87] © 2016 Elsevier.

Chitosan-based composite films with different fillers have found to be a practical solution in improving the shelf-life and quality of agricultural produce.[86] Hu et al. studied the ripening of bananas for several days. The fruit was coated with chitosan/carboxymethyl composite film which increases the antioxidant enzymes preventing membrane damage and ripening.[87] The ripening process of the banana fruit is shown in Figure 7.3.

7.8.2 ANTIMICROBIAL ACTIVITY

Bacteria, fungus, viruses, and protozoa are infectious microorganisms which are a major concern in food industries worldwide. The infection caused by these microorganisms can pose serious health issues. Chitosan and its derivatives can be of great interest due to their biological and physiological behavior.[88] Chitosan is deacetylated from chitin, which is a linear cationic polysaccharide. The polycationic nature of chitosan can kill bacteria, fungi, and yeast.[89] The interaction between negatively charged microbial membranes and polycationic nature of chitosan produces intracellular constituents, creates an osmatic imbalance, and inhibits the growth of microorganisms.[90] Chitosan-based films used for food packaging show strong antimicrobial activity against pathogens and bacterial growth.[91]

7.9 CONCLUSIONS

The current food packaging strategies use the most popular synthetic plastics. The use of these petroleum-based materials creates huge pressure on the consumption of nonrenewable sources. The nondegradable nature of these plastics is also polluting the various natural resources on earth. This is where we find a possible solution to address this environmental issue with necessary alternatives, which are biodegradable, biocompatible, stronger, cost-effective, and immune to various microbes and bacteria. Numerous research works have been published on the use of various biopolymers in the food packaging industry. Chitosan is among the widely appreciated and studied biopolymers, which has been used to improve the shelf life of agricultural produce. The reinforced chitosan composites films also find their place in food packaging applications, chitosan with nanofillers, cross-linkers, and metal oxides show improved mechanical and barrier properties with the improvement in the quality and life of the food. Hence, the use of chitosan for food packaging increases life and protects perishables from various pathogens.

KEYWORDS

- **food packaging**
- **chitosan**
- **biodegradability**
- **biocompatibility**
- **antimicrobial**

REFERENCES

1. Realini, C. E.; Marcos, B. Active and Intelligent Packaging Systems for a Modern Society. *Meat Sci.* **2014,** *98*, 404–419.
2. Nitschke, J.; Altenbach, H.-J.; Malolepszy, T.; Mölleken, H. A New Method for the Quantification of Chitin and Chitosan in Edible Mushrooms. *Carbohydr. Res.* **2011,** *346*, 1307–1310.
3. Cooper, T. A. Developments in Bioplastic Materials for Packaging Food, Beverages, and Other Fast-moving Consumer Goods. *Trends in Packaging of Food, Beverages and Other Fast-Moving Consumer Goods (FMCG)* **2013,** *5*, 108–152.
4. Lopez-Carballo, G.; Gomez-Estaca, J.; Catala, R.; Hernandez-Munoz, P.; Gavara, R. Active Antimicrobial Food and Beverage Packaging. *Emerg. Food Pack. Tech.* **2012,** *3*, 27–54.
5. Huang, W.; Xu, H.; Xue, Y.; Huang, R.; Deng, H.; Pan, S. Layer-by-layer Immobilization of Lysozymechitosan–Organic Rectorite Composites on Electrospun Nanofibrous Mats for Pork Preservation. *Food Res. Int.* **2012,** *48*, 784–791.
6. Zemljic, L. F.; Tkavc, T.; Vesel, A.; Sauperl, O. Chitosan Coatings Onto Polyethylene Terephthalate for the Development of Potential Active Packaging Material. *Appl. Surf. Sci.* **2013,** *265*, 697–703.
7. Abdollahi, M.; Rezaei, M.; Farzi, G. A Novel Active Bionanocomposite Film Incorporating Rosemary Essential Oil and Nanoclay into Chitosan. *J. Food Eng.* **2012,** *111*, 343–350.
8. Abdou, E. S.; Nagy, K. S. A.; Elsabee, M. Z. Extraction and Characterization of Chitin and Chitosan from Local Sources. *Bioresour. Technol.* **2008,** *99*, 1359–1367.
9. Luiz Dotto, G.; de Souza, V. C.; de Moura, J. M.; de Moura, C. M.; de Almeida Pinto, L. A. Influence of Drying Techniques on the Characteristics of Chitosan and the Quality of Biopolymer Films. *Dry. Technol.* **2011,** *29*, 1784–1791.
10. Arvanitoyannis, I. S. Waste Management for Polymers in Food Packaging Industries. *Plastic Films Food Pack.* **2013,** *14*, 249–310.
11. Rinaudo, M. Chitin and Chitosan: Properties and Applications. *Prog. Polym. Sci.* **2006,** *31*, 603–632.
12. Ruocco, N.; Costantini, S.; Guariniello, S.; Costantini, M. Polysaccharides from the Marine Environment with Pharmacological, Cosmeceutical and Nutraceutical Potential. *Molecules* **2016,** *21* (5).

13. Vázquez, J. A.; Rodríguez-Amado, I.; Montemayor, M. I.; Fraguas, J.; del Pilar González, M.; Murado, M. A. Chondroitin Sulfate, Hyaluronic Acid and Chitin/Chitosan Production Using Marine Waste Characteristics, Applications and Eco-friendly Processes: A Review. *Mar. Drugs* **2013,** *11*, 747–774.
14. Karthik, R.; Manigandan, V.; Saravanan, R.; Rajesh, R. P.; Chandrika, B. Structural Characterization and In Vitro Biomedical Activities of Sulfated Chitosan from *Sepia pharaonis*. *Int. J. Biol. Macromol.* **2016,** *84*, 319–328.
15. Manigandan, V.; Karthik, R.; Saravanan, R. Marine Carbohydrate-based Therapeutics for Alzheimer's Disease: Mini Review. *J. Neuro. Neurosci.*, **2015,** 1–6 (Special issue).
16. Younes, I.; Rinaudo, M. Chitin and Chitosan Preparation from Marine Sources: Structure, Properties and Applications. *Mar. Drugs* **2015,** *13*, 1133–1174.
17. Hajji, S.; Younes, I.; Ghorbel-Bellaaj, O.; Hajji, R.; Rinaudo, M.; Nasri, M.; Jellouli, K. Structural Differences between Chitin and Chitosan Extracted from Three Different Marine Sources. *Int. J. Biol. Macromolec.* **2014,** *65*, 298–306.
18. Aytekin, O.; EliboL, M. Cocultivation of *Lactococcus lactis* and *Teredinobacter turnirae* for Biological Chitin Extraction from Prawn Waste. *Bioprocess Biosyst. Eng.* **2010,** *33*, 393–399.
19. Cardenas, L.; Dominguez, J.; Quinto, C.; Lopez-Lara, I. M.; Lugtenberg, B. J. J.; Spaink, H. P.; Rademaker, G. J.; Haverkamp, J.; Thomas-Oates, Isolation, J. E. Chemical Structures and Biological Activity of the Lipo-chitin Oligosaccharide Nodulation Signals from *Rhizobium etli*. *Plant Mol. Biol.* **1995,** *29*, 453–464.
20. Percot, A.; Viton, C.; Domard, A. Optimization of Chitin Extraction from Shrimp Shells. *Biomacromolecules* **2003,** *4*, 12–18.
21. Manni, L.; Ghorbel-Bellaaj, O.; Jellouli, K.; Younes, I.; Nasri, M. Extraction and Characterization of Chitin, Chitosan, and Protein Hydrolysates Prepared from Shrimp Waste by Treatment with Crude Protease from *Bacillus cereus* SV1. *Applied Biochem. Biotechnol.* **2010,** *162*, 345–357.
22. Kato, Y.; Onishi, H.; Machida, Y. Application of Chitin and Chitosan Derivatives in the Pharmaceutical Field. *Curr. Pharm.Biotechnol.* **2003,** *4*, 303–309.
23. Varshosaz, J. The Promise of Chitosan Microspheres in Drug Delivery Systems. *Drug Deliv.* **2007,** *4*, 263–273.
24. Galed, G.; Miralles, B.; Ines Panos, I.; Santiago, A.; Heras, A. N-Deacetylation and Depolymerization Reactions of Chitin/Chitosan: Influence of the Source of Chitin. *Carbohydr. Polym.* **2005,** *62*, 316–320.
25. Huang, M.; Khor, E.; Lim, L. Y. Uptake and Cytotoxicity of Chitosan Molecules and Nanoparticles: Effects of Molecular Weight and Degree of Deacetylation. *Pharm. Res.* **2004,** *21*, 344–353.
26. Chien, R.; Yen, M.; Mau, J. Antimicrobial and Antitumor Activities of Chitosan from Shiitake Stipes, Compared to Commercial Chitosan from Crab Shells. *Carbohydr. Polym.* **2016,** *138*, 259–264.
27. Aiping, Z.; Tian, C.; Lanhua, Y.; Hao, W.; Ping, L. Synthesis and Characterization of N-succinyl-chitosan and Its Self-assembly of Nanospheres. *Carbohydr. Polym.* **2006,** *66*, 274–279.
28. Yan, C.; Gu, J.; Hou, D.; Jing, H.; Wang, J.; Guo, Y.; Katsumi, H.; Sakane, T.; Yamamoto, A. Synthesis of Tat Tagged and Folate Modified N-succinyl-chitosan Self-assembly Nanoparticles as a Novel Gene Vector. *Int. J. Biol. Macromol.* **2015,** *72*, 751–756.

29. Goy, R. C.; Britto, D. D.; Assis Oiii, B. G. A Review of the Antimicrobial Activity of Chitosan. *Polímeros* **2009,** *19*, 241–247.
30. Younes, I.; Sellimi, S.; Rinaudo, M.; Jellouli, K.; Nasri, M. Influence of Acetylation Degree and Molecular Weight of Homogeneous Chitosan on Antibacterial and Antifungal Activities. *Int. J. Food Microbiol.* **2014,** *185*, 57–63.
31. Kong, M., Chen, X. G.; Xing, K.; Park, H. J. Antimicrobial Properties of Chitosan and Mode of Action: A State of the Art Review. *Int. J. Food Microbiol.* **2010,** *144*, 51–63.
32. Ngo, D. H.; Kim, S. K. Antioxidant Effects of Chitin, Chitosan, and Their Derivatives. *Adv. Food Nutr. Res.* **2014,** *73*, 15–31.
33. Park, P. J.; Je, J. Y.; Kim, S. K. Free Radical Scavenging Activity of Chitooligosaccharides by Electron Spin Resonance Spectrometry. *J. Agric. Food Chem.* **2003,** *51*, 4624–4627.
34. Islam, N.; Ferro, V. Recent Advances in Chitosan-based Nanoparticulate Pulmonary Drug Delivery. *Nanoscale* **2016,** *8*, 14341–14358.
35. Cai, J.; Dang, Q.; Liu, C.; Fan, B.; Yan, J.; Xu, Y.; Li, J. Preparation and Characterization of N-benzoyl-O-acetyl-chitosan. *Int. J. Biol. Macromol*. **2015,** *77*, 52–58.
36. Kurita, Y.; Isogai, A. N-Alkylations of Chitosan Promoted with Sodium Hydrogen Carbonate Under Aqueous Conditions. *Int. J. Biol. Macromol.* **2012,** *50*, 741–746.
37. Ma, G.; Yang, D.; Zhou, Y.; Xiao, M.; Kennedy, J. F.; Nie, J. Preparation and Characterization of Water-Soluble N-alkylated Chitosan. *Carbohydr. Polym.* **2008,** *74*, 121–126.
38. Yang, T. C.; Chou, C. C.; Li, C. F. Antibacterial Activity of N-alkylated Disaccharide Chitosan Derivatives. *Int. J. Food Microbiol.* **2005,** *97*, 237–245.
39. Burr, S. J.; Williams, P. A.; Ratclie, I. Synthesis of Cationic Alkylated Chitosans and An Investigation of their Rheological Properties and Interaction with Anionic Surfactant. *Carbohydr. Polym.* **2018,** *201*, 615–623.
40. Onésippe, C.; Lagerge, S. Studies of the Association of Chitosan and Alkylated Chitosan with Oppositely Charged Sodium Dodecyl Sulfate. *Colloids Surf. A* **2008,** *330*, 201–206.
41. Mohammadi, E.; Daraei, H.; Ghanbari, R.; Dehestani Athar, S.; Zandsalimi, Y.; Ziaee, A.; Maleki, A.; Yetilmezsoy, K. Synthesis of Carboxylated Chitosan Modified with Ferromagnetic Nanoparticles for Adsorptive Removal of Fluoride, Nitrate, and Phosphate Anions from Aqueous Solutions. *J. Mol. Liq.* **2019,** *273*, 116–124.
42. Kurniasih, M.; Cahyati, T.; Dewi, R. S. Carboxymethyl Chitosan As An Antifungal Agent on Gauze. *Int. J. Biol. Macromol.* **2018,** *119*, 166–171.
43. Zhang, A.; Zhang, Y.; Pan, G.; Xu, J.; Yan, H.; Liu, Y. In Situ Formation of Copper Nanoparticles in Carboxylated Chitosan Layer: Preparation and Characterization of Surface Modified TFC Membrane with Protein Fouling Resistance and Long-lasting Antibacterial Properties. *Sep. Purif. Technol.* **2017,** *176*, 164–172.
44. Chen, K.; Guo, B.; Luo, J. Quaternized Carboxymethyl Chitosan/Organic Montmorillonite Nanocomposite as a Novel Cosmetic Ingredient Against Skin Aging. *Carbohydr. Polym.* **2017,** *173*, 100–106.
45. Huang, X.; Xu, C.; Li, Y.; Cheng, H.; Wang, X.; Sun, R. Quaternized Chitosan-stabilized Copper sulfide Nanoparticles for Cancer Therapy. *Mater. Sci. Eng. C* **2019,** *96*, 129–137.
46. Jang, S. C.; Tsen, W. C.; Chuang, F. S.; Gong, C. Simultaneously Enhanced Hydroxide Conductivity and Mechanical Properties of Quaternized Chitosan/Functionalized Carbon Nanotubes Composite Anion Exchange Membranes. *Int. J. Hydrogen Energy* **2019,** *44*, 18134–18144.

47. Rahimi, M.; Ahmadi, R.; Samadi Kafil, H.; Shafiei-Irannejad, V. A Novel Bioactive Quaternized Chitosan and Its Silver-containing Nanocomposites as a Potent Antimicrobial Wound Dressing: Structural and Biological Properties. *Mater. Sci. Eng. C* **2019,** *101*, 360–369.
48. Senra, T. D. A.; Campana-Filho, S. P.; Desbrières, J. Surfactant-polysaccharide Complexes Based on Quaternized Chitosan. Characterization and application to emulsion stability. *Eur. Polym. J.* **2018,** *104*, 128–135.
49. Ramasamy, P.; Subhapradha, N.; Thinesh, T.; Selvin, J.; Selvan, K. M.; Shanmugam, V.; Shanmugam, A. Characterization of Bioactive Chitosan and Sulfated Chitosan from Doryteuthis Singhalensis. *Int. J. Biol. Macromol.* **2017,** *99*, 682–691.
50. Yang, Y.; Xing, R.; Liu, S.; Qin, Y.; Li, K.; Yu, H.; Li, P. Immunostimulatory e_ects of Sulfated Chitosans on RAW 264.7 Mouse Macrophages Via the Activation of PI3K/Akt Signaling Pathway. *Int. J. Biol. Macromol.* **2018,** *108*, 1310–1321.
51. Dutta, P. K.; Tripathi, S.; Mehrotra, G. K.; Dutta, J. Perspectives for Chitosan Based Antimicrobial Films in Food Applications. *Food Chem.* **2009,** *114*, 1173–1182. https://doi.org/10.1016/j. foodchem.2008.11.047.
52. Elsabee, M. Z.; Abdou, E. S. Chitosan Based Edible Films and Coatings: A Review. *Mater. Sci. Eng. C* **2013,** *33*, 1819–1841. https://doi.org/10.1016/j.msec.2013.01.010.
53. Fernandez-Saiz, P.; Lagaron, J. M.; Ocio, M. J. Optimization of the Biocide Properties of Chitosan for Its Application in the Design of Active Films of Interest in the Food Area. *Food Hydrocoll.* **2009,** *23*, 913–921. https://doi.org/10.1016/j.foodhyd.2008.06.001.
54. Pelissari, F. M.; Yamashita, F.; Grossmann, M. V. E. Extrusion Parameters Related to Starch/Chitosan Active Films Properties: Extrusion Variables _ Films Characteristics. *Int. J. Food Sci. Technol.* **2011,** *46*, 702–710. https://doi.org/10.1111/j.1365-2621.2010.02533.x.
55. van den Broek, L. A. M.; Knoop, R. J. I.; Kappen, F. H. J.; Boeriu, C. G. Chitosan Films and Blends for Packaging Material. *Carbohydr. Polym.* **2015,** *116*, 237–242. https://doi.org/10.1016/j.carbpol.2014.07.039
56. Park, S. Y.; Marsh, K. S.; Rhim, J. W. Characteristics of Different Molecular Weight Chitosan Films Affected by the Type of Organic Solvents. *J. Food Sci.* **2002,** *67*, 194–197.
57. Thakhiew, W.; Champahom, M.; Devahastin, S.; Soponronnarit, S. Improvement of Mechanical Properties of Chitosan-based Films via Physical Treatment of Film-forming Solution. *J. Food Eng.* **2015,** *158*, 66–72. https://doi.org/10.1016/j.jfoodeng.2015.02.027
58. Siripatrawan, U. Active Food Packaging from Chitosan Incorporated with Plant Polyphenols. In *Novel Approaches of Nanotechnology in Food*, Grumezescu, A. M. Ed.; Elsevier: Amsterdam, Netherlands, 2016; 465–507.
59. Aider, M. Chitosan Application for Active Bio-based Films Production and Potential in the Food Industry: Review. *LWT Food Sci. Technol.* **2010,** 43, 837–842. https://doi.org/10.1016/j.lwt.2010.01.021
60. Kim, K. M.; Son, J. H.; Kim, S.-K.; Weller, C. L.; Hanna, M. A. Properties of Chitosan Films as a Function of pH and Solvent Type. *J. Food Sci.* **2006,** *71* (3), E119–E124.
61. Butler, B. L.; Vergano, P. J.; Testin, R. F.; Bunn, J. M.; Wiles, J. L. Mechanical and Barrier Properties of Edible Chitosan Films as Affected by Composition and Storage. *J. Food Sci.* **1996,** *61* (5), 953–956.

62. Sharma, R. K.; Tahiliani, S.; Jain, N.; Priyadarshi, R.; Chhangani, S.; Purohit, S. D.; Joshi, P. Cynodon Dactylon Leaf Extract Assisted Green Synthesis of Silver Nanoparticles and Their Anti-Microbial Activity. *Adv. Sci. Eng. Med.* **2013,** *5* (8), 858–863.
63. Tripathi, S.; Mehrotra, G. K.; Dutta, P. K. Chitosan–silver Oxide Nanocomposite Film: Preparation and Antimicrobial Activity. *Bull. Mater. Sci.* **2011,** *34* (1), 29–35.
64. Vimala, K.; Yallapu, M. M.; Varaprasad, K.; Reddy, N. N.; Ravindra, S.; Naidu, N. S.; Raju, K. M. Fabrication of Curcumin Encapsulated Chitosan-PVA silver Nanocomposite Films for Improved Antimicrobial Activity. *J. Biomater. Nanobiotechnol.* **2011,** *2* (1), 55–64.
65. Priyadarshi, R.; Negi, Y. S. Effect of Varying Filler Concentration on Zinc Oxide Nanoparticle Embedded Chitosan Films as Potential Food Packaging Material. *J. Polym. Environ.* **2017,** *25* (4), 1087–1098. https://doi.org/10.1007/s10924-016-0890-4.
66. Rodrigues, C.; de Mello, J. M. M.; Dalcanton, F.; Macuvele, D. L. P.; Padoin, N.; Fiori, M. A.; ... Riella, H. G. Mechanical, Thermal and Antimicrobial Properties of Chitosan-based Nanocomposite with Potential Applications for Food Packaging. *J. Polym. Environ.* **2020,** *28* (4), 1216–1236.
67. Youssef, A. M.; Abou-Yousef, H.; El-Sayed, S. M.; Kamel, S. Mechanical and Antibacterial Properties of Novel High Performance Chitosan/Nanocomposite Films. *Int. J. Biol. Macromol.* **2015,** *76*, 25–32.
68. Zhang, L.; Wang, H.; Jin, C.; Zhang, R.; Li, L.; Li, X.; Jiang, S. Sodium Lactate Loaded Chitosan-polyvinyl Alcohol/Montmorillonite Composite Film Towards Active Food Packaging. *Innov. Food Sci. Emerg. Technol.* **2017,** *42*, 101–108.
69. Cárdenas, G.; Díaz, V. J.; Meléndrez, M. F.; Cruzat, C. C.; García Cancino, A. Colloidal Cu Nanoparticles/Chitosan Composite Film Obtained by Microwave Heating for Food Package Applications. *Polym. Bull.* **2009,** *62* (4), 511–524. https://doi.org/10.1007/s00289-008-0031-x.
70. Ahmed, J.; Mulla, M.; Arfat, Y. A.; Thai, T. L. A. Mechanical, Thermal, Structural and Barrier Properties of Crab Shell Chitosan/Graphene Oxide Composite Films. *Food Hydrocoll.* **2017,** *71*, 141–148.
71. Sun, F.; Cha, H. R.; Bae, K. E.; Hong, S.; Kim, J. M.; Kim, S. H.; ... Lee, D. Mechanical Properties of Multilayered Chitosan/CNT Nanocomposite Films. *Mater. Sci. Eng. A* **2011,** *528* (21), 6636–6641.
72. Casariego, A.; Souza, B. W. S.; Cerqueira, M. A.; Teixeira, J. A.; Cruz, L.; Díaz, R.; Vicente, A. A. Chitosan/Clay Films' Properties as Affected by Biopolymer and Clay Micro/Nanoparticles' Concentrations. *Food Hydrocoll.* **2009,** *23* (7), 1895–1902.
73. Zhang, X.; Xiao, G.; Wang, Y.; Zhao, Y.; Su, H.; Tan, T. Preparation of Chitosan-TiO2 Composite Film with Efficient Antimicrobial Activities Under Visible Light for Food Packaging Applications. *Carbohydr. Polym.* **2017,** *169*, 101–107.
74. Rhim, J. W. Effect of Clay Contents on Mechanical and Water Vapor Barrier Properties of Agar-based Nanocomposite Films. *Carbohydr. Polym.* **2011,** *86* (2), 691–699.
75. Kusmono, K.; Abdurrahim, I. Water Sorption, Antimicrobial Activity, and Thermal and Mechanical Properties of Chitosan/Clay/Glycerol Nanocomposite Films. *Heliyon* **2019,** *5* (8), e02342.
76. Swain, S. K.; Kisku, S. K.; Sahoo, G. Preparation of Thermal Resistant Gas Barrier Chitosan Nanobiocomposites. *Polym. Compos.* **2014,** *35* (12), 2324–2328.
77. Lee, Y.; Hwang, K. T. Changes in Physicochemical Properties of Mulberry Fruits (Morus alba L.) During Ripening. *Sci. Hortic.* **2017,** *217*, 189–196.

78. Perotti, V. E.; Moreno, A. S.; Podestá, F. E. Physiological Aspects of Fruit Ripening: The Mitochondrial Connection. *Mitochondrion* **2014,** *17*, 1–6.
79. Ali, S.; Nawaz, A.; Ejaz, S.; Haider, S. T.-A.; Alam, M. W.; Javed, H. U. Effects of Hydrogen Sulfide on Postharvest Physiology of Fruits and Vegetables: An Overview. *Sci. Hortic.* **2019,** *243*, 290–299.
80. Martínez, K.; Ortiz, M.; Albis, A.; Gilma Gutiérrez Castañeda, C.; Valencia, M. E.; Grande Tovar, C. D. The Effect of Edible Chitosan Coatings Incorporated with Thymus Capitatus Essential Oil on the Shelf-life of Strawberry (Fragaria x Ananassa) During Cold Storage. *Biomolecules* **2018,** *8* (4), 155.
81. Xu, D. A. N.; Qin, H.-R.; Ren, D. A. N.; Yu, Y.-L. Influence of Coating Time on the Preservation Performance of Chitosan/Montmorillonite Composite Coating on Tangerine Fruits. The 21st IAPRI world Conference on packaging). DEStech Publications, Inc, 2018.
82. Zhang, X.; Xiao, G.; Wang, Y.; Zhao, Y.; Su, H.; Tan, T. Preparation of Chitosan-TiO2 Composite Film with Efficient Antimicrobial Activities Under Visible Light for Food Packaging Applications. *Carbohydr. Polym.* **2017,** *169*, 101–107.
83. Al-Naamani, L.; Dutta, J.; Dobretsov, S. Nanocomposite Zinc Oxide-chitosan Coatings on Polyethylene Films for Extending Storage Life of Okra (Abelmoschus Esculentus). *Nanomaterials* **2018,** *8* (7).
84. Harish Prashanth, K. V.; Tharanathan, R. N. Chitin/chitosan: Modifications and Their Unlimited Application Potential—An Overview. *Trends Food Sci. Technol.* **2007,** *18* (3), 117–131.
85. Kumar, S.; Ye, F.; Dobretsov, S.; Dutta, J. Chitosan Nanocomposite Coatings for Food, Paints, and Water Treatment Applications. *Appl. Sci.* **2019,** *9* (12).
86. Xing, Y.; Li, W.; Wang, Q.; Li, X.; Xu, Q.; Guo, X.; et al. Antimicrobial Nanoparticles Incorporated In Edible Coatings and Films for the Preservation of Fruits and Vegetables. *Molecules* **2019,** *24* (9), 1695.
87. Hu, D.; Wang, H.; Wang, L. Physical Properties and Antibacterial Activity of Quaternized Chitosan/Carboxymethyl Cellulose Blend Films. *LWT-Food Sci. Technol.* **2016,** *65*, 398–405.
88. Karthik, R.; Manigandan, V.; Saravanan, R.; Rajesh, R. P.; Chandrika, B. Structural Characterization and In Vitro Biomedical Activities of Sulfated Chitosan from *Sepia pharaonis*. *Int. J. Biol. Macromol.* **2016,** *84*, 319–328.
89. Sayari, N.; Sila, A.; Abdelmalek, B. E.; Abdallah, R. B.; Ellouz-Chaabouni, S.; Bougatef, A.; Balti, R. Chitin and Chitosan from the Norway Lobster by-products: Antimicrobial and Anti-proliferative Activities. *Int. J. Biol. Macromol.* **2016,** *87*, 163–171.
90. Pieniz, S.; Andreazza, R.; Okeke, B. C.; Camargo, F. A.; Brandelli, A. Antimicrobial and Antioxidant Activities of *Enterococcus* Species Isolated from Meat and Dairy Products. *Braz. J. Biol.* **2015,** *75*, 923–931.
91. Monu, E. A.; David, J. R.; Schmidt, M.; Davidson, P. M. Effect of White Mustard Essential Oil on the Growth of Foodborne Pathogens and Spoilage Microorganisms and the Effect of Food Components on Its Efficacy. *J. Food. Prot.* **2014,** *77*, 2062–2068.

CHAPTER 8

Biopolymer Matriccs: Current Trends in Tissue Engineering

AARTHI S. BHATT[1] and G. SANTHOSH[2]

[1]*Department of Chemistry, NMAM Institute of Technology (Deemed to be University), Udupi, Karnataka, India*

[2]*Department of Mechanical Engineering, NMAM Institute of Technology (Deemed to be University), Udupi, Karnataka, India*

ABSTRACT

Biopolymers are naturally occurring polymers, which are extracted largely from animal and plant sources. These biopolymers have been found to be bioactive and biocompatible with the extracellular matrix system of mammalian cells, thereby making them ideal substitutes for medical applications such as tissue engineering. Biomaterials such as scaffolds and hydrogels derived from biopolymers are popular for their biocompatibility, biodegradability, and abundance. The structure and biochemistry of biopolymer-based scaffolds are excellent matrices for cell adhesion, proliferation, and differentiation. In this work, a brief review on the structural and morphological properties of some of the most promising naturally derived polysaccharides and proteins have been portrayed. In particular, the effective-specific functional applications of chitosan, hyaluronic acid, alginate (polysaccharides), collagen, gelatin, and silk fibroin (proteins) in the repair and reconstruction of damaged cell lines have been outlined. Since naturally derived polymers have eminence in tissue engineering over synthetic polymers, the future perspectives of biopolymers have also been discussed.

Tailored Functional Materials for Clean and Sustainable Development. Tamara Tatrishvili, PhD, Neha Kanwar Rawat, PhD, Ann Rose Abraham, PhD & A. K. Haghi, PhD (Eds.)

8.1 INTRODUCTION

Biopolymers are polymers originating from living organisms or chemically produced from a natural material. Unlike commercial polymers obtained from crude petroleum, biopolymers derived from renewable natural sources are likely to be biodegradable. As a matter of fact, biopolymers are found in all living matter in the form of proteins, polysaccharides, metabolites, genetic molecules, and other polystructural materials. This makes them more compatible and sustainable for biomedical applications. It is not surprising that the present technological advancements in the field of biomedical applications are aimed at designing biopolymers with biocompatible structural and functional properties.

The progression in the field of reconstructive medicine has led to the employment of biocompatible materials that can repair and rebuild the damaged tissues without any side effects. At the same time, the material must possess sufficient mechanical stamina to hold up during the entire restoration period synchronized with the material's biodegradation. In brief, the desirable features for a biopolymer to be employed as matrices are (1) cellular adhesion, (2) cellular proliferation, (3) integration with the local tissues, (4) structural integrity, and (5) efficient biodegradation. Synthetic polymers have been extensively used as a support system for tissue engineering. Nevertheless, bio-derived polymers can naturally interact with cells and stimulate cell growth. Moreover, these organic polymers have the ability to control and optimize enzymatic metabolism, all the while regulating the moisture content essential for wound healing and providing reliable structural support. The diversity rendered by the biopolymers in terms of their structure and chemistry is far ahead and superior when compared to their inert synthetic counterparts. Hence, it is not surprising to come across novel biopolymer scaffolds with various designs and integrations.

The most commonly employed bio-derived materials explored for tissue engineering, are polysaccharides, proteins, and polyhydroxyalkanoates. Most of the time, they act as monomers and are subjected to chemical modifications so as to attain the desired structural and functional features.[1] For instance, thiolated arabinoxylan and cellulose acetate have been extensively used in drug delivery, wound healing, and skin/bone tissue engineering.[2,3] A brief overview of the current applications of biomaterials in tissue engineering has been provided in Table 8.1.[4,5]

TABLE 8.1 A Brief Overview of the Current Applications of Bio-derived Polymers.

Natural polymer	Applications as biomaterials	Applications in tissue engineering
Polysaccharides and derivatives		
Alginate	Immobilization matrix for cells and enzymes; controlled release of bioactive substances; injectable microspheres	Cartilage engineering[5]
Agarose	Immobilization matrix	Load-sensitive construction for in vitro cartilage development[6]
Cellulose	Cell immobilization, drug delivery, dialysis membranes	Injection-molded porous scaffolds for regeneration of load-bearing tissues[7]
Chitosan	Controlled delivery systems	Matrix support for cell attachment and proliferation[5]
Dextran	Plasma expander, drug carrier	Amine cross-linkage for cell delivery[8]
Heparin	Anti-thrombotic, anti-coagulant	Photocurable scaffolds[9]
Hyaluronic acid	Lubricant, visco-supplement for knee and eye surgery, post-surgical adhesion prevention	Attributing high solubility and water-binding capacity to the synovial fluid and cartilage[5]
Proteins and protein-based polymers		
Collagen	Absorbable sutures, sponge wound and burn dressing, drug-delivery microspheres, soft-tissue augmentation,	Phenotype with elastic-like behavior[10]
Fibrin/Fibrinogen	Glues/sealants as soft tissue adhesives; wound dressing	Gels for cartilage or vascular reconstruction[11]
Fibronectin	Coatings on synthetic materials to improve cell adhesion	Wound repair[12]
Silk	Absorbable sutures	Fiber with low density, high strength, and excellent elasticity[5]
Polyhydroxyalkanoates		
Polyhydroxy-butyrate	Controlled drug release, sutures, artificial skin	Nerve regeneration conduits,[13] heart valves,[14] vessel augmentation, vascular grafts[15]

Source: Adapted from Ref. [4, 5]

Other than the general structural attributes, these biopolymers are biocompatible and possess negligible cytotoxicity and predictable degradation. Besides, they contribute to biological signaling in terms of host tissue integration without creating any histological changes. The host-friendly nature and structural polymorphism of biopolymers have led to their applications in tissue repair and reconstruction (Fig. 8.1). In this chapter, we discuss the recent employment of some of the naturally derived polysaccharides and proteins for tissue engineering applications; with a special emphasis on their biological aspects.

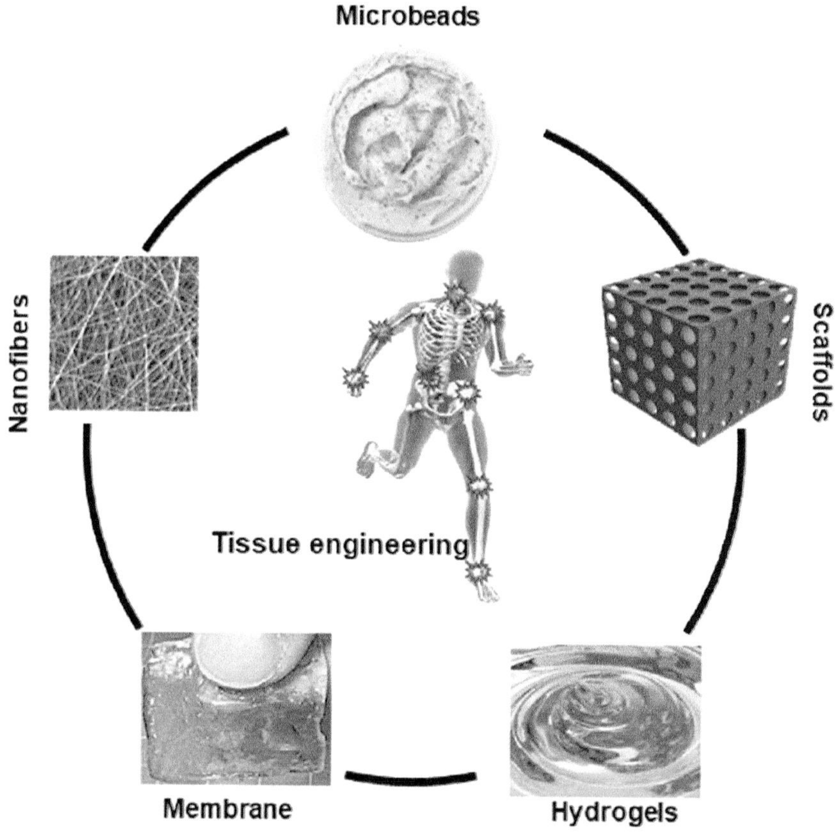

FIGURE 8.1 Various forms of biopolymers in tissue engineering applications.[16]

Source: Reprinted from Ref. [16]. https://creativecommons.org/licenses/by/4.0/

8.2 POLYSACCHARIDES

The polysaccharides majorly comprise proteoglycans, glycosaminoglycans, glycoproteins, and glycolipids which are close analogs to the extracellular matrix. At the same time, their inherent feature of readily forming a visco-elastic gel with aqueous systems provides ample opportunity to attain desirable mechanical properties. Moreover, their abundance and renewability make them suitable for tissue engineering applications.

8.2.1 *CHITOSAN*

Chitosan is the second most abundant biopolymer (after cellulose) which is found in the exoskeleton of crustaceans. It can be obtained from partial/complete deacetylation of chitin with a preferable deacetylation degree (fraction of D-glucosamine units) of more than 60%. The molecular weight and deacetylation degree has a high influence on the biocompatibility, hydrophilicity, and solubility of chitosan in an aqueous environment.[17]

Although, in solitude, chitosan fails to provide an efficient mechanical support, chitosan cross-linked with different functional groups like maleic,[18] azido[19], and amino[20] groups help in tailoring morphology, rheology, and compressive modulus of the hybrid material. Such cross-linked structures perform as excellent scaffold materials with antimicrobial properties.[21,22] The polycationic nature of chitosan creates a favorable physicochemical environment for electrostatic interactions with other biomolecules which are largely anionic.[20,22] Other than cross-linking, chitosan-based porous scaffolds can also be prepared by freeze-drying process. Madihally and Matthew have derived porous chitosan structures by nucleation of frozen chitosan–acetic acid solutions.[23] They showed that with variation in the cooling rate, freezing temperature, and rate of ice crystal growth, it is possible to obtain scaffolds with pore diameter 1–250 μm. Additionally, the introduction of $CaCO_3$ through the internal bubbling process results in achieving a pore size 100–1000 μm.[24] High efficacy biomaterial can be obtained by blending chitosan with other natural polymers such as glutaraldehyde,[25] genipin,[25,26] cellulose,[27] gelatin,[28] collagen[29], and even synthetic polymers like polyvinyl acetate,[30] polyacrylamide.[31] Such derived chitosan-based scaffolds have been employed for tissue repair governing skin,[32] cartilage,[27] bone,[33] fibroblasts,[21] nerve cells[34], and hepatocytes.[35]

8.2.2 HYALURONIC ACID

Hyaluronic acid (HA) is a linear polysaccharide that is largely extracted from rooster combs and also produced from bacterial sources such as *Streptococcus equi* and *Streptococcus zooepidemicus*. HA forms a major component of the extracellular matrix which makes this polymer biocompatible with the human body. It is widely used in medicine, surgery, and cosmetics industry since it displays contradictory biological features based on its molecular weight. Low-molecular-weight HA manifests proinflammatory and proangiogenic activities while with increase in molecular weight, it tends to be anti-inflammatory and anti-angiogenic in nature.[36,37] The other advantages of HA are high moisture retention, easy biodegradation, viscoelasticity, and hydrophilicity.

Similar to chitosan, HA and its derivatives have also been modified by various methods like cross-linking,[38–40] freeze drying,[41] cold plasma treatment[42], and even dip coating.[43] Ranga and co-workers[39] have carried out FXIIIa-mediated cross-linking to produce *in situ* HA-based hydrogels. The reaction was catalyzed by transglutaminase peptide and the resultant hydrogels exhibited superior cell adhesion and gelling kinetics. In another work, high-molecular-weight HA was immobilized on polylactic acid scaffolds via cold plasma method and the resulting scaffold was found to be more biocompatible with proangiogenic properties.[42] In an attempt to understand the gel swelling capacity of HA, porous elastomeric scaffolds were prepared with a template leaching procedure and coated with HA solutions.[44] It was observed that depending on the coating process parameters, the HA deposition on the pores resulted in either aggregates or a thick continuous layer (Fig. 8.2). Moreover, hybrid biomaterials of HA with other polymers like collagen, heparin, and other peptides have also been carried out to improve its structural and biological functionalities.[39,40,45]

8.2.3 ALGINATE

Alginate is a natural polysaccharide extracted from brown algae, kelp, and certain bacteria. Similar to HA, alginate has a polyanionic structure with either consecutively or alternatively arranged blocks of α-L-guluronic acid (G) and β-D-mannuronic acid (M), linked via α-(1-4) and β-(1-4) glycosidic bond, respectively. Alginate has a major drawback of poor cell adherence, which restricts its application in the medical field.[46] However, the structural and biological properties of alginate can be tailored to the desired level by varying the composition, sequence pattern, and M and G block length.

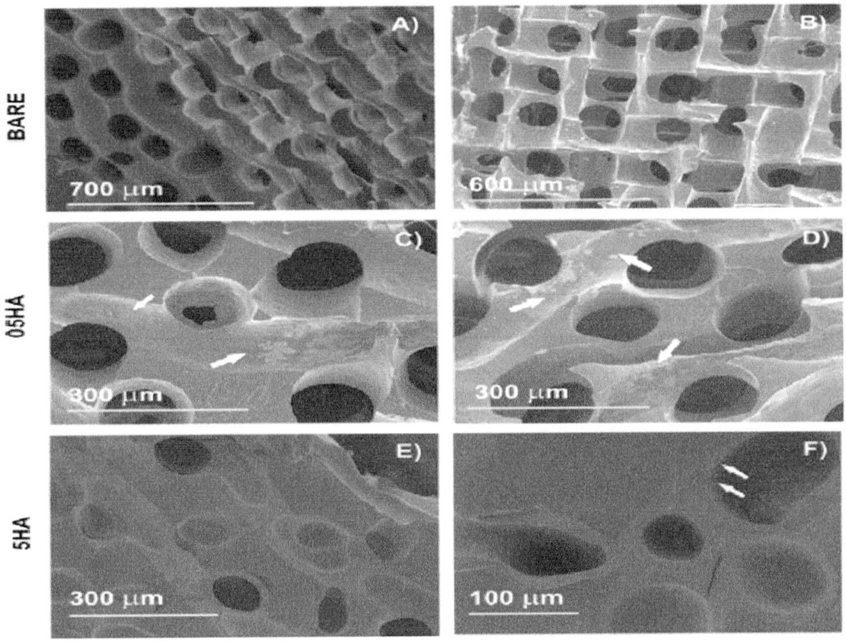

FIGURE 8.2 Scanning electron microscopic images of the dry poly(ethyl acrylate) (PEA) scaffolds: (A and B) Bare scaffold, cross-section, and frontal view, respectively; (C) 05HA1; (D) 05HA5; (E) 5HA2; (F) 5HA5. The arrowheads point at the adsorbed HA. With one coating cycle, 0.5 wt% HA solution produced aggregates on the pore surface (shown in C). These aggregates become more distinct as the number of cycles increases, but a uniform layer is not obtained with 0.5wt% even after five cycles (D). In contrast, coating with 5 wt% HA produces a uniform continuous layer after the first coating cycle. The effect of further cycles is to achieve the layer thickness. This is accompanied with a decreased pore diameter and the clogging of some pores (E). After the fifth cycle (F), the channels are filled with HA to a high degree. (05HA# and 5HA# designate the scaffolds coated, respectively, with 0.5 wt% and 5 wt% HA solutions, # being the number of cycles).

Source: Reproduced and modified from R ef. [44] © 2011 . Elsevier.

The problem of poor cell attachment by alginate has been addressed by employing RGD peptides. Several works have been carried out to show the crucial role of RGD in the linkage of the intracellular skeleton with the extracellular matrix.[47–50] This enhanced cell adhesion; while in some cases it resulted in rapid tissue regeneration without cell apoptosis.[47] The nonadhesive and microporous structure of alginate has been shown to be favorable for the stabilization of hepatocytes.[51,52] To improvise on its medical applications, alginate has been frequently blended with other polysaccharides such as chitosan,[32] peptides,[53] gelatin[54], and carbon nanotubes.[55] The incorporation of these additives is vital for the enhancement of mechanical and biological activities.

8.3 PEPTIDES

Like polysaccharides, proteins are another class of natural polymers that are currently being employed as bioactive materials for tissue engineering. As discussed before, scaffolds that mimic the extracellular matrix are the most suitable candidates for tissue repair. Proteins form a major proportion of extracellular matrix, making them more biocompatible. The major drawback of biopolymers is low mechanical strength. However, fibrous proteins (such as collagen, gelatin, and silk fibroin), which have a complex macro structure consisting of secondary and tertiary peptide structures, possess high mechanical stability and have promising applications in tissue engineering.

8.3.1 COLLAGEN

Collagen is the most abundant protein found in mammals and is known to exist in around 29 different forms with different supra structures. Out of these different types, only few (Type I, II, III, V, XI, XXIV, and XXVII) exist in fiber form. Each of the types is distributed in different tissue cells of mammals. For instance, Type I is abundant in the connective tissues of tendons, bone, and skin while Type II is invariably found in cartilages. Furthermore, collagen can be extracted from various animal sources such as bovine skin and tendons, porcine skin, rat tail[56,57] and even aquatic flora and fauna (fish, sponges, and jellyfish).[57–59] Nevertheless, the collagen derived from each of these sources differs, making the investigation on collagen fibers all the more challenging and interesting.

A majority of the collagen-based biomaterials have glutaraldehyde cross-linking to its structure which enhances their mechanical characteristics and enzymatic response behavior.[60] The currently available commercial Zimmer® Collagen Repair Patch biomaterial for tissue repair employs isocyanate cross-linkage. However, a demerit of this method is the generation of toxic by-products which limits its potential applications.[61] Cross-linking is basically carried out to modify the amine and carboxyl functional groups present on the polymer structure so as to generate covalent bonds. However, recently it has been found that integrating chitosan with collagen leads to the formation of ionic bonds between the amine and carboxyl functional groups of the two polymers. These ionic bonds are strong enough to develop sufficient mechanical stability besides rendering the biomaterial nontoxic.[62] Collagen-based scaffolds have been extensively used for bone and cartilage reconstruction,[62] vascular diseases,[63] skin and cornea tissue repair,[64,65] peripheral nerve regeneration[66], and wound dressing.[64]

8.3.2 GELATIN

Gelatin is a biocompatible protein which is derived from collagen. The derivation involves the denaturation of collagen either through heat treatment or enzymatic hydrolysis. The denaturation results in the conversion of the triple helical structure of collagen to a randomly arranged unorganized structure of gelatin. However, this does not reduce the impact of gelatin. Instead, gelatin is considered to be more bioactive than collagen in terms of solubility and reactivity. Moreover, gelatin exhibits a unique reversible gelation phenomenon with respect to temperature. At 40–50°C, gelatin exists in a solution form with a random coil arrangement.[67] At temperature below 37°C, a slow reversion of the random coil to the triple helical structure takes place to convert the morphology to a gel.[68]

There are numerous routes of cross-linking gelatin viz. physical,[67,68] chemical (basically with glutaraldehyde),[69] natural compounds (basically with genipin)[70], and enzymatic.[71] Many times, cross-linking is carried out in the presence of other agents to achieve different purposes. For instance, the hybrid structure of chitosan-gelatin has been reported to be more bioactive and mechanically stable than its individual counterparts.[72] Meanwhile, the employment of hydroxyapatite with gelatin is found to be an excellent route for bone regeneration with the osteoblast cells showing enhanced cell proliferation both *in vitro* and *in vivo*.[73] Gelatin is also the most sought polymer for the preparation of hydrogels as it provides the required 3D environment for the attachment and proliferation of the cells of damaged tissues. Gelatin hydrogels are substantially used for skin[74] and bone tissue repair.[75,76]

8.3.3 SILK FIBROIN

Silk fibroin (SF) has been traditionally used in textiles and surgical sutures for a long time. SF is a natural polymer found in the cocoon produced by silkworms. Although spider silk (obtained from the ampullate gland of spider) is also commonly used, it is tedious to get a high yield when compared to silkworm cocoons. The SF structure basically comprises two chains with different molecular weights (Fig. 8.3). The heavy (H) chain (~390 KDa) and light (L) chain (~25 KDa) are linked to each other via disulfide bonds.[77,78] A glycoprotein, referred to as P25, is also linked to the H–L complex via hydrophobic interactions.[79] The SF structure can be either crystalline or amorphous with antiparallel β-sheets (formed by the interactions of hydrophobic peptide residues) representing crystalline structure while randomly arranged α-turns and helices representing amorphous structure.[80] The crystalline SF is dominant with a strong hydrogen bonding among the adjacent peptide fragments to achieve high rigidity and tensile strength.[81]

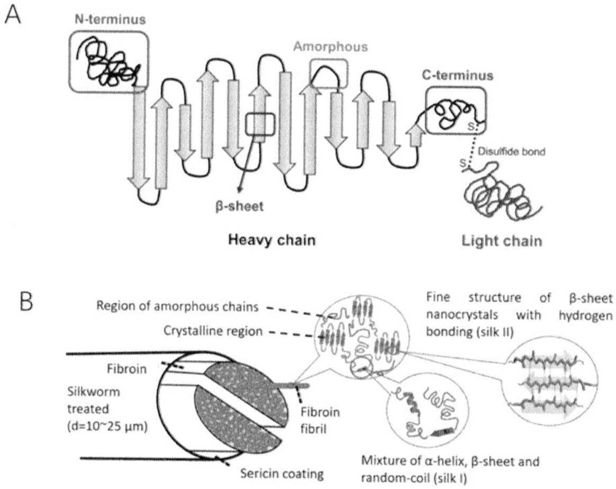

FIGURE 8.3 (A) Heavy chain (i.e., N-terminus, β-sheets, Amorphous and C-terminus) and light chain which are linked via disulfide bonds.
Source: Reproduced with permission from Ref. [78]. https://creativecommons.org/licenses/by/4.0/ (B) silkworm thread, fibril overall structure and silk fibroin polypeptide chains. Reproduced with permission from Ref. [77]. https://creativecommons.org/licenses/by/4.0/

The superior mechanical strength of SF makes SF-derived scaffolds preferential for tissue regeneration. Regenerated silk fibroin (RSF) is integrated with hydroxyapatite, to prepare composite scaffolds with high osteoconductivity.[82] RSF has also been blended with chitosan[83,84] and collagen[85] for cartilage reconstruction while its integration with gelatin and hyaluronic acid is known to enhance the cellular activity of the ligament and tendon connective tissues.[86,87] RSF film, in itself, has the ability to fasten the wound-healing process and repair the damaged skin tissue complex. This has been proven on rabbit and porcine models.[88] Other than this, RSF and chitosan composite scaffolds have been studied in a large scale for skin repair with the former providing the mechanical support and the latter imparting antimicrobial property.[89–91]

8.4 CONCLUSION AND FUTURE PROSPECTS

It is highly desirable that the impaired/damaged tissues are effectively repaired via tissue regeneration with no adverse effects or toxic residues. It is also recommended having the wound healing in a brief span, thereby alleviating postoperative care. In this direction, it becomes all the more crucial to understand the complex molecular interactions between the local cells and the bioactive scaffolds so as to design suitable and precise 2D and/or 3D biomaterials for tissue engineering. Naturally derived polymers like polysaccharides and proteins, because of their biocompatibility, biodegradability and abundance, are most suitable for tissue regeneration. However, progressive research needs to be undertaken in analyzing various parameters such as molecular weight, stereochemistry, sequential arrangement of the monomer units and their interactions with aqueous media for their impact on cell adhesion, proliferation, and differentiation in the local environment. With the advancement in technology, state-of-the-art and ultra-modern equipment and techniques have made it possible to have an in-depth understanding of the interactions taking place at the cellular level. Taking advantage of the upcoming technologies, future investigations must be channeled toward the development of compatible bioactive materials for tissue engineering applications.

KEYWORDS

- **biopolymers**
- **polysaccharides**
- **proteins**
- **biocompatible**
- **tissue engineering**

REFERENCES

1. Gasperini, L.; Mano, J. F.; Reis, R. L. Natural Polymers for the Microencapsulation of Cells. *J. R. Soc. Interface* **2014,** *11*, 20140817.
2. Thomas, B.; Raj, M. C.; Joy, J.; Moores, A.; Drisko, G. L.; Sanchez, C. M. Nanocellulose, A Versatile Green Platform: From Biosources to Materials and Their Applications. *Chem. Rev.* **2018,** *118*, 11575–11625.
3. N. Morin-Crini, Lichtfouse, E.; Torri, G.; Crini, G. Applications of Chitosan in Food, Pharmaceuticals, Medicine, Cosmetics, Agriculture, Textiles, Pulp and Paper, Biotechnology, and Environmental Chemistry. *Environ. Chem. Lett.* **2019,** *17*, 1667–1692.
4. Angelova, N.; Hunkeler, D. Rationalizing the Design of Polymeric Biomaterials. *Trends Biotechnol.* **1999,** *17*, 409–4521.
5. Velema, J.; Kaplan, D. Biopolymer-based Biomaterials as Scaffolds for Tissue Engineering. *Adv. Biochem. Eng. Biotechnol.* **2006,** *102*, 187–238.
6. Mauck, R. L.; Nicoll, S. B.; Seyhan, S. L.; Ateshian, G. A.; Hung, C. T. Synergistic Action of Growth Factors and Dynamic Loading for Articular Cartilage Tissue Engineering. *Tissue Eng.* **2003,** *9*, 597–611.
7. Gomes, M. E.; Ribeiro, A. S.; Malayfaya, P. B.; Reis, R. L.; Cunha, A. M. A New Approach Based on Injection Molding to Produce Biodegradable Starch-based Polymeric Scaffolds: Morphology, Mechanical, and Degradation Behaviour. *Biomaterials* **2001,** *22*, 883–889.
8. Ehrenfreund-Kleinman, T.; Domb, A. J.; Golenser, J. Polysaccharide Scaffolds Prepared by Crosslinking of Polysaccharides with Chitosan or Proteins for Cell Growth. *J. Bioact. Compat. Polym.* **2003,** *18*, 323–338.
9. Matsuda, T.; Magoshi, T. Preparation of Vinylated Polysaccharides and Photofabrication of Tubular Scaffolds as Potential Use in Tissue Engineering. *Biomacromolecules* **2002,** *3*, 942–950.
10. Kim, B. S.; Mooney, D. J. Scaffolds for Engineering Smooth Muscle Under Cyclic Mechanical Strain Conditions. *J. Biomech. Eng.* **2000,** *122*, 210–215.
11. Ye, Q.; Zund, G.; Benedikt, P.; Jockenhoevel, S.; Hoerstup, S. P.; Sakyama, S.; Hubbell, J. A.; Turina, M. Fibrin Gel as a Three-dimensional Matrix in Cardiovascular Tissue Engineering. *Eur. J. Cardiothor. Surg.* **2000,** *17*, 587–591.
12. Underwood, S.; Afoke, A.; Brown, R. A.; MacLeod, A. J.; Shamlou, P. A.; Dunnill, P. Wet Extrusion of Fibronectin–Fibrinogen Cables for Application in Tissue Engineering. *Biotechnol. Bioeng.* **2001,** *73*, 295–305.
13. Stock, U. A.; Nagashima, M.; Khalil, P. N.; Nollert, G. D.; Herden, T.; Sperling, J. S.; Moran, A.; Lien, J.; Martin, D. P.; Schoen, F. J.; Vacanti, J. P.; Mayer, J. E. Tissue Engineered Valve Conduits in Pulmonary Circulation. *J. Thorac. Cardiovasc. Surg.* **2000,** *119*, 732–740.
14. Sodian, R.; Sperling, J. S.; Martin, D. P.; Egozy, A.; Stock, U.; Mayer, J. E.; Vacanti, J. P. Fabrication of a Trileaflet Heart Valve Scaffold from a Polyhydroxyalkanoate Biopolyester for Use in Tissue Engineering. *Tissue Eng.* **2000,** *6*, 183–188.
15. Stock, U. A.; Sakamoto, T.; Hatsuoka, S.; Martin, D. P.; Nagashima, M.; Moran, A. M.; Moses, M. A.; Khalil, P. N.; Schoen, F. J.; Vacanti, J. P.; Mayer, J. R. Patch Augmentation of the Pulmonary Artery with Bioabsorbable Polymers and Autologous Cell Seeding. *J. Thorac. Cardiovasc. Surg.* **2000,** *120*, 1158–1167.

16. Khan, M. U. A.; Razak, S. I. A.; Al Arjan, W. S.; Nazir, S.; Anand, T. J. S.; Mehboob, H.; Amin, R. Recent Advances in Biopolymeric Composite Materials for Tissue Engineering and Regenerative Medicines: A Review. *Molecules* **2021,** *26*, 619.
17. Huang, M.; Khor, E.; Lim, L. Y. Uptake and Cytotoxicity of Chitosan Molecules and Nanoparticles: Effects of Molecular Weight and Degree of Deacetylation. *Pharm. Res.* **2004,** *21*, 344–353.
18. Zhou, Y.; Zhao, S.; Zhang, C.; Liang, K.; Li, J.; Yang, H.; Gu, S.; Bai, Z.; Ye, D.; Xu, W. Photopolymerized Maleilated Chitosan/Thiol-terminated Poly (Vinyl Alcohol) Hydrogels as Potential Tissue Engineering Scaffolds. *Carbohydr. Polym.* **2018,** *184*, 383–389.
19. Fan, M.; Ma, Y.; Mao, J.; Zhang, Z.; Tan, H. Cytocompatible In Situ Forming Chitosan/Hyaluronan Hydrogels via a Metal-free Click Chemistry for Soft Tissue Engineering. *Acta Biomater*. **2015,** *20*, 60–68.
20. J. Nilsen-Nygaard, Strand, S.; Vårum, K.; Draget K.; Nordgård, C. Chitosan: Gels and Interfacial Properties. *Polymers* **2015,** *7*, 552–579.
21. de Souza, R. F. B.; de Souza, F. C. B.; Rodrigues, C.; Drouin, B.; Popat, K.; Mantovani, D. Mechanically-enhanced Polysaccharide-based Scaffolds for Tissue Engineering of Soft Tissues. *Mater. Sci. Eng. C* **2019,** *94*, 364–375.
22. Kutlusoy, T.; Oktay, B.; Apohan, N. K.; Süleymano, G.; Kuruca, M. Chitosan-co-hyaluronic Acid Porous Cryogels and Their Application in Tissue Engineering. *Int. J. Biol. Macromol.* **2017,** *103*, 366–378.
23. Madihally, S. V.; Matthew, H. W. Porous Chitosan Scaffolds for Tissue Engineering. *Biomaterials* **1999,** *20*, 1133–1142.
24. Khor, E.; Lim, L. Y. Implantable Applications of Chitin and Chitosan. *Biomaterials* **2003,** *24*, 2339–2349.
25. Kil'deeva, N. R.; Kasatkina, M. A.; Drozdova, M. G.; Demina, T. S.; Uspenskii, S. A.; Mikhailov, S. N.; Markvicheva, E. A. Biodegradable Scaffolds Based on Chitosan: Preparation, Properties and Use for the Cultivation of Animal Cells. *Appl. Biochem. Microbiol.* **2016,** *52*, 515–524.
26. Silva, J. M.; Custódio, C. A.; Reis, R. L.; Mano, J. F. Multilayered Hollow Tubes as Blood Vessel Substitutes. *ACS Biomater. Sci. Eng.* **2016,** *2*, 2304–2314.
27. Belluzo, M. S.; Medina, L. F.; Cortizo, A. M.; Cortizo, M. S. Ultrasonic Compatibilization of Polyelectrolyte Complex Based on Polysaccharides for Biomedical Applications. *Ultrason. Sonochem.* **2016,** *30*, 1–8.
28. Ng, W. L.; Yeong, W. Y.; Naing, M. W. Development of Polyelectrolyte Chitosan-gelatin Hydrogels for Skin Bioprinting. *Procedia CIRP* **2016,** *49*, 105–112.
29. Sanchez-Sanchez, R.; Martínez-Arredondo, E.; Martínez-López, V.; Melgarejo-Ramírez, Y.; Brena-Molina, A.; Lugo-Martínez, H.; et al. Development of Hydrogel with Anti-Inflammatory Properties Permissive for the Growth of Human Adipose Mesenchymal Stem Cells. *J. Nanomater.* **2016,** *2016*, 1–8.
30. Zeng, Q.; Qin, J.; Yin, X.; Liu, H.; Zhu, L.; Dong, W.; Zhang, S. Preparation and Hemocompatibility of Electrospun O-carboxymethyl Chitosan/PVA Nanofibers. *J. Appl. Polym. Sci.* **2016,** *133*.
31. Yang, Y.; Wang, X.; Yang, F.; Shen, H.; Wu, D. A Universal Soaking Strategy to Convert Composite Hydrogels into Extremely Tough and Rapidly Recoverable Double-Network Hydrogels. *Adv. Mater.* **2016,** *28*, 7178–7184.

32. Bierhalz, A. C. K.; Moraes, A. M. Tuning the Properties of Alginate-chitosan Membranes by Varying the Viscosity and the Proportions of Polymers. *J. Appl. Polym. Sci.* **2016,** *133*.
33. Przekora, A.; Benko, A.; Blazewicz, M.; Ginalska, G. Hybrid Chitosan/β-1,3-glucan Matrix of Bone Scaffold Enhances Osteoblast Adhesion, Spreading and Proliferation via Promotion of Serum Protein Adsorption. *Biomed. Mater.* **2016,** *11*, 45001.
34. Sadeghi, A.; Moztarzadeh, F.; Mohandesi, J. A. Investigating the Effect of Chitosan on Hydrophilicity and Bioactivity of Conductive Electrospun Composite Scaffold for Neural Tissue Engineering. *Int. J. Biol. Macromol.* **2018,** *121*, 625–632.
35. Elcin, Y. M.; Dixit, V.; Gitnick, G. Hepatocyte Attachment on Biodegradable Modified Chitosan Membranes: In Vitro Evaluation for the Development of Liver Organoids. *Artif. Organs* **1998,** *22,* 837–846.
36. Maharjan, A. S.; Pilling, D.; Gomer, R. H. High and Low Molecular Weight Hyaluronic Acid Differentially Regulate Human Fibrocyte Differentiation. *PLoS One* **2011,** *6*, e26078.
37. Kobayashi, H.; Terao, T. Hyaluronic Acid-specific Regulation of Cytokines by Human Uterine Fibroblasts. *Am. J. Physiol.* **1997,** *273*, C1151–C1159.
38. Puperi, D. S.; O'Connell, R. W.; Punske, Z. E.; Wu, Y.; West, J. L.; Grande-Allen, K. J. Hyaluronan Hydrogels for a Biomimetic Spongiosa Layer of Tissue Engineered Heart Valve Scaffolds. *Biomacromolecules* **2016,** *17*, 1766–1775.
39. Ranga, A.; Lutolf, M. P.; Hilborn, J.; Ossipov, D. A. Hyaluronic Acid Hydrogels Formed in Situ by Transglutaminase-Catalyzed Reaction. *Biomacromolecules* **2016,** *17*, 1553–1560.
40. Jha, A. K.; Tharp, K. M.; Browne, S.; Ye, J.; Stahl, A.; Yeghiazarians, Y.; Healy, K. E. Matrix Metalloproteinase-13 Mediated Degradation of Hyaluronic Acid-based Matrices Orchestrates Stem Cell Engraftment Through Vascular Integration. *Biomaterials* **2016,** *89*, 136–147.
41. Hortensius, R. A.; Ebens, J. H.; Harley, B. A. C. Immunomodulatory Effects of Amniotic Membrane Matrix Incorporated into Collagen Scaffolds. *J. Biomed. Mater. Res. A* **2016,** *104*, 1332–1342.
42. Kudryavtseva, V.; Stankevich, K.; Gudima, A.; Kibler, E.; Zhukov, Y.; Bolbasov, E.; Malashicheva, A.; Zhuravlev, M.; Riabov, V.; Liu, T.; et al. Atmospheric Pressure Plasma Assisted Immobilization of Hyaluronic Acid on Tissue Engineering PLA-based Scaffolds and Its Effect on Primary Human Macrophages. *Mater. Des.* **2017,** *127*, 261–271.
43. Zamboni, F.; Keays, M.; Hayes, S.; Albadarin, A. B.; Walker, G. M.; Kiely, P. A.; Collins, M. N. Enhanced Cell Viability in Hyaluronic Acid Coated Poly(lactic-co-glycolic acid) Porous Scaffolds Within Microfluidic Channels. *Int. J. Pharm.* **2017,** *532*, 595–602.
44. Arnal-Pastor, M.; Valles-Lluch, A.; Keicher, M.; Pradas, M. M. Coating Typologies and Constrained Swelling of Hyaluronic Acid Gels within Scaffold Pores. *J. Colloid Interface Sci.* **2011,** *361*, 361–369.
45. Hortensius, R. A.; Ebens, J. H.; and Harley, B. A. C. Immunomodulatory Effects of Amniotic Membrane Matrix Incorporated into Collagen Scaffolds. *J. Biomed. Mater. Res. A* **2016,** *104*, 1332–1342.
46. Sarker, B.; Singh, R.; Silva, R.; Roether, J. A.; Kaschta, J.; Detsch, R.; Schubert, D. W.; Cicha, I.; Boccaccini, A. R. Evaluation of Fibroblasts Adhesion and Proliferation on Alginate-gelatin Crosslinked Hydrogel. *PLoS One* **2014,** *9*, e107952.

47. Shachar, M.; Tsur-Gang, O.; Dvir, T.; Leor, J.; Cohen, S. The Effect of Immobilized RGD Peptide in Alginate Scaffolds on Cardiac Tissue Engineering. *Acta Biomater.* **2011,** *7*, 152–162.
48. Bernstein-Levi, O.; Ochbaum, G.; Bitton, R. The Effect of Covalently Linked RGD Peptide on the Conformation of Polysaccharides in Aqueous Solutions. *Colloids Surf. B* **2016,** *137*, 214–220.
49. Giammanco, G. E.; Carrion, B.; Coleman, R. M.; Ostrowski, A. D. Photoresponsive Polysaccharide-Based Hydrogels with Tunable Mechanical Properties for Cartilage Tissue Engineering. *ACS Appl. Mater. Interfaces* **2016,** *8*, 14423–14429.
50. Ning, L.; Xu, Y.; Chen, X.; Schreyer, D. J. Influence of Mechanical Properties of Alginate-based Substrates on the Performance of Schwann Cells in Culture. *J. Biomater. Sci. Polym. Ed.* **2016,** *27*, 898–915.
51. Dvir-Ginzberg, M.; Elkayam, T.; Cohen, S. Induced Differentiation and Maturation of Newborn Liver Cells into Functional Hepatic Tissue in Macroporous Alginate Scaffolds. *FASEB J.* **2008,** *22*, 1440–1449.
52. Shteyer, E.; Ben, Y. A.; Zolotaryova, L.; Sinai, A.; Lichtenstein, Y.; Pappo, O.; Kryukov, O.; Elkayam, T.; Cohen, S.; Ilan, Y. Reduced Liver Cell Death Using an Alginate Scaffold Bandage: A Novel Approach for Liver Reconstruction After Extended Partial Hepatectomy. *Acta Biomater.* **2014,** *10*, 3209–3216.
53. Celik, E.; Bayram, C.; Akçapınar, R.; Türk, M.; Denkbaş, E. B. The Effect of Calcium Chloride Concentration on Alginate/Fmoc-diphenylalanine Hydrogel Networks. *Mater. Sci. Eng. C* **2016,** *66*, 221–229.
54. Pan, T.; Song, W.; Cao, X.; Wang, Y. 3D Bioplotting of Gelatin/Alginate Scaffolds for Tissue Engineering: Influence of Crosslinking Degree and Pore Architecture on Physicochemical Properties. *J. Mater. Sci. Technol.* **2016,** *32*, 889–900.
55. Yildirim, E. D.; Yin, X.; Nair, K.; Sun, W. Fabrication, Characterization, and Biocompatibility of Single-walled Carbon Nanotube-reinforced Alginate Composite Scaffolds Manufactured Using Freeform Fabrication Technique. *J. Biomed. Mater. Res. B Appl. Biomater.* **2008,** *87*, 406–414.
56. Copes, F.; Pien, N.; Van Vlierberghe, S.; Boccafoschi, F.; Mantovani, D. Collagen-based Tissue Engineering Strategies for Vascular Medicine. *Front. Bioeng. Biotechnol.* **2019,** *7*, 1–15.
57. Muthukumar, T.; Sreekumar, G.; Sastry, T. P.; Chamundeeswari, M. Collagen as a Potential Biomaterial in Biomedical Applications. *Rev. Adv. Mater. Sci.* **2018,** *53*, 29–39.
58. Exposito, J. Y.; Le Guellec, D.; Lu, Q.; Garrone, R. Short Chain Collagens in Sponges are Encoded by a Family of Closely Related Genes. *J. Biol. Chem.* **1991,** *266*, 21923–21928.
59. Song, E.; Kim, S. Y.; Chun, T.; Byun, H. J.; Lee, Y. M. Collagen scaffolds derived from a marine source and their biocompatibility. *Biomaterials* **2006,** *27*, 2951–2961.
60. Wu, X.; Black, L.; Santacana-Laffitte G.; Patrick C. W. Jr. Preparation and Assessment of Glutaraldehyde-Crosslinked Collagen-chitosan Hydrogels for Adipose Tissue Engineering. *J. Biomed. Mater. Res. A* **2007,** *81*, 59–65.
61. Sung, H. W.; Huang, R. N.; Huang, L. L.; Tsai, C. C. In Vitro Evaluation of Cytotoxicity of a Naturally Occurring Cross-linking Reagent for Biological Tissue Fixation. *J. Biomater. Sci. Polym. Ed.* **1999,** *10*, 63–78.
62. Arakawa, C.; Ng, R.; Tan, S.; Kim, S.; Wu, B.; Lee, M. Photopolymerizable Chitosan-collagen Hydrogels for Bone Tissue Engineering: Photopolymerizable Chitosan-collagen Hydrogels. *J. Tissue Eng. Regen. Med.* **2017,** *11*, 164–74.

63. Elomaa, L.; Yang, Y. P. Additive Manufacturing of Vascular Grafts and Vascularized Tissue Constructs. *Tissue Eng. B Rev.* **2017,** *23*, 436–50.
64. Song, M.; Wang, W.; Ye, Q.; Bu, S.; Shen, Z.; Zhu, Y. The Repairing of Full-thickness Skin Deficiency and Its Biological Mechanism using Decellularized Human Amniotic Membrane as the Wound Dressing. *Mater. Sci. Eng. C* **2017,** *77*, 739–47.
65. Yamauchi, M.; Sricholpech, M. Lysine Post-translational Modifcations of Collagen. *Essays Biochem.* **2012,** *52*, 113–33.
66. Sun, W.; Lin, H.; Chen, B.; Zhao, W.; Zhao, Y.; Dai, J. Promotion of Peripheral Nerve Growth by Collagen Scaffolds Loaded with Collagen-Targeting Human Nerve Growth Factor-beta. *J. Biomed. Mater. Res. A* **2007,** *83*, 1054–1061.
67. Boedtker, H.; Doty, P. A Study of Gelatin Molecules, Aggregates and Gels. *J. Phys. Chem.* **1954,** *58*, 968–983.
68. Harrington, W. F.; Rao, N. V. Collagen Structure in Solution I. Kinetics of Helix Regeneration in Single-chain Gelatins. *Biochemistry* **1970,** *9*, 3714–3724.
69. Bigi, A.; Cojazzi, G.; Panzavolta, S.; Rubini, K.; Roveri, N. Mechanical and Thermal Properties of Gelatin Films at Different Degrees of Glutaraldehyde Crosslinking. *Biomaterials* **2001,** *22*, 763–768.
70. Cheng, N. C.; Estes, B. T.; Young, T. H.; Guilak, F. Genipin-crosslinked Cartilage-derived Matrix as a Scaffold for Human Adipose-derived Stem Cell Chondrogenesis. *Tissue Eng. Part A* **2013,** *19*, 484–496.
71. Taddei, P.; Chiono, V.; Anghileri, A.; Vozzi, G.; Freddi, G.; Ciardelli, G. Silk Fibroin/Gelatin Blend Films Crosslinked with Enzymes for Biomedical Applications. *Macromol. Biosci.* **2013,** 13, 1492–510.
72. Kozlowska, J.; Stachowiak, N.; Sionkowska, A. Collagen/Gelatin/Hydroxyethyl Cellulose Composites Containing Microspheres Based on Collagen and Gelatin: Design and Evaluation. *Polymers* **2018,** *10*, 456.
73. Chao, S. C.; Wang, M. J.; Pai, N. S.; Yen, S. K. Preparation and Characterization of Gelatin–hydroxyapatite Composite Microspheres for Hard Tissue Repair. *Mat. Sci. Eng.: C* **2015,** *57*, 113–122.
74. Zhao, X.; Lang, Q.; Yildirimer, L.; Lin, Z. Y.; Cui, W.; Annabi, N.; Ng, K. W.; Dokmeci, M. R.; Ghaemmaghami, A. M.; Khademhosseini, A. Photocrosslinkable Gelatin Hydrogel for Epidermal Tissue Engineering. *Adv. Healthc. Mater.* **2016,** *5*, 108–118.
75. Kuo, Z. K.; Lai, P. L.; Toh, E. K. W.; Weng, C. H.; Tseng, H. W.; Chang, P. Z.; Chen, C. C.; Cheng, C. M. Osteogenic Differentiation of Preosteoblasts on a Hemostatic Gelatin Sponge. *Sci. Rep.* **2016,** *6*, 32884.
76. Kim, G.; Park, Y. S.; Lee, Y.; Jin, Y. M.; Choi, D. H.; Ryu, K.-H.; Park, Y. J.; Park, K. D.; Jo, I. Tonsil-derived Mesenchymal Stem Cell-embedded In Situ Crosslinkable Gelatin Hydrogel Therapy Recovers Postmenopausal Osteoporosis Through Bone Regeneration. *PLOS One* **2018,** *13*, e0200111.
77. Qi, Y.; Wang, H.; Wei, K.; Yang, Y.; Zheng, R. Y.; Kim, I. S.; Zhang, K. Q. A Review of Structure Construction of Silk Fibroin Biomaterials from Single Structures to Multi-Level Structures. *Int. J. Mol. Sci.* **2017,** *18*, 237.
78. Tomeh, M. A.; Hadianamrei, R.; Zhao, X. Silk Fibroin as a Functional Biomaterial for Drug and Gene Delivery. *Pharmaceutics* **2019,** *11*, 494.
79. Tanaka, K.; Inoue, S.; Mizuno, S. Hydrophobic Interaction of P25, Containing Asn-linked Oligosaccharide Chains, with the H-L Complex of Silk Fibroin Produced by Bombyx Mori. *Insect Biochem. Mol. Biol.* **1999,** *29*, 269–276.

80. Jaramillo-Quiceno, N.; Álvarez-López C.; Restrepo-Osorio, A. Structural and Thermal Properties of Silk Fibroin Films Obtained from Cocoon and Waste Silk Fibers as Raw Materials. *Procedia Eng.* **2017,** *200*, 384–388.
81. Cebe, P.; Partlow, B. P.; Kaplan, D. L.; Wurm, A.; Zhuravlev, E.; Schick, C. Silk I and Silk II Studied by Fast Scanning Calorimetry. *Acta Biomater.* **2017,** *55*, 323–332.
82. Jiang, J.; Hao, W.; Li, Y.; Yao, J.; Shao, Z.; Li, H.; Yang, J.; Chen, S. Hydroxyapatite/Regenerated Silk Fibroin Scaffold-enhanced Osteoinductivity and Osteoconductivity of Bone Marrow-derived Mesenchymal Stromal Cells. *Biotechnol. Lett.* **2013,** *35*, 657–661.
83. Bhardwaj, N.; Nguyen, Q. T.; Chen, A. C.; Kaplan, D. L.; Sah, R. L.; Kundu, S. C. Potential of 3-D Tissue Constructs Engineered from Bovine Chondrocytes/Silk Fibroin-chitosan for In Vitro Cartilage Tissue Engineering. *Biomaterials* **2011,** *32*, 5773–5781.
84. Silva, S. S.; Motta, A.; Rodrigues, M. T.; Pinheiro, A. F. M.; Gomes, M. E.; Mano, J. F.; Reis, R. L.; Migliaresi, C. Novel Genipin-CrossLinked Chitosan/Silk Fibroin Sponges for Cartilage Engineering Strategies. *Biomacromolecules* **2008,** *9*, 2764–2774.
85. Lien, S. M.; Ko, L. Y.; Huang, T. J. Effect of Pore Size on ECM Secretion and Cell Growth in Gelatin Scaffold for Articular Cartilage Tissue Engineering. *Acta Biomater.* **2009,** *5*, 670–679.
86. Seo, Y. K.; Yoon, H. H.; Song, K. Y.; Kwon, S. Y.; Lee, H. S.; Park, Y. S.; Park, J. K. Increase in Cell Migration and Angiogenesis in a Composite Silk Scaffold for Tissue-engineered Ligaments. *J. Orthop. Res.* **2009,** *27*, 495–503.
87. Sahoo, S.; Toh, S. L.; Goh, J. C. H. A bFGF-releasing Silk/PLGA-based Biohybrid Scaffold for Ligament/Tendon Tissue Engineering Using Mesenchymal Progenitor Cells. *Biomaterials* **2010,** *31*, 2990–2998.
88. Zhang, W.; Chen, L.; Chen, J.; Wang, L.; Gui, X.; Ran, J.; Xu, G.; Zhao, H.; Zeng, M.; Ji, J.; et al. Silk Fibroin Biomaterial Shows Safe and Effective Wound Healing in Animal Models and a Randomized Controlled Clinical Trial. *Adv. Healthc. Mater.* **2017,** *6*, 1700121.
89. Xu, Y.; Li, L.; Yu, X.; Gu, Z.; Zhang, X. Feasibility Study of a Novel Crosslinking Reagent (Alginate Dialdehyde) for Biological Tissue Fixation. *Carbohydr. Polym.* **2012,** *87*, 1589–1595.
90. Gu, Z.; Xie, H.; Huang, C.; Li, L.; Yu, X. Preparation of Chitosan/Silk Fibroin Blending Membrane Fixed with Alginate Dialdehyde for Wound Dressing. *Int. J. Biol. Macromol.* **2013,** *58*, 121–126.
91. Guang, S.; An, Y.; Ke, F.; Zhao, D.; Shen, Y.; Xu, H. Chitosan/Silk Fibroin Composite Scaffolds for Wound Dressing. *J. Appl. Polym. Sci.* **2015,** *132*, 42503.

CHAPTER 9

Biopolymer Sustainable Films: Mechanical and Biological Properties

HARSHITHA M JATHANNA[1] and G. SANTHOSH[2]

[1]*Department of Biotechnology, NMAM Institute of Technology (Deemed to be University), Karnataka, India*

[2]*Department of Mechanical Engineering, NMAM Institute of Technology (Deemed to be University), Karnataka, India*

ABSTRACT

The current research trend focusing on biopolymers is getting a lot of attention with the aim of developing biocomposites which are nontoxic, biodegradable, and of low cost. The environmental-friendly nature of biopolymers is expected to replace many synthetic-petroleum-based polymers in various engineering applications. The major source of energy is fossil fuels, which are nonrenewable and are depleting at an alarming rate due to their heavy consumption in various industrial sectors, such as transportation, power, petrochemicals, and food industry. Biopolymer composites have been considered promising candidates for food industrial applications due to their amazing biological and physiological properties. The desired properties of the biocomposites for food packaging can be achieved by reinforcing fillers and additives into biopolymer. The polymer–filler interaction has a major effect on the mechanical, chemical, and biological properties of the composites. The present chapter is focused on the fundamentals of biopolymers in food packaging applications and their advancement in mechanical and biological properties.

Tailored Functional Materials for Clean and Sustainable Development. Tamara Tatrishvili, PhD, Neha Kanwar Rawat, PhD, Ann Rose Abraham, PhD & A. K. Haghi, PhD (Eds.)

9.1 INTRODUCTION

A continuous surge in the global energy consumption has led to a global energy crisis. The major source of energy is fossil fuels, which are nonrenewable and are depleting at an alarming rate due to their heavy consumption in various industrial sectors, such as transportation, power, and petrochemicals, etc.. The food industry is another major consumer of fossil-fuel-based products, at various stages, starting from agricultural practices to the marketing of the food product in the form of food packaging films. On the one hand, there is concern and fear that these sources will be depleted in the near future and on the other hand, the use of these sources is associated with unhealthy and undesirable environmental effects. Because of this, there have been continuous efforts throughout various industrial sectors to de-promote their use and switch to sources that are alternatives to fossil fuels. Similarly, in the food industry, conventional petroleum-based food packaging films have been attempted to be replaced with natural polymer-based films, or simply biopolymer films, due to their sustainability and environmental friendliness. The fact that biopolymer-based films are not only derived from natural sources, such as proteins and polysaccharides[1] but also from wastes, such as agricultural residues, food wastes, etc.[2] contributes to their sustainability. Eco-friendliness stems from the fact that as opposed to conventional plastic packaging films that are unsafe, toxic, nondegradable, and a great concern for environmental pollution of both terrestrial and marine litter[1] due to heaps of them being dumped in landfills or in incineration plants[3] resulting in their bioconcentration and bioaccumulation; biopolymer films are safe and highly degradable. More importantly, biopolymer films are cost-effective, thus attracting the global market. Furthermore, food-borne diseases are one of the leading causes of illness and death worldwide; with approximately 600 million cases of illness and 420,000 deaths reported each year as a result of food-borne conditions,[3] signaling the need to prevent food spoilage, which occurs mainly due to microbial contamination. Microbial attacks on food, in addition to causing harmful health conditions, reduce the freshness of the food. Thus, along with other health, environmental, and economic benefits offered by bio-based films, these films have been attracting customers because they are generally loaded with active ingredients that ensure the maintenance of food quality. It has been reported that the global use of biopolymer films is currently surging, with a market size estimated to reach about $7.5 billion by the year 2027. A major chunk of this is to be contributed by packaging industries.[4]

To use biopolymers for food packaging applications, it is essential to increase the mechanical properties of the polymer by modifying their composition.[5] Further, organic and inorganic nanofillers can also be used to enhance the properties of the composite films. The incorporation of nanofillers, with a high surface-to-volume ratio, into the polymers is considered an alternative, novel, and cost-effective method for the fabrication of bio-nanocomposites.[6–9] This chapter's main focus is to brief the biological and mechanical properties of biopolymer composites as food packaging films.

9.2 BIOMATERIALS USED IN FILM PACKAGING

The most commonly used biomaterials in film packaging are natural polymers, such as polysaccharides and proteins,[5,10] which are isolated from various sources, such as microbes, plants, and animals.[11] They are abundantly found in nature and are thus low cost, which combined with other properties, such as excellent biocompatibility and stable physicochemical film properties, make them the ideal choices for bio-based packaging films. However, forming films from single polymeric sources is a challenge due to processing difficulties,[10] which results in poor mechanical properties as well as lack of required biological effects—a reason which has led to the development of biocomposite films, which are films produced by combining two or more biopolymers. Generally, biocomposite films are produced by combining a polymer with another copolymer which is of low cost[2] so that the entire process of biopolymer film production becomes sustainable. Some of the polysaccharides that are commonly reported in the production of films for food packaging include starch, cellulose, pectin, pullulan, alginate, and chitosan.[4] Chitin is the most exploited polymer in edible film packaging due to its abundance because of the fact that it is the second largest polysaccharide in nature and shows nontoxicity and biodegradable properties.[12] It is a major structural component of fungi, planktons, insects, and crustacean shells, making it abundantly available even in food wastes, such as crab shells, facilitating the sustained development of biopolymer films.[13] Chitosan, a linear polysaccharide and a derivative of chitin, has also been widely used to produce films for food and pharmaceutical applications. In addition to possessing exceptional film-forming capacities, it has also been shown to have antimicrobial effects, adding to the functional properties of films produced. However, to enhance the mechanical properties of pure chitosan films, they are regularly conjugated with copolymers such as pectin, and so on.[14]

Starch is a polysaccharide of interest in biopolymer films due to its favorable features, such as renewability, accessibility, low cost, safety, biocompatibility, and high tensile strength.[10] Carboxymethyl cellulose (CMC) is another polymer with high viscosity and excellent tensile strength that is generally combined with other polymeric materials to improve the mechanical properties of the composite film. Furthermore, because of its nontoxic and nonallergic properties, it has additional advantages of using in biofilm production for food or pharmaceutical applications.[10]

Carrageenan, a charged polysaccharide extracted from seaweeds, and gelatin, a fibrous protein obtained from the skins of bovine, porcine, poultry, and fish, are two natural polymers with good oxygen-bearing capacities being used in biopolymer film production.[3] Alginate that is isolated from sources such as seaweeds finds regular applications in food industries as a thickener and stabilizer; it is also exploited as a low-cost film-forming material.[11] Because of its charged behavior that allows it to interact with oppositely charged polymers used to make composite films, xanthan gum, a bacterial exopolysaccharide that is commonly used as a food additive and in drug delivery, is used in the production of biopolymer films.[13] Polysaccharides composed of branched fructose units known as levans isolated from microorganisms, such as bacteria, yeasts, and fungi, as well as plants, have also demonstrated film-forming abilities. Aside from its gas barrier properties, its neutral taste and odor are considered as added benefits for its use in edible films[15]; however, the cost is a barrier. Some research works have also attempted to incorporate nanomaterials into the polymeric matrix such that the final mechanical and physicochemical properties of the film are not compromised. In one study,[4] hemicellulose matrix was combined with cellulose nanocrystal and cellulose nanofiber to enhance the film-forming capacities of hemicellulose. When seaweed-based carrageenan films were combined with nanocellulose fillers, their tensile strength increased significantly.[16] In order to ensure sustainability in production and to ensure the safety of the final films produced, these nanomaterials have been isolated from biological sources, for example, nanocellulose has been extracted from agricultural wastes, such as sugarcane bagasse and bamboo,[17] and nanoparticles of natural polymers, such as chitosan and cellulose have been used.[18]

9.3 ACTIVE COMPONENTS IN PACKAGING FILMS

Consumer interest toward biopolymer food packaging films may be enhanced by incorporating the films with active ingredients that can monitor, and in turn, maintain the freshness of food and prolong its shelf life; such packaging systems are referred to as "active packaging" or "smart packaging." The enhancement of the functional properties of packaging films, especially the ones used for food applications, involves combining additives that either have antibacterial or antimicrobial effects so that the microbial activity resulting in food spoilage and adverse health effects is thereby prevented,[3] as well as those that have antioxidant properties, which prevent food deterioration upon storage. In addition to maintaining food quality, these ingredients may also improve the quality of food by acting as moisture scavengers and regulators, light screeners, texture modifiers, and so on.[11] These additives are generally natural ingredients with high phenolic content, extracted from sources, such as plants and food wastes.[1,14,19] Many previous studies have attempted to combine chitosan with additives that are mostly natural ingredients, particularly those with high phenolic content to enhance the antimicrobial activities of chitosan-based films. However, limitations associated with such substances, such as heat sensitivity, processing difficulties, high cost have necessitated the search for alternate solutions.[19]

In addition to these natural extracts, nanoparticles of metals and metal oxides are being used; however, to avoid any ill effects from the consumption of these nanoparticles that are released into the food from the packaging films, as well as to overcome the limitations associated with natural extracts, particularly in terms of volume requirements, nanoparticles of natural origin, such as those derived from plants, are in use currently.[3] Essential oils, such as those isolated from thyme, tea tree, cinnamon, and other plants, are another class of natural products that are used in the production of functional films because in addition to their antimicrobial and antioxidant activities, they also enhance the physicochemical properties of the film, such as providing a water barrier due to their hydrophobicity.[11,17,20] There have also been reports of organic acids and antimicrobial peptides like LL-37 being used as active ingredients.[15,20] Overall, these natural active ingredients are preferred not only for the beneficial properties they provide but also because they are safe, nontoxic, biocompatible, and readily degradable.

9.4 SUSTAINABLE SOLUTIONS TO CONVENTIONAL FILM MATERIALS

This section summarizes some of the natural polymers and extracts that have been exploited in the production of sustainable biopolymer films:

As a mode of circular economy, chitosan, a derivative of a food industry residue, that is, chitin, was used in the production of biopolymer films due to its excellent antibacterial and antioxidant properties, as well as favorable mechanical properties.[2] These films were also incorporated with oak extracts, which have high polyphenol content, in order to improve both physical and biological properties of the film. Because chitosan degradation had previously only been studied in the human body and aquatic environments, the study primarily focused on producing chitosan-based films and evaluating their degradation capacity and rates in the soil. Degradation rates in soils with varying compositions and moisture contents were studied. The film's mechanical properties were found to be comparable to those of conventional plastic wraps. Even after the addition of oak extract, there was no considerable increase in the moisture content. However, in different soil types tested, namely, garden soil, vineyard soil, and industrial compost, the rate of degradation in chitosan films without oak extracts was higher than that in films with extract, which was attributed to the higher content of polyphenols in the extract, which leads to antimicrobial properties. Nevertheless, complete degradation was observed for few days, which is a significant environmental benefit compared with the conventional plastic packages, which remain in the soil for years.

In the case of perishable foods and foods that must be stored for an extended period, one of the prime ways to improve the shelf life of the food is to use packaging materials that possess antimicrobial capacities. In this regard, chitosan-based films with inherent antimicrobial properties were developed as a replacement for plastic films, which fail to provide long-term food storage; however, because chitosan has high water vapor permeability (WVP), chitosan matrix was integrated with nanofillers to enhance the mechanical and also the biological properties of the film, resulting in nanocomposite films, that is, the low-cost urea-derived graphitic carbon nitride (g-C_3N_4) assembled in chitosan matrix.[19] This composite film was generated as a result of self-assembly caused by opposite charge interactions between positively charged chitosan and negatively charged g-C_3N_4. This nanomaterial-based film was expected to exhibit antibacterial activity that was stimulated by visible light, thereby assisting in long-term food storage. In terms of biological activity, it was observed that in the absence of light, even films not integrated with g-C_3N_4 nanoparticles had the same antibacterial activity as the bio-nanocomposite films. In the absence of visible light, increased g-C_3N_4 addition resulted in lower antibacterial activity, which was presumed to be due to an increase in the hydrophobicity of the film caused by g-C_3N_4 addition. The antibacterial activity of the nanocomposite film increased significantly when exposed to visible light, in contrast to the film containing only chitosan, which showed no improvement in antibacterial activity upon visible light irradiation. The obtained results could demonstrate the composite films' sustained antibacterial action when compared with the neat chitosan films with limited antibacterial activity. The films produced were wrapped around the tangerines at the end of all the preliminary assays. While the unpacked tangerines and the tangerines wrapped in chitosan films showed bacterial infection after about 10 days of storage and were completely rotted on the 18th day, the fruits wrapped in plastic films remaining fresh during the first few days of storage, did spoil on approaching day 18. However, the bionanocomposite films could retain the fruits in a good condition even till the 24th day of storage, as evidenced by the lowest change in weight and pH, as well as the highest stiffness observed for this film on the 24th day when compared with the tangerines with other types of packing. In addition to having very good antibacterial activities, the films also had least cytotoxicity, demonstrating their suitability for use in food packaging.

In another attempt to improve both physical and biological properties by using heterogenous mixtures rather than pure polymers in the synthesis of edible packaging films, black carob extract, which is used as a food source in some parts of the world like Argentina, was used in the synthesis of edible food packing films in conjugation with red seaweed-derived polysaccharide, κ- carrageenan.[1] These active ingredients were chosen because of the known medicinal properties of the extracts from black carob, such as the antioxidant and anti-inflammatory activities, as well as its effectiveness against metabolic syndrome and the film-forming and gas barrier properties of κ-carrageenan. The biological effects of the films produced from black carob extract, which primarily contains proteins, polyphenols and sugars, and κ- carrageenan films were evaluated by assessing the extent of oxidation that occurred when these films were used to store cheese slices at refrigerated conditions. In comparison to the uncoated slices used as controls, the coated samples showed lower peroxide values (PV), confirming the antioxidant effects of the coating films. Based on the observation that films containing mixtures of black carob extract and κ- carrageenan had lower PV values than films produced from the extracts alone, it was concluded that the total antioxidant effect exhibited by the film is dependent on both the chemical composition of the film, that is, the polyphenol content and physical properties such as the WVP.

Bulgur bran, a by-product of bulgur production rich in the amply available natural polymers, cellulose, and hemicellulose, was used as another low-cost alternative for the production of biopolymer films in order to add value to agro-industrial waste.[4] However, to enhance the physical and mechanical properties, the plasticizing effect, and the biodegradability of the films, they were reinforced with nanocellulose in the form of cellulose nanocrystals and cellulose nanofibers, as well as citric acid. Besides, because hemicelluloses are highly hydrophilic, different weight ratios of the nanomaterials were incorporated into the film and statistical analysis has determined that these nanomaterials in 10% weight ratios were optimal for imparting favorable water solubility properties to the film. On the whole, because all of the components used in the production of films, such as bulgur bran, nanocellulose, and citric acid are nontoxic, the film provided environmental benefits. Soil biodegradation occurs as a result of various factors, such as enzyme load, water content, and so on. The nanocellulose-integrated hemicellulose films showed a slight delay in biodegradation compared with the films made of hemicellulose alone due to an increase in film crystallinity after the addition of cellulose nanomaterials to the hemicellulose matrix. However, the results of both hemicellulose and nanocellulose-based hemicellulose films were positive when compared with nondegradable conventional LDPE films. This work, in addition to revealing the compatibility of the two biomaterials, demonstrated value addition to food industry waste, avoiding competition with food production in sustainable film manufacturing.

In a related study,[3] biopolymer films reinforced with berberine–cinnamic acid nanoparticles were used to develop biodegradable films based on gelatin and carrageenan matrices with high antibacterial properties that aid in the prevention of food spoilage. Berberine, a natural plant extract, was added to impart antimicrobial properties to the edible films. This was then conjugated with cinnamic acid obtained from the spice *Cinnamomi cortex*, which has known properties of medicinal relevance, such as antibacterial, anti-inflammatory, and antivirulence properties, to further enhance the antimicrobial properties and to obtain berberine–cinnamic acid composite in nanodimensions to increase the surface area, as well as berberine in the aggregated state to display aggregation-induced emissivity, that is, light-induced production of more reactive oxygen species toward further improvement in antibacterial activity. Because the film was made entirely of natural ingredients, it exhibited excellent biocompatibility, with a hemolysis rate of less than 5%. The film also exhibited excellent bacteriostatic activity against common food-borne pathogens, *Staphylococcus aureus* and *Escherichia coli*, as demonstrated by agar plate tests, particularly when exposed to light; additionally, it retained a bacterial inhibition rate of up to 97% even on the 7th day of storing 2 g of fresh chicken wrapped in these films. For the noninvasive release of an anti-inflammatory drug, Piroxicam, which is primarily used in the treatment of bone-related conditions, topical films based on the natural polysaccharides, such as chitosan and xanthan gum were developed.[13] These polymers were used in combination because their opposite charges (chitosan has a positive charge, whereas xanthan gum has a negative charge) form a polyelectrolyte complex (PEC) due to electrostatic interaction, imparting better physical and biological film properties. However, in order to have a better control over drug release by increasing xanthan gum solubility in aqueous systems, xanthan gum in its native form was converted into its carboxymethyl derivatives. To maximize the therapeutic effect of the active ingredient, the combination of chitosan and xanthan gum has been extensively explored for controlled drug release using various routes of administration[21,22]; the combination has also been used in tissue engineering and enzyme immobilization. The feasibility of using babassu coconut mesocarp, alginate, and glycerol as natural polymeric materials in the preparation of composite films was attempted.[23] Different weight combinations of these polymers were investigated in order to obtain films with the most desirable physical and mechanical properties. In this study, coconut mesocarp was chosen with the goal of increasing the economic value of this source. While alginate, a polysaccharide extracted from algae, is used for a variety of commercial and pharmaceutical applications, glycerol has very good film-forming capacities and is also generated as a by-product of biodiesel, making it available as a low-cost feedstock for biopolymer film production. Thus, the work was based on the development of biopolymer films from agricultural and industrial residues. The mechanical and physicochemical properties, as well as the biodegradability of the biopolymer film based on chitin extracted from crab shells, were further enhanced by the addition of an agricultural residue, coconut shell powder, and castor oil, where the optimum conditions of the composite film's synthesis was analyzed statistically using response surface methodology (RSM).[12] Similarly, the possibility of using furcelleran, a type of polysaccharide extracted from sea weeds as a novel source in the production of biopolymer-based food packaging films, was reported due to its abundance, excellent film-forming capacities, nontoxicity, and biodegradability.[20] The possibility of producing smart films containing antioxidant and antimicrobial additives was also reported.

Active food packaging films are those that contain additives, such as an antioxidant or an antimicrobial agent, that are released consistently over a specified period and can improve food quality by inhibiting food spoilage and thus extending the shelf life of foods.[24] In this regard, biopolymer films based on cassava starch and CMC to be used as a coating for kraft paper were further coated with turmeric oil to improve the shelf life of food products due to the presence of active ingredients with excellent antimicrobial activities, such as turmerone, curcumene, and so on.[10] This study confirms that the volume of turmeric oil and its coating thickness affect the migration rate of the antimicrobial agent, and as a result, the extent of antimicrobial activity as well as the extent of biodegradation, indicating that selecting the proper volume of the antimicrobial agent and film thickness is essential to ensure better release of these agents from the film.

Calcium alginate biopolymer films with high film-forming capacities containing antimicrobial nanoemulsions based on thyme oil as active ingredients were developed.[11] The microemulsion film also showed the highest antimicrobial activity against a variety of microbial populations, including bacteria, yeasts, and molds when compared with control films that did not contain the active ingredient, indicating its potential use in storing highly perishable food items, such as fish and meat products. Levan/sodium bentonite clay films were incorporated with glycerol, and antimicrobial peptides as active agents, and their mechanical, physicochemical, and antimicrobial activities against *E. coli* were assessed.[15] In the study, levan was produced from the bacterium *Pseudomonas mandelii*, and the resulting film was studied for its medicinal application due to the nontoxicity and high biocompatibility of the polysaccharide, which synergized with the antibacterial and antifungal activities of the active agent in the film. Because of their volatility, the active compounds incorporated into edible films do not generally exhibit sustained release. These volatile compounds typically show high release at first, followed by unstable release. They also have weak interactions with the polymer matrix. To overcome these limitations, a CMC-based biopolymer film was combined with stearic acid for the purpose of transporting and releasing carvacrol, which has both antibacterial and antifungal activities.[25] The rate of release was regulated by homogenizing the lipid polymer emulsion at various intensities and for varying lengths of time. Similarly, chitosan and polyvinyl alcohol composite films were developed to ensure acid resistivity and thus the sustained release of the antimicrobial agent procyanidin.[19] When used to preserve salmon, it had a storage capacity of up to 10 days.

9.5 ASSESSING MECHANICAL PROPERTIES OF BIOPOLYMER COMPOSITE FILMS

Biopolymers are nontoxic, biodegradable, biocompatible, and environmental-friendly materials; they possess high mechanical strength and impact resistance[26–32] which make them ideal candidates for tissue engineering and food packaging applications over traditional/synthetic materials.[33–40] The incorporation of various filler materials into the biopolymer system improves the mechanical properties, such as tensile and impact strength of the composites.[41] The biopolymers reinforced with naturally derived nanomaterials enhance young's modulus, tensile strength, thermal, and barrier properties of the composites.[35,42] In recent years, biopolymers, such as chitosan, starch, polyvinyl alcohol (PVA), polylactic acid (PLA), and polycaprolactone (PCL) have gained significant attention.[43]

The layered silicates with a large surface area used as reinforcements in biopolymers enhance the mechanical properties.[44–47] The montmorillonite (MMT) is a layered clay mineral, which is widely used to improve the mechanical property of the polymer matrix, and chitosan and MMT are the two components considered most effective and are considered perfectly intercalated nanocomposite used in food packaging. The nanocomposite showcases superior properties as polymeric chain diffuses through the clay layers, interrupting the van der Waals force of attraction between MMT layers. The favorable interaction between the negatively charged clay minerals and polycationic chitosan is established through the cationic exchange process.[48]

The evaluation of mechanical behavior is necessary to determine the feasibility of the composites. The polymer matrix mixed with nanoclay exhibits refined mechanical properties; the tensile strength and young's modulus are the key parameters achieved by the clay-reinforced composites.[49] Many researchers have studied the effect of MMT in chitosan with the varying content of MMT. Hong et al. fabricated MMT–chitosan composite films; the composite containing 3 wt% of MMT exhibited the highest tensile strength.[50] Potarniche et al. fabricated chitosan–MMT nanocomposite films at different temperatures and pH values; it was found that the composite films prepared at higher temperatures with pH 5 showed better polymer–clay interaction and exhibited higher mechanical properties.[51]

9.6 ASSESSING BIOLOGICAL ACTIVITY OF BIOPOLYMER FILMS

One of the important studies conducted to evaluate the biological effect of biopolymer films is the antimicrobial assay to assess the antimicrobial effects of the film. The following are the general methods for carrying out the test: (1) the conventional CFU (colony forming units)[19] method of counting the viable bacterial colonies, (2) the zone of inhibition assay method, which involves measuring the zone of bacterial inhibition on agar plates containing bacterial colonies caused by the active ingredient (cut pieces of the film) loaded on the plate,[10,11,15] (3) UV spectrophotometric method to estimate the antibacterial activities of the films incubated with suitable bacterial cultures,[14] (4) fluorescent assay: to distinguish between the live and dead bacterial cells using fluorescent dyes.[19] The antimicrobial activity of the films is also assessed in terms of the film's biofilm inhibition capacity.[14]

The safety of edible biopolymer films is also evaluated using preliminary assays such as the hemolysis test, which generally involves the in vitro treatment of red blood cells with the biopolymer film material.[3,19] In certain studies, the bacteriostatic activity of the film materials was further confirmed by wrapping fresh food material with the films produced and monitoring the freshness of the food for a given storage period in terms of microbial contamination, volatile gas release, textile profile analysis, and so on.[3,11,14] Biopolymer films were prepared for pharmaceutical applications in one study,[13] and their bioadhesive properties were studied by attaching them to excised rat abdominal skin and measuring the force of adhesion. On the other hand, the biodegradability of the films is assessed by cutting the films produced into given size ranges, burying them in soil, and measuring the loss of weight after the burial period, which corresponds to the rate of degradation.[10] In another study,[12] the biodegradability of the chitosan-based films was evaluated by incubating the film with protease solution for a certain number of days and measuring the reduction in weight of the sample at regular intervals during incubation, as protease is known to have chitosan-degrading capacity.

9.7 CONCLUSIONS

Biopolymers are the safest materials for packaging film applications, both in terms of human safety and environmental conservation. Furthermore, when compared with conventional plastic packaging films, they are more sustainable and cost-effective options. Advances in smart packaging film technologies allow for further improvements in meeting consumer needs. However, maintaining the overall film quality in terms of physical, chemical, mechanical, and biological properties, as well as preserving the sensory qualities of the food after the addition of active ingredients are some of the challenges that must be addressed.

KEYWORDS

- **biopolymers**
- **petrochemicals**
- **food industry**
- **mechanical and biological properties**
- **polysaccharides**
- **proteins**
- **tissue engineering**

REFERENCES

1. Pérez, M. J.; Moreno, M. A.; Martínez-Abad, A.; Cattaneo, F.; Zampini, C.; Isla, M. I.; López-Rubio, A.; Fabra, M. J. Interest of Black Carob Extract for the Development of Active Biopolymer Films for Cheese Preservation. *Food Hydrocoll.* **2021**, *113*.
2. Oberlintner, A.; Bajić, M.; Kalčíková, G.; Likozar, B.; Novak, U. Biodegradability Study of Active Chitosan Biopolymer Films Enriched with Quercus Polyphenol Extract in Different Soil Types. *Environ. Technol. Innov.* **2021,** *21*.
3. Ma, K.; Zhe, T.; Li, F.; Zhang, Y.; Yu, M.; Li, R.; Wang, L. Sustainable Films Containing AIE-Active Berberine-Based Nanoparticles: A Promising Antibacterial Food Packaging. *Food Hydrocoll.* **2022,** *123*, 107147.
4. Sutay Kocabaş, D.; Erkoç Akçelik, M.; Bahçegül, E.; Özbek, H. N. Bulgur Bran as a Biopolymer Source: Production and Characterization of Nanocellulose-Reinforced Hemicellulose-Based Biodegradable Films with Decreased Water Solubility. *Ind. Crops Prod.* **2021,** *171*.
5. Jamróz, E.; Kopel, P.; Tkaczewska, J.; Dordevic, D.; Jancikova, S.; Kulawik, P.; Milosavljevic, V.; Dolezelikova, K.; Smerkova, K.; Svec, P.; et al. Nanocomposite Furcellaran Films—the Influence of Nanofillers on Functional Properties of Furcellaran Films and Effect on Linseed Oil Preservation. *Polymers* **2019,** *11*, 2046.
6. Fortunati, E.; Luzi, F.; Yang, W.; Kenny, J. M.; Torre, L.; Puglia, D. Bio-Based Nanocomposites in Food Packaging. In *Nanomaterials for Food Packaging*; Elsevier: Amsterdam, The Netherlands, 2018; pp 71–110. ISBN 9780323512718.

7. Pires, J. R. A.; de Souza, V. G. L.; Fernando, A. L. Chitosan/Montmorillonite Bionanocomposites Incorporated with Rosemary and Ginger Essential Oil as Packaging for Fresh Poultry Meat. *Food Packag. Shelf Life* **2018,** *17*, 142–149.
8. Pires, J. R. A.; Souza, V. G. L.; Fernando, A. L. Production of Nanocellulose from Lignocellulosic Biomass Wastes: Prospects and Limitations. In *Innovation, Engineering and Entrepreneurship*; Machado, J., Soares, F., Veiga, G., Eds.; Lecture Notes in Electrical Engineering; Springer International Publishing: Cham, Switzerland, 2019; vol 505, pp 719–725. ISBN 978-3-319-91333-9.
9. Souza, V. G. L.; Pires, J. R. A.; Rodrigues, C.; Rodrigues, P. F.; Lopes, A.; Silva, R. J.; Caldeira, J.; Duarte, M. P.; Fernandes, F. B.; Coelhoso, I. M.; et al. Physical and Morphological Characterization of Chitosan/Montmorillonite Films Incorporated with Ginger Essential Oil. *Coatings* **2019,** *9*, 700.
10. Mustapha, F. A.; Jai, J.; Nik Raikhan, N. H.; Sharif, Z. I. M.; Yusof, N. M. Response Surface Methodology Analysis Towards Biodegradability and Antimicrobial Activity of Biopolymer Film Containing Turmeric Oil Against Aspergillus Niger. *Food Control* **2019,** *99*, 106–113.
11. Almasi, L.; Radi, M.; Amiri, S.; McClements, D. J. Fabrication and Characterization of Antimicrobial Biopolymer Films Containing Essential Oil-Loaded Microemulsions or Nanoemulsions. *Food Hydrocoll.* **2021,** *117*, 106733.
12. Seenuvasan, M.; Malar, C. G.; Growther, L. Production of a Biopolymer Film from Biological Wastes and its Statistical Analysis. *Bioresour. Technol. Rep.* **2021,** *13*, 100610.
13. Salma, H.; Melha, Y. M.; Sonia, L.; Hamza, H.; Salim, N. Efficient Prediction of In Vitro Piroxicam Release and Diffusion from Topical Films Based on Biopolymers Using Deep Learning Models and Generative Adversarial Networks. *J. Pharm. Sci.* **2021,** *110*, 2531–2543.
14. Bi, J.; Tian, C.; Zhang, G. L.; Hao, H.; Hou, H. M. Novel Procyanidins-Loaded Chitosan-Graft-Polyvinyl Alcohol Film with Sustained Antibacterial Activity for Food Packaging. *Food Chem.* **2021,** *365*, 130534.
15. Ağçeli, G. K.; Cihangir, N. Synthesis, Characterization and Antimicrobial Performance of Novel Nanostructured Biopolymer Film Based on Levan/clay/LL-37 Antimicrobial Peptide. Biocatal. Agric. Biotechnol. **2020,** *23*, 101421.
16. Surya, I.; Chong, E. W. N.; Abdul Khalil, H. P. S.; Funmilayo, O. G.; Abdullah, C. K.; Sri Aprilia, N. A.; Olaiya, N. G.; Lai, T. K.; Oyekanmi, A. A. Augmentation of Physico-Mechanical, Thermal and Biodegradability Performances of Bio-Precipitated Material Reinforced in Eucheuma Cottonii Biopolymer Films. *J. Mater. Res. Technol.* **2021,** *12*, 1673–1688.
17. Oyekanmi, A. A.; Abdul Khalil, H. P. S.; Rahman, A. A.; Mistar, E. M.; Olaiya, N. G.; Alfatah, T.; Yahya, E. B.; Mariana, M.; Hazwan, C. M.; Abdullah, C. K. Extracted Supercritical CO2 Cinnamon Oil Functional Properties Enhancement in Cellulose Nanofibre Reinforced Euchema Cottoni Biopolymer Films. *J. Mater. Res. Technol.* **2021,** *15*, 4293–4308.
18. Alinaqi, Z.; Khezri, A.; Rezaeinia, H. Sustained Release Modeling of Clove Essential Oil from the Structure of Starch-Based Bio-Nanocomposite Film Reinforced by Electrosprayed Zein Nanoparticles. *Int. J. Biol. Macromol.* **2021,** *173*, 193–202.

19. Ni, Y.; Shi, S.; Li, M.; Zhang, L.; Yang, C.; Du, T.; Wang, S.; Nie, H.; Sun, J.; Zhang, W.; Wang, J. Visible Light Responsive, Self-Activated Bionanocomposite Films with Sustained Antimicrobial Activity for Food Packaging. *Food Chem.* **2021,** *362*, 130201.
20. Marangoni Júnior, L.; Vieira, R. P.; Jamróz, E.; Anjos, C. A. R. Furcellaran: An Innovative Biopolymer in the Production of Films and Coatings. *Carbohydr. Polym.* **2021,** *252*.
21. Ćirić, A.; Medarević, Đ.; Čalija, B.; Dobričić, V.; Mitrić, M.; Djekic, L. Study of Chitosan/Xanthan Gum Polyelectrolyte Complexes Formation, Solid State and Influence on Ibuprofen Release Kinetics. *Int. J. Biol. Macromol.* **2020,** *148*, 942–955.
22. Faralli, A.; Shekarforoush, E.; Ajalloueian, F.; Mendes, A. C.; Chronakis, I. S. In Vitro Permeability Enhancement of Curcumin Across Caco-2 Cells Monolayers Using Electrospun Xanthan-Chitosan Nanofibers. *Carbohydr. Polymers* **2019,** *206*, 38–47.
23. Alves Lopes, I.; Coelho Paixão, L.; Souza da Silva, L. J.; Almeida Rocha, A.; Allan, A. K. Amorim Santana, A. Elaboration and Characterization of Biopolymer Films with Alginate and Babassu Coconut Mesocarp. *Carbohydr. Polymers* **2020,** *234*, 115747.
24. Sani, M. A.; Azizi-Lalabadi, M.; Tavassoli, M.; Mohammadi, K.; McClements, D. J. Recent Advances in the Development of Smart and Active Biodegradable Packaging Materials. *Nanomaterials* **2021,** *11*, 1–34.
25. Zhong, T.; Yu, X. ; Xiao, Y.; Wang, L.; Tan, L.; Huang, R. Incorporating Lipid Microparticles into Carboxymethyl Cellulose Films as Secondary Carriers for Stable Loading and Sustained Releasing of Carvacrol. *Results Mater.* **2021,** *11*.
26. Ilyas, R. A.; Sapuan, S. M.; Ibrahim, R.; Abral, H.; Ishak, M. R.; Zainuddin, E. S.; Asrofi, M.; Atikah, M. S. N.; Huzaifah, M. R. M.; Radzi, A. M.; Azammi, A. M. N.; Shaharuzaman, M. A.; Nurazzi, N. M.; Syafri, E. Sari, N. H.; Norrrahim, M. N. F.; Jumaidin, R. *J. Mater. Res. Technol.* **2019,** *8*, 2753.
27. Ilyas, R. A.; Sapuan, S. M.; Ishak, M. R.; Zainuddin, E. S. *Int. J. Biol. Macromol.* **2019,** *123*, 379.
28. Ponnamma, D.; Sadasivuni, K. K.; Cabibihan, J. J.; Yoon, W. J.; Kumar, B. *Appl. Phys. Lett.* **2016,** *108*, 171906.
29. Ilyas, R. A.; Sapuan, S. M.; Ishak, M. R.; Zainuddin, E. S. *Bioresources* **2017,** *12*, 8734.
30. Ilyas, R. A.; Sapuan, S. M.; Ishak, M. R.; Zainuddin, E. S. *J. Adv. Res. Fluid. Mech. Therm. Sci.* **2018,** *51*, 234.
31. Abral, H.; Basri, A.; Muhammad, F.; Fernando, Y.; Hafizulhaq, F.; Mahardika, M.; Sugiarti, E.; Sapuan, S. M.; Ilyas, R. A.; Stephane, I. *Food Hydrocoll.* **2019,** *93*, 276.
32. Edin, B. B.; Ascari, L.; Beccai, L.; Roccella, S.; Cabibihan, J. J.; Carrozza, M. C. *Brain Res. Bull.* **2008,** *15*, 785.
33. Ilyas, R. A.; Sapuan, S. M.; Ishak, M. R.; Zainuddin, E. S. *Carbohydr. Polym.* **2018,** *202*, 186.
34. Sadasivuni, K. K.; Saiter, A.; Gautier, N.; Thomas, S.; Grohens, Y. *Colloid Polym. Sci.* **2013,** *291*, 1729.
35. Halimatul, M. J.; Sapuan, S. M.; Jawaid, M.; Ishak, M. R.; Ilyas, R. A. *Polymery* **2019,** *64*, 422.
36. Sadasivuni, K. K.; Ponnamma, D.; Thomas, S.; Grohens, Y. *Prog. Polym. Sci.* **2014,** *39*, 749.
37. Ponnamma, D.; Sung, S. H.; Hong, J. S.; Ahn, K. H.; Varughese, K. T.; Thomas, S. *Eur. Polym. J.* **2014,** *53*, 147.

38. Ponnamma, D.; Sadasivuni, K. K.; Strankowski, M.; Moldenaers, P.; Thomas, S.; Grohens, Y. *RSC Adv.* **2013,** *3*, 16068.
39. Sadasivuni, K. K.; Ponnamma, D.; Kumar, B.; Strankowskie, M.; Cardinaels, R.; Moldenaers, P.; Thomas, S.; Grohens, Y. *Compos. Sci. Technol.* **2014,** *104*, 18.
40. Zhang, Y. Z.; Lim, C. T. *J. Min. Met. Mater. Soc.* **2008,** *60*, 45.
41. Mohammad, J.; Sarat, K. S. Bionanocomposites for Packaging Applications; Springer International Publishing: Cham, 2018; p 125.
42. Ilyas, R. A.; Sapuan, S. M.; Ishak, M. R.; Zainuddin, E. S.; Atikah, M. S. N. Sugar Palm Biofibers, Biopolymers, and Biocomposites, 1st ed.; CRC Press, Boca Raton: FL, 2018; p 189.
43. John, M. J.; Thomas, S. *Carbohydr. Polym.* **2008,** *71*, 343.
44. Xu, Y.; Ren, X.; Hanna, M. A. Chitosan/Clay Nanocomposite Film Preparation and Characterization. *J. Appl. Polym. Sci.* **2006,** *99*, 1684–1691.
45. Lavorgna, M.; Piscitelli, F.; Mangiacapra, P.; Buonocore, G. G. Study of the Combined Effect of Both Clay and Glycerol Plasticizer on the Properties of Chitosan Films. *Carbohydr. Polym.* **2010,** *82*, 291–298.
46. Günister, E.; Pestreli, D.; .nlü, C.H.; Atıcı, O.; Güng.r, N. Synthesis and Characterization of Chitosan-MMT Biocomposite Systems. *Carbohydr. Polym.* **2007,** *67*, 358–365.
47. Lewandowska, K.; Sionkowska, A.; Kaczmarek, B.; Furtos, G. Characterization of Chitosan Composites with Various Clays. *Int. J. Biol. Macromol.* **2014,** *65*, 534–541.
48. Dias, M. V.; Machado Azevedo, V.; Borges, S. V.; Soares, N. D. F. F.; de Barros Fernandes, R.V.; Marques, J. J.; Medeiros, É. A. A. Development of Chitosan/Montmorillonite Nanocomposites with Encapsulated α-tocopherol. *Food Chem.* **2014,** *165*, 323–329.
49. Kumar, S.; Mukherjee, A.; Dutta, J. Chitosan Based Nanocomposite Films and Coatings: Emerging Antimicrobial Food Packaging Alternatives. *Trends Food Sci. Technol.* **2020,** *97*, 196–209.
50. Hong, S. I.; Lee, J. H.; Bae, H. J.; Koo, S. Y.; Lee, H. S.; Choi, J. H.; Kim, D. H.; Park, S.; Park, H. J. Effect of Shear Rate on Structural, Mechanical, and Barrier Properties of Chitosan/Montmorillonite Nanocomposite Film. *J. Appl. Polym. Sci.* **2009,** *119*, 2742–2749.
51. Potarniche, C. G.; Vuluga, Z.; Donescu, D.; Christiansen, J. D. C.; Eugeniu, V.; Radovici, C.; Serban, S.; Ghiurea, M.; Somoghi, R.; Beckmann, S. Morphology Study of Layered Silicate/Chitosan Nanohybrids. *Surf. Interface Anal.* **2011,** *44*, 200–207.

PART III

Solid Materials

CHAPTER 10

Computational Analysis of X_3O_4 (X= Mn, Fe, and Co) Nanofibers

SHAYERI DAS[1], PRABHAT RANJAN[1], and TANMOY CHAKRABORTY[2]

[1]*Department of Mechatronics Engineering, Manipal University Jaipur, Jaipur, Rajasthan, India*

[2]*Department of Chemistry and Biochemistry, School of Basic Sciences and Research, Sharda University, Greater Noida, India*

ABSTRACT

The vast contrast between the properties of nanomaterials with their bulk counterparts has led to the emergence of nanofibers. The noteworthy impression of the nanofiber technology has the potential to be sketched up to the broad spectrum of essential materials that might be utilized for the amalgamation of nanofibers. The materials vary from pure and man-made polymers to composite materials. The increasing areas of application have resulted in the study of X_3O_4 (X- Mn, Fe, Co) by using a computational approach. The results show that these systems have HOMO–LUMO gap in the range of 1.238 eV–2.700 eV. The bond lengths and DFT-based global descriptors, such as molecular hardness, softness, electronegativity, electrophilicity index, and dipole moment of the most stable configurations of the nanofibers are calculated. The bond lengths found for the lowest energy structure of X_3O_4 are in agreement with the experimental data. The dipole moment furnishes an idea about the nature of the bonds formed in the clusters. The present study opens a new domain of research for experimental scientists to explore more about nanofiber.

Tailored Functional Materials for Clean and Sustainable Development. Tamara Tatrishvili, PhD, Neha Kanwar Rawat, PhD, Ann Rose Abraham, PhD & A. K. Haghi, PhD (Eds.)

10.1 INTRODUCTION

Nanomaterials possess unique physical properties and characteristics, which are different from their bulk states. In this aspect, nanofibers are the new torchbearers as a 1D material for a diverse continuum of exploration and commercialization.[1–3] Nanofibers are a sensational genre of fibers bearing cross-sectional diameters of a huge range of nanometers and these retain an exceptionally high bulk volume. Nanofibers are proficient in developing grids of a vastly porous lattice with significant interconnectivity among the apertures, leading the nanofibers to be an impeccable prime material for a congregation of implementations. Nanofibers are fabricated from a range of materials, namely, pure polymers,[4] artificial polymers,[5,6] organic nanomaterials,[7] semiconductor nanomaterials,[8,9] and composite nanomaterials.[10–16] Besides, swift advancement in the creation and categorization of these clusters over some time, remarkable exertions have been engrossed on investigating the prospect of nanofibers in various operative uses, comprising of energy production and storage, aquatic and other environmental management, healthcare, and biomedical engineering.

Nanofibers have found applications in microchip technology, photonics, and regenerative medicine applications.[17–19] The core-shell structures have drawn broad attention and turned out to be one of the most favorable contenders for catalysis due to their interesting activity, selectivity, stability, and resistance to toxicity as compared with monometallic counterparts.[20–22] Chu et al.[20] synthesized AuPt nanoparticles and studied the features of polymer-reinforced metallic nanoparticles of couturier configuration for optimal performance in catalysis and further utilizations. Lee et al.[21] investigated the Au–Pd alloy nanodendrite experimentally and found an improvement in their geometry as well as catalytic activities for probable applications in fuel cells and fabrication of nanodevices. Dong et al.[22] created a fibrous nanosilica-based nanocatalyst (Ni@Pd/KCC-1) and studied the features to comprehend the potential in catalytic applications.

Koo et al.[23] found that iron-doped oxide indium-based core-shell nanofibers with upgraded magnetic characteristics can be fused aided by co-axial electrospinning. Co-electrospinning is implemented to create monolayer and bilayer nanofibers. However, the arrangements have restricted assembling alignments and relevance. Yet, by amplifying the amount of films of nanofibers, the subsequent connectivity of the complexes and their potential can be heightened. Starr et al.[24] investigated multifunctional nanofibers with multiple layers which are fabricated using co-electrospinning. Kayaci et al.[25]

reported that homo-assembled polyacrylonitrile nanofibrous mat embezzled with zinc oxide nanostructures are created with the aid of electrospinning. As an add-on to the electrospinning procedure, nanofibers are presently manufactured through a self-assembly technique in solution.

Owing to these several research works and applications on nanofibers, we have performed a computational study on X_3O_4 (X= Mn, Fe, Co) by using the density functional theory (DFT) technique. This report has been presented in such a way that Sections 10.2 and 10.3 provide a brief account of the computational methodology, results, and discussion, respectively.

10.2 COMPUTATIONAL DETAILS

Among several computational techniques, DFT has gained importance for being computational-friendly and having high accuracy. Various fields of study, such as electronics, solid-state physics, molecular and nuclear physics, life sciences have adopted this technique successfully.[26–31] DFT methods are efficaciously used in various cases.[32–35] In the past few years, our group has executed various analyses on metallic clusters[36–43] based on DFT.

In this report, a theoretical investigation on nanofiber clusters, X_3O_4 (X= Mn, Fe, Co) has been conducted. Optimization of the clusters has been done by Gaussian 16 program.[44] The hybrid functional parameter (B3LYP) along with the basis set LANL2DZ has been opted and is intended for optimization.

Ionization energy (I) and electron affinity (A) of all the clusters are determined via the subsequent equations[45]:

$$I = -\varepsilon_{HOMO} \tag{10.1}$$

$$A = -\varepsilon_{LUMO} \tag{10.2}$$

By using eqs (10.1) and (10.2), the conceptual descriptors: molecular hardness (η), molecular softness (S), electronegativity (χ), and electrophilicity index (ω) are computed as under:

$$\chi = -\mu = \frac{I + A}{2} \tag{10.3}$$

Here, μ implies the chemical potential of the cluster.

$$\eta = \frac{I - A}{2} \tag{10.4}$$

10.3 RESULTS AND DISCUSSION

10.3.1 OPTIMIZED GEOMETRY

The ground state of nanofiber X_3O_4 (X=Mn, Fe, Co) is presented in Figure 10.1, along with the point groups and spin multiplicity. Bond length is interpreted as the distance between the centers of two atoms that are covalently bonded atoms. The length of the bond is confirmed using the summation of bonded electrons. Overall, the span of the bond amid two adjacent atoms is almost identical to the summation of the covalent radii of the atoms.

The geometry of Fe_3O_4 is a triangular structure presented in 10.1-a with three Fe atoms linked with four O atoms. The orientation has the symmetry group C_{2V} and singlet spin multiplicity. On every side, the Fe atom is placed in between two O atoms. The bond lengths of Fe–O are 1.653Å to 1.718 Å, 1.917Å, and 2.086 Å. The reported experimental value of the bond length of Fe–O for compound Fe_3O_4 is 2.04 Å.[46]

Mn_3O_4 has a pentagonal-shaped ground-state structure, presented in Figure 10.1-b, which has two Mn placed at two of the vertices and one on the base side. The structure comprises the point group of C_{2V} and quartet spin multiplicity. All the Mn atoms are connected with three O atoms. The bond lengths of Mn–O are 1.552 Å, 1.614 Å, 1.893 Å, and 2.043 Å. Pomar et al. reported the bond length of Mn–O for Mn_3O_4 to be between 1.903 Å and 2.184 Å.[47]

Figure 10.1c shows the ground-state structure of Co_3O_4, which has point group C_{2V} and multiplicity doublet. The structure is similar in shape compared with Fe_3O_4. The Co-O bonds on the other sides have a bond length of 1.917 Å between the upper vertex and Co; the bond length with the lower vertex is 1.718 Å. Co-O bond length at the base side is 2.086 Å, and 1.653 Å with O atoms placed inside the triangular enclosed space. The computed bond length is in agreement with the experimental data.[48]

10.3.2 ELECTRONIC PROPERTIES AND DFT-BASED DESCRIPTORS

The theoretical study of the nanofibers is conducted using the electronic structure theory. The global DFT-based descriptors—molecular hardness, molecular softness, electronegativity, and electrophilicity index are evaluated and discussed in this section.

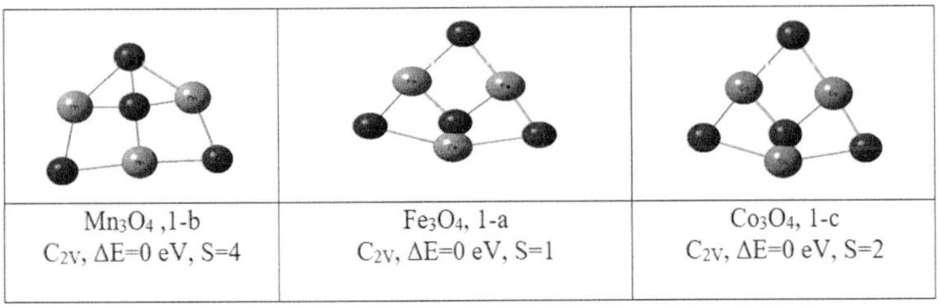

FIGURE 10.1 Lowest energy structure of X_3O_4 (X= Mn, Fe, Co).

The molecular orbitals of both energy bands have an imperative part in the transferral of charge and bond creation for the duration of the establishment of donor–acceptor system.[48–55] The energy alteration among orbitals HOMO and LUMO is an essential component to calculating the electronic properties of chemical species.[56,57] The kinetic stability of a cluster declines with the diminution of HOMO–LUMO gap. Among the nanofibers under study, Co_3O_4 exhibits the maximum HOMO–LUMO gap, whereas Fe_3O_4 has the lowest gap. There is a considerable increase in the HOMO–LUMO gap from Fe_3O_4. On the other hand, the remaining nanofibers fall in a compatible range in terms of HOMO–LUMO gap. Nanofibers have applications in various domains, including medicine, healthcare, and electronic devices. The range of HOMO–LUMO gap exhibited makes these nanofibers compatible with domains, such as optoelectronic devices, solar cells, as well as applications in life sciences.

In order to comprehend the molecular structure, stability, nature of bond, and dynamics of chemical species, molecular hardness is a vital factor.[58] Every molecule possesses the inherent tendency to configure itself with maximum hardness.[59] The movement of a molecule from a stable to an unstable state reduces its hardness value and vice versa.[51] As hardness is in direct variance with the HOMO–LUMO gap, consequently, Fe_3O_4 exhibits minimum hardness comparatively much lower than the other nanofibers and Co_3O_4 possesses maximum hardness.

The clusters that possess the peak HOMO–LUMO gap are most likely not to respond to any external disruption. The molecular softness has contrary relation with HOMO–LUMO gap of the cluster. The obtained data reveal that Co_3O_4 has the least value of molecular softness, whereas Fe_3O_4 has the maximum value of softness.

Electronegativity is the most vital factor to comprehend the transfer of charge from donor to acceptor in a chemical system.[60,61] In the present study, the electronegativity varies from 4.153 eV to 5.499 eV. The results show that nanofiber Fe_3O_4 has the maximum electronegativity, whereas Mn_3O_4 has the minimum electronegativity.

Both ionization energy and electron affinity influence the electrophilicity index, which is the constraint for computation of the expanse of energy reduction due to the additional flow of electrons for the duration of donor–acceptor interaction.[62] Fe_3O_4 with the lowest HOMO–LUMO gap exhibits the highest electrophilicity index.

The result reveals that Fe_3O_4 has an extremely high value of dipole moment. It indicates that the nanofiber has the presence of ionic bonds within itself. Both Mn_3O_4 and Co_3O_4 have dipole moments with a magnitude much less than 1 Debye.

TABLE 10.1 DFT-Based Global Descriptors of X_3O_4 (X- Mn, Fe, and Co).

Species	HOMO-LUMO gap (eV)	Hardness (eV)	Softness (eV)	Electronegativity (eV)	Electrophilicity index (eV)	Dipole moment (debye)
Mn_3O_4	2.379	1.189	0.420	4.153	7.249	0.738
Fe_3O_4	1.238	0.619	0.808	5.499	24.432	9.671
Co_3O_4	2.700	1.350	0.370	4.967	9.138	0.353

10.4 CONCLUSIONS

In the present study, we have investigated some compounds of nanofiber—X_3O_4 (X-Fe, Mn, Co) by using a computational approach. Recent advances in nanofibers have encouraged researchers to explore more about these compounds for probable uses in solar cells, optoelectronic, and nanodevices. The structure, electronic properties, and DFT-based descriptors of these compounds are computed. The result shows that Fe_3O_4 system is highly reactive, whereas Co_3O_4 is found as the least reactive compound. It is observed that the stability of these systems increases from Fe to Mn to Co. System Fe_3O_4 shows the maximum value of electronegativity, electrophilicity index, and dipole moment whereas hardness is minimum. The high value of the dipole moment for Fe_3O_4 indicates that the bond within the cluster is ionic in nature. The calculated data display that these nanofiber compounds can be utilized for optoelectronics, life sciences, and for building blocks for nanodevices.

ACKNOWLEDGMENTS

S. Das and P. Ranjan would like to acknowledge Manipal University Jaipur, Rajasthan, India for providing research facilities and computational resources.

T. Chakraborty would like to acknowledge Sharda University, Greater Noida, UP, India for providing research facilities and computational resources.

KEYWORDS

- **nanofiber**
- **DFT**
- **HOMO–LUMO gap**
- **Co_3O_4**
- **Fe_3O_4**

REFERENCES

1. Tiwari, J. N.; Tiwari, R. N., Kim, K. S. Zero-dimensional, One-dimensional, Two-dimensional and Three-dimensional Nanostructured Materials for Advanced Electrochemical Energy Devices. *Prog. Mater. Sci.* **2012,** *57* (4), 724–803.
2. Yong, K. T., Yu, S. F. AlN Nanowires: Synthesis, Physical Properties, and Nanoelectronics Applications. *J. Mater. Sci.* **2012,** *47* (14), 5341–5360.
3. Lim, C. T. Synthesis, Optical Properties, and Chemical–Biological Sensing Applications of One-Dimensional Inorganic Semiconductor Nanowires. *Prog. Mater. Sci.* **2013,** *58* (5), 705–748.
4. Hassanzadeh, P.; Kharaziha, M.; Nikkhah, M.; Shin, S. R.; Jin, J.; He, S.; Sun, W.; Zhong, C.; Dokmeci, M. R.; Khademhosseini, A.; Rolandi, M. Chitin Nanofiber Micropatterned Flexible Substrates for Tissue Engineering. *J. Mater. Chem. B* 2013, *1* (34), 4217–4224.
5. Behrens, A. M.; Casey, B. J.; Sikorski, M. J.; Wu, K. L.; Tutak, W.; Sandler, A. D.; Kofinas, P. In Situ Deposition of PLGA Nanofibers Via Solution Blow Spinning. *ACS Macro Lett.* **2014,** *3* (3), 249–254.

6. Shah, S.; Yin, P. T.; Uehara, T. M.; Chueng, S. T. D.; Yang, L.; Lee, K. B. Guiding Stem Cell Differentiation into Oligodendrocytes Using Graphene-Nanofiber Hybrid Scaffolds. *Adv. Mater.* **2014,** *26* (22), 3673–3680.
7. Yang, X.; Zou, W.; Su, Y.; Zhu, Y.; Jiang, H.; Shen, J.; Li, C. Activated Nitrogen-Doped Carbon Nanofibers with Hierarchical Pore as Efficient Oxygen Reduction Reaction Catalyst for Microbial Fuel Cells. *J. Power Sources* **2014,** *266*, 36–42.
8. Shang, M.; Wang, W.; Sun, S.; Gao, E.; Zhang, Z.; Zhang, L.; O'Hayre, R. The Design and Realization of a Large-Area Flexible Nanofiber-Based Mat for Pollutant Degradation: An Application in Photocatalysis. *Nanoscale* **2013,** *5* (11), 5036–5042.
9. Cheng, L.; Ma, S. Y.; Wang, T. T.; Li, X. B.; Luo, J.; Li, W. Q.; Mao, Y. Z.; Gz, D. J. Synthesis and Characterization of SnO2 Hollow Nanofibers by Electrospinning for Ethanol Sensing Properties. *Mater. Lett.* **2014,** *131*, 23–26.
10. Wu, Q.; Tran, T.; Lu, W.; Wu, J. Electrospun Silicon/Carbon/Titanium Oxide Composite Nanofibers for Lithium Ion Batteries. *J. Power Sources* **2014,** *258*, 39–45.
11. Liu, Y.; Zhao, L.; Li, M.; Guo, L.. TiO 2/CdSe Core–Shell Nanofiber Film for Photoelectrochemical Hydrogen Generation. *Nanoscale* **2014,** *6* (13), 7397–7404.
12. Shi, H.; Zhou, M.; Song, D.; Pan, X.; Fu, J.; Zhou, J.; Ma, S.; Wang, T. Highly Porous SnO2/TiO2 Electrospun Nanofibers with High Photocatalytic Activities. *Ceram. Int.* **2014,** *40* (7), 10383–10393.
13. Liao, Y.; Li, X. G.; Hoek, E. M.; Kaner, R. B. Carbon Nanotube/Polyaniline Nanofiber Ultrafiltration Membranes. *J. Mater. Chem. A* **2013,** *1* (48), 15390–15396.
14. Wang, Y.; Zhang, X.; He, X.; Zhang, W.; Zhang, X.; Lu, C. In Situ Synthesis of MnO2 Coated Cellulose Nanofibers Hybrid for Effective Removal of Methylene Blue. *Carbohydr. Polym.* **2014,** *110*, 302–308.
15. Ayad, M. M.; Salahuddin, N. A.; Minisy, I. M.; Amer, W. A. Chitosan/Polyaniline Nanofibers Coating on the Quartz Crystal Microbalance Electrode for Gas Sensing. *Sens. Actuators B* **2014,** *202*, 144–153.
16. Zhao, R.; Li, X.; Sun, B.; Zhang, Y.; Zhang, D.; Tang, Z.; Chen, X.; Wang, C. Electrospun Chitosan/Sericin Composite Nanofibers with Antibacterial Property as Potential Wound Dressings. *Int. J. Biol. Macromol.* **2014,** *68*, 92–97.
17. Agarwal, S.; Greiner, A.; Wendorff, J. H. Functional Materials by Electrospinning of Polymers. *Prog. Polym. Sci.* **2013,** *38* (6), 963–991.
18. Persano, L.; Camposeo, A.; Pisignano, D. Active Polymer NJanofibers for photonics, Electronics, Energy Generation and Micromechanics. *Prog. Polym. Sci.* **2015,** *43*, 48–95.
19. Jiang, T.; Carbone, E. J.; Lo, K. W. H.; Laurencin, C. T. Electrospinning of Polymer Nanofibers for Tissue Regeneration. *Prog. Polym. Sci.* **2015,** *46*, 1–24.
20. Chu, C.; Su, Z. Facile Synthesis of AuPt Alloy Nanoparticles in Polyelectrolyte Multilayers with Enhanced Catalytic Activity for Reduction of 4-nitrophenol. *Langmuir* **2014,** *30* (50), 15345–15350.
21. Lee, Y. W.; Kim, N. H.; Lee, K. Y.; Kwon, K.; Kim, M.; Han, S. W. Synthesis and Characterization of Flower-Shaped Porous Au– Pd Alloy Nanoparticles. *J. Phys. Chem. C* **2008,** *112* (17), 6717–6722.
22. Dong, Z.; Le, X.; Dong, C.; Zhang, W.; Li, X.; Ma, J. Ni@ Pd Core–Shell Nanoparticles Modified Fibrous Silica Nanospheres as Highly Efficient and Recoverable Catalyst for Reduction of 4-nitrophenol and Hydrodechlorination of 4-chlorophenol. *Appl. Catal. B* **2015,** *162*, 372–380.

23. Koo, B. R.; Park, I. K.; Ahn, H. J. Fe-doped In2O3/α-Fe2O3 Core/Shell Nanofibers Fabricated by Using a Co-electrospinning Method and its Magnetic Properties. *J. Alloys Compd.* **2014,** *603*, 52–56.
24. Starr, J. D.; Andrew, J. S. A Route to Synthesize Multifunctional Tri-phasic Nanofibers. *J. Mater. Chem. C* **2013,** *1* (14), 2529–2533.
25. Kayaci, F.; Vempati, S.; Ozgit-Akgun, C.; Biyikli, N.; Uyar, T. Enhanced Photocatalytic Activity of Homoassembled ZnO Nanostructures on Electrospun Polymeric Nanofibers: A Combination of Atomic Layer Deposition and Hydrothermal Growth. *Appl. Catal. B* **2014,** *156*, 173–183.
26. Illas, F.; Martin, R. L. Magnetic Coupling in Ionic Solids Studied by Density Functional Theory. *J. Chem. Phys.* **1998,** *108* (6), 2519–2527.
27. Gyorffy, B. L.; Staunton, J. B.; Stocks, G. M. Fluctuations in Density Functional Theory: Random Metallic Alloys and Itinerant Paramagnets. In *Density Functional Theory*; Springer: Boston, MA, **1995;** pp 461–484.
28. Kümmel, S.; Brack, M. Quantum Fluid Dynamics from Density-Functional Theory. *Phys. Rev. A* **2001,** *64* (2), 022506.
29. Car, R.; Parrinello, M. Unified Approach for Molecular Dynamics and Density-Functional Theory. *Phys. Rev. Lett.* **1985,** *55* (22), 2471.
30. Koskinen, M.; Lipas, P. O.; Manninen, M. Unrestricted Shapes of Light Nuclei in the Local-Density Approximation: Comparison with Jellium Clusters. *Nucl. Phys. A* **1995,** *591* (3), 421–434.
31. Schmid, R. N.; Engel, E.; Dreizler, R. M. Density Functional Approach to Quantum Hadrodynamics: Local Exchange Potential for Nuclear Structure Calculations. *Phys. Rev. C* **1995,** *52* (1), 164.
32. Chen, P.; Qin, M.; Chen, H.; Yang, C.; Wang, Y.; Huang, F. Cr Incorporation in Cu G a S 2 Chalcopyrite: A New Intermediate-Band Photovoltaic Material with Wide-Spectrum Solar Absorption. *Phys. Status Solidi A* **2013,** *210* (6), 1098–1102.
33. Palacios, P.; Sánchez, K.; Conesa, J. C.; Wahnón, P. First Principles Calculation of Isolated Intermediate Bands Formation in a Transition Metal-Doped Chalcopyrite-Type Semiconductor. *Phys. Status Solidi A* **2006,** *203* (6), 1395–1401.
34. Palacios, P.; Sánchez, K.; Conesa, J. C.; Fernández, J. J.; Wahnón, P. Theoretical Modelling of Intermediate Band Solar Cell Materials Based on Metal-Doped Chalcopyrite Compounds. *Thin Solid Films* **2007,** *515* (15), 6280–6284.
35. Palacios, P.; Wahnón, P.; Pizzinato, S.; Conesa, J. C. Energetics of Formation of Ti Ga 3 As 4 and Ti Ga 3 P 4 Intermediate Band Materials. *J. Chem. Phys.* **2006,** *124* (1), 014711.
36. Ranjan, P.; Chakraborty, T. Structure and Optical Properties of (CuAg) n (n= 1–6) Nanoalloy Clusters Within Density Functional Theory Framework. *J. Nanopart. Res.* **2020,** *22* (9), 1–11.
37. Ranjan, P.; Dhail, S.; Venigalla, S.; Kumar, A.; Ledwani, L.; Chakraborty, T. A Theoretical Analysis of Bi-metallic (Cu–Ag) Nano Alloy Clusters Invoking DFT Based Descriptors. *Mater. Sci.-Pol.* **2015,** *33* (4), 719–724.
38. Ranjan, P.; Venigalla, S.; Kumar, A.; Chakraborty, T. Theoretical Study Of Bi-Metallic Ag m Au n;(m+ n= 2-8) Nano Alloy Clusters In Terms Of DFT Based Descriptors. *New Front. Chem* **2014,** *23* (2), 111–122.
39. Ranjan, P.; Chakraborty, T. Density Functional Approach: To Study Copper Sulfide Nanoalloy Clusters. *Acta Chim. Slov.* **2019,** *66* (1), 173–181.

40. Ranjan, P.; Chakraborty, T.; Kumar, A. Density Functional Study of Structures, Stabilities and Electronic Properties of Clusters: Comparison with Pure Gold Clusters. *Mater. Sci. Pol.* **2020,** *38* (1), 97–107.
41. Das, S.; Ranjan, P.; Chakraborty, T. Computational Study of CunAgAu (n= 1–4) Clusters Invoking DFT Based Descriptors. *Phys. Sci. Rev.* 2022. DOI: https://doi.org/10.1515/psr-2021-0141
42. Ranjan, P.; Das, S.; Yadav, P.; Tandon, H.; Chaudhary, S.; Malik, B.; Rajak, A. K.; Suhag, V.; Chakraborty, T. Structure and Electronic Properties of [AunV] λ (n= 1–9; λ=0,±1) Nanoalloy Clusters Within Density Functional Theory Framework. *Theor. Chem. Acc.* **2021,** *140* (5), 1–12.
43. Das, S.; Chakraborty, T.; Ranjan, P. Theoretical Analysis of AgFen (n= 1–5) Clusters: A DFT Study. *Mater. Today Proc.* **2021,** *54* (3), 873–877.
44. Gaussian 16, Revision C.01, Frisch, M. J.; Trucks, G. W.; Schlegel, H. B.; Scuseria, G. E.; Robb, M. A.; Cheeseman, J. R.; Scalmani, G.; Barone, V.; Petersson, G. A.; Nakatsuji, H.; Li, X.; Caricato, M.; Marenich, A. V.; Bloino, J.; Janesko, B. G.; Gomperts, R.; Mennucci, B.; Hratchian, H. P.; Ortiz, J. V.; Izmaylov, A. F.; Sonnenberg, J. L.; Williams-Young, D.; Ding, F.; Lipparini, F.; Egidi, F.; Goings, J.; Peng, B.; Petrone, A.; Henderson, T.; Ranasinghe, D.; Zakrzewski, V. G.; Gao, J.; Rega, N.; Zheng, G.; Liang, W.; Hada, M.; Ehara, M.; Toyota, K.; Fukuda, R.; Hasegawa, J.; Ishida, M.; Nakajima, T.; Honda, Y.; Kitao, O.; Nakai, H.; Vreven, T.; Throssell, K.; Montgomery, J. A., Jr.; Peralta, J. E.; Ogliaro, F.; Bearpark, M. J.; Heyd, J. J.; Brothers, E. N.; Kudin, K. N.; Staroverov, V. N.; Keith, T. A.; Kobayashi, R.; Normand, J.; Raghavachari, K.; Rendell, A. P.; Burant, J. C.; Iyengar, S. S.; Tomasi, J.; Cossi, M.; Millam, J. M.; Klene, M.; Adamo, C.; Cammi, R.; Ochterski, J. W.; Martin, R. L.; Morokuma, K.; Farkas, O.; Foresman, J. B.; Fox, D. J. Gaussian, Inc., Wallingford CT, 2016.
45. Parr, R. G.; Yang, W. Density-Functional Theory of Atoms and Molecules; Oxford University Press: New York, Oxford, 1989.
46. Novotny, Z.; Mulakaluri, N.; Edes, Z.; Schmid, M.; Pentcheva, R.; Diebold, U.; Parkinson, G. S. Probing the Surface Phase Diagram of Fe3O4 (001) Towards the Fe-rich Limit: Evidence for Progressive Reduction of the Surface. *Phys. Rev. B* **2013,** *87* (19), 19541-1 (8 p) .
47. Pomar, A.; Konstantinovic, Z.; Bahues, N.; Roqueta, J.; Lopez-Mir, L.; et Al. Formation of Self-Organized Mn3O4 Nanoinclusions in LaMnO3 Films. *Front. Phys.* **2016,** *4* (41), 1–14.
48. Nayak, S.; Dasari, K.; Joshi, D. C.; Pramanik, P.; Palai, R.; Sathe, V.; Chauhan, R.; Tiwari, N.; Thota, S. Spectroscopic Studies of Co2TiO4 and Co3O4 Two-Phase Composites. *Phys. Status Solidi* **2016,** *253* (11), 2270–2282.
49. Fujimoto, H.; Kato, S.; Yamabe, S.; Fukui, K. Molecular Orbital Calculations of the Electronic Structure of Borazane. *J. Chem. Phys.* **1974,** *60*, 572–578.
50. Kato, S.; Fujimoto, H.; Yamabe, S.; Fukui, K. Molecular Orbital Calculation of the Electronic Structure of Borane Carbonyl. *J. Am. Chem. Soc.* **1974,** *96* (7), 2024–2029.
51. Ghosh, D. C.; Bhattacharyya, S. Molecular Orbital and Density Functional Study of the Formation, Charge Transfer, Bonding and the Conformational Isomerism of the Boron Trifluoride (BF3) and Ammonia (NH3) Donor-Acceptor Complex. *Int. J. Mol. Sci.* **2004,** *5* (8), 239–264.

52. Ghosh, D. C. A Comparative Cndo/2 and Cndo 2d Study of the Orbital Interaction, Charge-Transfer and Bond Formation in Ammonia-Borane. *Indian J. Pure Appl. Phys.* **1984,** *22* (6), 346–350.
53. Ghosh, D. C. A Comparative Cndo 2 and Cndo 2d Study of the Orbital Interaction, Charge-Transfer and Bond Formation in Borane-Adduct Molecules-. 2.-H3b-Co, H3b-N2,(H3b-Cn)-And (H3b-Nc)-Systems. *Indian J. Pure Appl. Phys.* **1989,** *27* (4), 160–166.
54. Xiao, H.; Tahir-Kheli, J.; Goddard III, W. A. Accurate Band Gaps for Semiconductors from Density Functional Theory. *J. Phys. Chem. Lett.* **2011,** *2* (3), 212–217.
55. Saravanan, S.; Balachandran, V. Quantum Chemical Studies, Natural Bond Orbital Analysis and Thermodynamic Function of 2, 5-Dichlorophenylisocyanate. *Spectrochim. Acta Part A* **2014,** *120*, 351–364.
56. Azam, F.; Alabdullah, N. H.; Ehmedat, H. M.; Abulifa, A. R.; Taban, I.; Upadhyayula, S. NSAIDs as Potential Treatment Option for Preventing Amyloid β Toxicity in Alzheimer's Disease: An Investigation by Docking, Molecular Dynamics, and DFT Studies. *J. Biomol. Struct. Dyn.* **2018,** *36* (8), 2099–2117.
57. Parr, R. G.; Zhou, Z. Absolute Hardness: Unifying Concept for Identifying Shells and Subshells in Nuclei, Atoms, Molecules, and Metallic Clusters. *Acc. Chem. Res.* **1993,** *26* (5), 256–258.
58. Chattaraj, P. K.; Sengupta, S. Chemical Hardness as a Possible Diagnostic of the Chaotic Dynamics of Rydberg Atoms in an External Field. *J. Phys. Chem. A* **1999,** *103* (31), 6122–6126.
59. Pearson, R. G. Recent Advances in the Concept of Hard and Soft Acids and Bases. *J. Chem. Educ.* **1987,** *64* (7), 561.
60. Sanderson, R. T. An Interpretation of Bond Lengths and a Classification of Bonds. *Science* **1951,** *114* (2973), 670–672.
61. Sanderson, R. T. Carbon—Carbon Bond Lengths. *Science* 1952, *116* (3002), 41–42.
62. Parr, R. G.; Szentpály, L. V.; Liu, S. Electrophilicity Index. *J. Am. Chem. Soc.* **1999,** *121* (9), 1922–1924.

CHAPTER 11

Solid Materials for Applications in Life Sciences: An Introduction

SHAYERI DAS[1,2], PRABHAT RANJAN[1], and TANMOY CHAKRABORTY[3]

[1]*Department of Mechatronics Engineering, Manipal University Jaipur, Jaipur, Rajasthan, India*

[2]*Department of Electrical Engineering, Ideal Institute of Engineering, Kalyani, Nadia, West Bengal, India*

[3]*Department of Chemistry and Biochemistry, School of Basic Sciences and Research, Sharda University, Greater Noida, India*

ABSTRACT

Metallic biomaterials have drawn a lot of interest, as they are capable of even replacing hard tissues. The materials and their alloys have been put into use in various compositions. Different fabrication processes have been used in order to compare the suitability of the materials in the field of application. Research and development of their alloys are also under process which will enable us to apply the materials in nanotech-based biomedical fields as well as targeted medications. The nontoxicity of these materials makes them even more biocompatible.

11.1 INTRODUCTION

Metals and some semiconducting elements possess a unique physio-chemical feature, which allows them to contribute to various areas of research related to life sciences, energy storage, and solar cells, etc.[1–6]

Tailored Functional Materials for Clean and Sustainable Development. Tamara Tatrishvili, PhD, Neha Kanwar Rawat, PhD, Ann Rose Abraham, PhD & A. K. Haghi, PhD (Eds.)

Metallic elements date back in history and they have been used for biomedical purposes. The entire surgical field would be obsolete without the implementation of stainless steel.[7] The establishment of aseptic surgery further leads to the implementation of these metal-based components. Vitallium which is a cobalt-based alloy was put to real-world use. Among all elements, titanium is the latest metallic biomaterial among the three main metallic biomaterials which also include stainless steel and cobalt-based alloys, and these remain the most prevalent metallic alloys. Various metallic materials are used as biomaterials, but among them, titanium and its alloys are getting additional consideration in both medical as well as dental fields due to its outstanding biocompatibility, lightweight, attractive balance of mechanical properties, exceptional corrosion resistance, and other properties. Titanium has been predominantly used for implant devices replacing failed hard tissue, for example, artificial hip joints, artificial knee joints, bone plates, dental implants, and other applications. Various dental products, such as crowns, bridges, and dentures, etc., have been also fabricated using titanium using the precision casting method.

One of the traditional methods to evaluate the biocompatibility of any material is by analyzing the osteogenesis pattern. Takao[8] reported the patterns of osteogenesis after analyzing the results. It has been found that titanium and its alloys exhibit the finest biocompatibility among all the materials applicable as biomaterials. Titanium is not only antitoxic but also one of the recognized anti-allergens. This makes titanium a strong contender keeping in mind the allergy issues. Pure titanium and a few of its alloys are more commonly used for biomedical purposes in comparison to other alloys. These alloys are synthesized as structural materials for building aerospace structures. This has led to fewer alloys being left for targeted medicine. Titanium is one of the elements considered a contender to heal soft tissues.[9] Cobalt-based alloys have also shown high potential in this domain.

The main prospect of the metallic biomaterial counterpart or replacement for soft tissues is among senior citizens. As per the population ratio, the percentage of senior citizens has been increasing rapidly in all countries, especially before 2000. This age group requires multiple replacements of failed soft tissues. Based on all these aspects, some promising developments have been made with special emphasis on metallic elements and their alloys for life sciences and biomedical applications.

11.2 BIOMEDICAL APPLICATIONS OF ELEMENTS AND THEIR ALLOYS

11.2.1 APPLICATION OF TITANIUM

Niinomi[10] reported all the biocompatible titanium alloys introduced to date. The alloys that are licensed and approved for human testing and application as well the ones that are still under study have been reported.[10] Pure titanium and a few titanium alloys, in conjunction with the Ti-6Al-4V ELI and Ti-6Al-4V, were initially designed for implementation as general structural materials, particularly for aerospace structures, and only later were approved for biomedical applications. The toxicity of β-stabilizing element of V has been later reported.[11,12] Hence, in both alloys, different stabilizing elements have been used. Both iron and niobium are considered to be human-safe and have been used as a replacement. It can be deduced that the development of titanium alloys for living organisms was initiated with the inception of these alloys. Based on the same approach, other alloys were fabricated. Subsequently, a 1b-type titanium alloy that does not contain V or Al, such as the Ti-15Zr-based and Ti-15Sn-based alloys have been found.[13] It has been established that the elastic moduli of biomaterials are equal to that of the cortical bone because, if the elastic moduli of biomaterials are much greater than that of the cortical bone, bone resorption occurs. The elastic moduli of the titanium alloy Ti-6Al-4V are much lower than the counterparts of stainless steel and cobalt-based alloy.[14] Despite the extensive research, the elastic modulus of Ti-6Al-4V remains much greater than that of the cortical bone. The elastic moduli of β -type titanium alloys are known to be smaller than those of α- or α+β-type titanium alloys. Due to this aspect, β-type titanium alloys have been introduced in the application field.[15–24] The latest biocompatible titanium alloys contain a huge amount of tantalum and niobium. Both elements have higher melting points as well as specific gravity than titanium.

The most important aspect of implementing any material for biomedical purposes is nontoxicity. Due to the same reason, nontoxic elements are selected for amalgamation with titanium to form the alloys. On the other hand, at present, metal allergy has become a substantial problem. In Europe, about 20% of young females and 4% of young males suffer from nickel allergy. This tendency has amplified vividly for both sexes.[25] Allergy to metals is generated by the metallic ions that are released from an alloy over body fluids. In odontology, cobalt, chromium, and nickel have been indicated to be most supplementary with metal allergy, and the application of nickel is quickly being avoided. The percentage of metal allergy caused by each metallic element is unsteady.[26] It is essential to exclude metallic elements that may activate metal allergy.

It has been observed that new titanium alloys developed for biomedical uses comprise alloying elements that have been deemed to be nontoxic and nonallergic. Niobium, tantalum, and zirconium are known to be the safest alloying elements for biocompatible titanium alloy. Previously, the alloy, Ti-Ni, was implemented as an implant material.[27] On the other hand, since Ti-Ni alloy comprises a large amount of Ni, the usage of Ti–Ni shape-memory alloys is constrained. The shape-memory titanium alloy, which comprises nontoxic and nonallergic elements has been implemented for implants.[28]

The main issue with titanium alloy is that the machining process is very expensive. Recent low-modulus titanium alloys designed for biomedical uses comprise a large amount of alloying elements with considerably high melting temperatures and gravities, such as niobium and tantalum in comparison with titanium. In these cases, the powder metallurgy process is also more important to make homogeneous alloys. The mechanical properties of the alloys produced through the ingot melting process and powder metallurgy process are extremely different.[29] The mechanical properties fabricated through the powder metallurgy process are similar in the case of alloys containing niobium and zirconium but proof stress is much lower in powder metallurgy than that in ingot melting.

The dental application also demands materials with excellent biocompatibility. Few of the alloys used have very different chemical compositions than the alloys which are normally used for implants. Pure titanium and Ti-6Al4V are the focal materials in odontology and in the surgical field. Ti-6Al-7Nb, which is being used for surgical implants, is also potent for dental applications.[30,31] Very recently, Ti-40Zr,[32] Ti-5Al-13Ta,[33] and Ti-43.1Zr-10.2Al-3.6V (at. pct)[34] have been researched and development are in progress. In applications related to odontology, casting is a dominant process. Methods of enhancing the elongation without dropping strength are being explored in cast titanium alloys. Hydrogenation and dehydrogenation processing, which are thermochemical processing followed by postheat treatment,[31] have proven to be more effective for elongation purposes without any degradation of strength. Titanium alloys are extremely reactive and possess moderately higher melting points in comparison to other alloys used in the dental field, such as gold-based and silver-based alloys. Consequently, the low melting point titanium alloys and low-reactive mold materials are mostly selected for dental precision castings. In dental precision casting, magnesia-based mold and alumina-based mold materials are predominant.[35]

11.2.2 APPLICATION OF COBALT AND CHROMIUM

Cobalt and chromium-based alloys are mainly cobalt representatives for biomedical applications. In comparison to stainless steel and titanium alloys, they have much higher wear and tear resistance. This property makes them extremely suitable for artificial hip joints as hip prostheses. Co–Cr-based alloys are congregated into cast and wrought alloys. The Co–Cr-based alloys have been standardized for biomedical applications. Wrought Co–Cr-based alloys are used for implant devices, which require much higher strength compared with cast Co–Cr-based alloys. Wrought Co–Cr-based alloys carry a great amount of Ni, which is a great risk element for allergy.

Concerning the alloy incorporating cobalt, chromium, and molybdenum, low-carbon and high-nitrogen variants are implemented for forging and machining applications. Consequently, in the case of cast alloy, both high and low nitrogen contents have been mass-produced. The beads-sintering process, when applied to the low-carbon and high-nitrogen material, necessitates the use of high temperature, which causes a decline in the tensile strength and fatigue strength. Therefore, there rises a demand for a forged version of the high-carbon alloy containing cobalt, chromium, and molybdenum that would provide not only enhanced wear resistance but also the facility for using lower beads-sintering temperature, thereby decreasing the reduction of fatigue strength.[36] A process route using vacuum induction melting, electroslag remelting, press and precision forging, and rolling can cause higher carbon alloy with cobalt, chromium, and molybdenum composition.

The design of cobalt–chromium alloy is also done to meet the necessities of both corrosion resistance and mechanical properties. The metal injection molding process is lucrative for cost reduction of orthopedic implant devices, and it has been tried to apply to cobalt, chromium, and molybdenum alloy. The static mechanical properties and fatigue properties of the metal injection molding process are shown to be superior. The results of a metal injection molding process study on cobalt, chromium, and molybdenum alloy indicate the validity of the metal injection molding process for manufacturing cobalt, chromium, and molybdenum alloy implants.

11.2.3 APPLICATION OF STAINLESS STEEL

The chemical composition of conventional stainless steel is different from that which is being used today. The stainless steel which is currently used is austenitic stainless steel. There are two grades in ASTM F138. The carbon content of grade 2 is lesser than that of grade 1. Others are nitrogen-strengthened high chromium and manganese stainless steels with lower nitrogen content. Numerous variations of austenitic stainless steels, martensitic stainless steels, ferritic stainless steels, and precipitation-hardening stainless steels for surgical instruments are registered in ASTM F899-95.

Nickel is one of the main metallic elements that is highly envisioned to cause allergic problems. The representative stainless steel for biomedical use, SUS 316 L (nearly equivalent to ASTM F138 Grade 2), which is an austenitic stainless steel, contains a large amount of nickel. Meanwhile, localized corrosion problems, such as pitting, crevice, and stress corrosion, have been detected to occur in implants of SUS 316 L.[37] Reducing the nonmetallic inclusion contents leads to the minimization of pitting and general corrosion. Nevertheless, crevice corrosion may still occur. Nonmagnetic nature is also necessary due to magnetic resonance imaging, applied to get tomographic images of tissue and is influenced by ferromagnetic materials. As a result, high nitrogen is alloyed with reducing nickel content in order to make superior corrosion-resistant austenitic stainless steel for surgical implants. Consequently, the high-nitrogen austenitic stainless steel with a slight variation in composition has been developed.[37]

Additionally, in order to avoid nickel allergy problems, more high-nitrogen austenitic stainless steel, that is, nickel-free stainless steel, has been proposed by Speidel and Uggowitzer.[38] Their nickel-free austenitic stainless steel has the chemical composition of chromium viable for biomedical applications. The biomedical fulfillment of the similar compositional nickel-free austenitic stainless steel has been investigated.[39,40] Low-nickel and high-nitrogen stainless steel have been also developed for application as the implanted stainless steel electrode for applied electrical stimulation.[41]

11.2.4 APPLICATION OF METALS FOR SOFT TISSUES

Metallic biomaterials are mostly used to restore failed hard tissue. However, there are cases for metallic biomaterials to be used for repairing soft tissue. Demonstrative cases are vascular support structures considered stents. Stent devices[42] are located in diseased arteries to preserve healthy blood flow where plaque has partly or totally prevented the flow. The stents are grouped into the following two types:

(a) Balloon expandable
(b) Self-expandable ones.

The main materials used for stent structures are given below:

- Stainless steel
- Nitinol
- Elgiloy

It should be noted that nitinol is a Ti–Ni alloy. However, elgiloy is considered a Co-based alloy. Ti–Ni is expected to be used more widely because of its shape-memory and highly restored properties. The stent is preshaped to fit inside a delivery catheter. On introduction into the body and release from the delivery system, the stent transforms from martensite to austenite phase and could then expand to support the artery.

11.2.5 APPLICATION OF NOBLE METALS

Although noble metals generally refer to eight elements (ruthenium, rhodium, palladium, silver, osmium, iridium, platinum, and gold), clusters of gold,[43–47] silver,[48,49] as well as their alloys[50–52] are the most studied, while several clusters of copper,[53–55] platinum, [56–58] and palladium[59,60] are also reported. Over the past two decades, metallic clusters of various core sizes and ligands with unique properties have been utilized in diverse applications of energy, environment, and biomedicine.

Unique properties of clusters combined with their robust nature enable them to be developed into useful materials. Energy, environment, and biology are the three main domains in which clusters have shown significant promise. Recent research shows great promise of clusters in providing solutions to several environmentally relevant issues, which include pollution control and access to clean water. Noble metals have huge benefits in the medical field which have been in use for decades back.[61] Many of these noble metal preparations have withstood the test of time and are still prevalent and practiced by many. The optical properties of metal nanoclusters, especially silver and gold, exhibit properties, such as absorption, scattering, and their surface-enhancing characteristics offered promising results in medical diagnosis and treatment. Due to their insignificant size, these materials possess the potential to be used in target-specific areas inside cells previously inaccessible to large-sized nanoparticles. Whereas good photostability, lesser toxicity, and smaller size make clusters advantageous in biomedical applications over semiconductor quantum dots; large Stokes shift, simplicity of functionalization, and better stability against photobleaching are the auxiliary advantages of clusters in contrast with traditional fluorescent dyes. Clusters incorporated with drug molecules can work for multiple purposes-(1) they can be used as resourceful drug carriers due to their miniature size that gave access to easy cell penetration, (2) luminescence from the clusters can be used to instantaneously track the consignment and thus can ensure their delivery in targeted areas, and (3) release of drugs in cell organelles and subsequent changes can be scrutinized using changes in the luminescence of the clusters. Accordingly, fluorescent metal clusters having superior biocompatibility, high quantum yields, excellent photostability, and NIR luminescence are better contenders for multiple applications in molecular biotechnology and biomedical engineering.

Sensing biomolecules and targeted areas are imperative in biomedical diagnosis as it can provide vital information on the stages of many diseases including cancer. Florescent properties of noble metal clusters have also been utilized for exposure to biomolecules, such as biothiols, nanoparticles, aminoacids, proteins, DNA, nucleic acids. Fluorescent gold clusters protected by the enzyme horseradish peroxidase (HRP) were used to detect the presence of hydrogen peroxide in solutions.[62]

Cysteine and GSH commonly seen in biological systems have numerous functions in cellular metabolism and redox reactions.[63] Deficiency of cysteine can lead to various medical infirmities, such as edema, depigmentation of hair, and liver damage. Thus, a facile analysis of cysteine levels in our body is important for the early diagnosis and treatment. Yuan et al.[64] established a simple method of detection of cysteine using GSH-protected silver clusters coalescing the thiol-silver chemistry and steric hindrance of the ligand shell protecting the cluster surface. A dual optical signal alteration for the detection of cysteine, both fluorometrically and colorimetrically, avoids false-positives. The change is attributed to the decomposition of the cluster into smaller species. Silver cluster-based sensors for biothiols[65–67] containing luminescence quenching of the cluster as a consequence of thiol-adsorption-accelerated oxidation are also available. Detection of biopolymers, such as proteins and DNA is yet another fascinating application of these nanoclusters. Sensors engaged for such applications stereotypically use antibody or aptamer-functionalized clusters to selectively conjugate with the analyte. Dual-detection strategies, such as colorimetry and fluorescence were used to quantitatively determine the presence of Human IgG using an optical immunosensor employing fluorescent Au cluster as the signal-transducing agent.[68] Adenosine triphosphate (ATP), regularly labeled as the "molecular unit of currency" for intracellular energy transfer, is an important constituent of the various metabolic processes taking place inside cells. A simple route using DNA-protected Ag clusters as fluorescent molecular inspirations for discerning recognition of ATP was reported by Ye et al.[69]

The unique photophysical properties of noble metal clusters give upsurge to innumerable applications in imaging techniques extending from cellular staining to imaging specific proteins inside live cells. Size, biocompatibility, fluorescence quantum yield, and stability against photobleaching are essential components for any fluorescence-based optical probe. Makarava et al.[70] demonstrated the use of silver clusters in bioimaging in amalgamation with a fluorophore, thioflavin T. The altered clusters were used to stain amyloid fibrils. Yu et al.[71] used intracellularly synthesized silver clusters using argyrophilic proteins and imaged living cells using cluster emission. Conjugation with organically active molecules that can preserve activity posttreatment is yet another way to target precise locations. Due to their ease of functionalization, luminescence tunability, and specific binding capabilities, the clusters are implemented for cellular labeling, nuclei staining, and tumor detection. Yu et al.[72] utilized DNA-encapsulated silver clusters fused with proteins, such as avidin for surface labeling in live cells by exploiting the precise avidin–biotin interactions. Lin et al.[73] validated that dihydrolipoic acid (DHLA)-protected water-soluble gold clusters can be conjugated effectively to biomolecules such as polyethylene glycol (PEG). While the use of several biopolymers for the synthesis and subsequent imaging of clusters has been established by manipulating its luminescence; the use of a biologically active protein such as insulin, for the synthesis of clusters and retaining its biological activity, was advantageous as it adds another modality toward diagnosis and imaging of such clusters.

Studies demonstrated the detection of cancerous cells for the first time using Au clusters. Later, the potential of various Au and Ag clusters for simultaneous cancer cell targeting and imaging was demonstrated.[74,75] The effectiveness of such materials for therapy was limited due to the lack of effective nuclear drug delivery vehicles to succor the transport of the drug from the cytoplasm to the nucleus. The passage of drug into the nucleus can expressively enhance its therapeutic capability. Wang et al.[76] reported the nuclear localization and targeting capability of fluorescent BSA—protected gold clusters conjugated with Herceptin. Combined use of fluorescent correlation spectroscopy and fluorescence lifetime imaging microscopy was performed to track the dynamics of the fluorescent cluster probes inside the nucleus of live cells with single-particle sensitivity. Biosynthesis of fluorescent clusters by cancerous cells has been monitored by fluorescence imaging.[75]

MRI is a prevailing and widely used imaging technique owing to its fast scan rate, exceptional spatial resolution, and deep tissue penetration. However, the requirement of large doses of gadolinium-based agents for adequate image contrast is a cause of concern. The use of gold-based BSA clusters as bimodal MRI or optical nanoprobes was demonstrated by covalently grafting gadolinium complex of diethylenetriaminepentacetic acid (DTPA), denoted as Gd-DTPA, onto the gold clusters.[77] Liang et al.[78] used contemporary strategy for multimodal imaging using gadolinium-functionalized gold clusters sheltered by cyclodecapeptide for dual model (fluorescence and magnetic resonance) imaging. X-ray computer tomography signals received from gold clusters are stabilized by insulin.[79] The clusters exhibited a dependence on dose which got enhanced in contrast when tested for imaging.

Yet, it remains the main concern for any biomedical application is the issue of toxicity. Internal applications of noble metal nanoparticles as well as clusters are still sternly hampered primarily by their slow renal clearance and high nonspecific accumulation in the organs of the reticuloendothelial system, such as the liver and spleen, after systematic administration. In this perspective, sub-nanometer-sized clusters are advantageous compared with NPs for biomedical applications due to their extremely small size. Both the nature of the ligand and the particle size are crucial for efficient renal clearance. More than 50% of the cluster was excreted out of the body through urine within 24 h and up to 65% was observed in the urine, 72 h of post-injection. The prominence of choice of the surface capping ligands during in vivo claims of clusters was exposed in experiments conducted with various ligand-protected clusters. Simpson et al.[80] demonstrated that the histological damage to the renal tubules caused by the use of tiopronin-passivated Au clusters during in vivo applications can be eliminated by the incorporation of PEG into the clusters.

While water-soluble clusters have undoubtedly proven to have immense potential in biological and biomedical applications, caution should be implemented in the choice of appropriate ligand and their functionalization and core size in order to avoid additional effects.

11.3 CONCLUSIONS

The unique capabilities and biocompatibility make these materials promising candidates for application in biomedicine and other medical fields for disease diagnosis and treatment. As the properties of many of the materials have been studied only to a limited extent, many favorable avenues in terms of their medical and materials science applications are yet to be explored. With more experimental and theoretical achievements toward understanding these materials and solving demands in their synthesis, a nearly unlimited field of applications could be anticipated. At present, it is necessary to have a suitable terminology for labeling such materials, to utilize as many of them as possible, correlate structures with associated properties, come up with a general rule or rules for the fabrication of these materials, and expand their field of application. In this, the inputs of computational studies will be highly beneficial.

KEYWORDS

- **biomaterials**
- **life sciences**
- **cobalt**
- **chromium**
- **titanium**

REFERENCES

1. Boisselier, E.; Astruc, D. *Chem. Soc. Rev.* **2009,** *38*, 1759–1782.
2. Daniel, M. C.; Astruc, D. *Chem. Rev.* **2004,** *104*, 293.
3. Sajanlal, P. R.; Pradeep, T. In *Kirk-Othmer Encyclopedia of Chemical Technology*; John Wiley & Sons, Inc.: New York, 2011.
4. Sapsford, K. E.; Algar, W. R.; Berti, L.; Gemmill, K. B.; Casey, B. J.; Oh, E.; Stewart, M. H.; Medintz, I. L. *Chem. Rev.* **2013,** *113*, 1904.
5. Amelia, M.; Lincheneau, C.; Silvi, S.; Credi, A. *Chem. Soc. Rev.* **2012,** *41*, 5728.
6. Wu, C.; Chiu, D. T. *Angew. Chem. Int. Ed.* **2013,** *52*, 3086.
7. Park, J. B. In *The Biomedical Engineering Handbook Volume 1*; Bronzino, J. D., Ed.; CRC Press LLC: Boca Raton, FL, 2000; pp IV-1–IV-8.
8. Takao *J. Jpn. Soc. Biomater.* **1989,** *7*, 19–23.
9. Whitcher, F. D. *Comput. Struct.* **1997,** *64*, 1005–1011.

10. Niinomi, M. *Mater. Sci. Eng.* **1998,** *A243*, 231–236.
11. Zwicker, R.; Buheler, K.; Mueller, R.; Beck, H.; Schmid, H. J. In *Titanium '80: Science and Technology*; Kimura, H., Izumi, O., Eds.; TMS-AIME: Warrendale, PA, 1980; vol. 2, pp 505–514.
12. Semlitsch, M.; Staub, F.; Webber, H. *Biomed. Technol.* **1985,** *30*, 334–339.
13. Okazaki, Y.; Ito, Y.; Ito, A.; Tateishi, T. In *Medical Applications of Titanium and Its Alloys*; Brown, S. A., Lemons, J. E., Eds.; ASTM STP 1272, ASTM: Philadelphia, PA, 1996; pp 45–59.
14. Lemons, J. E. In *Perspectives on Biomaterials*; Lin, O. C. C., Chao, E. Y. S., Eds.; Elsevier: New York, NY, 1986; pp 1–13.
15. Mishra, A. K.; Davidson, J. A.; Kovacs, P.; Poggie, R. A. In *Beta Titanium in the 1990's*; Eylon, D.; Boyer, R. R., Koss, D. A., Eds.; TMS: Warrendale, PA, 1993; pp 61–72.
16. Wang, K.; Gustavson, L.; Dumbleton, J. In *Beta Titanium in the 1990's*; Eylon, D., Boyer, R. R., Koss, D. A., Eds.; 1993; pp 49–60.
17. Zardiackas, L. D.; Mitchell, D. W.; Diesegi, J. A. In *Medical Applications of Titanium and Its Alloys*; Brown, S. A., Lemons, J. E., Eds.; ASTM STP 1272, ASTM: Philadelphia, PA, 1996; pp 60–74.
18. Wang, K. *Mater. Sci. Eng. A* **1996,** *A213*, 134–137.
19. Nishimura, T. In *Metallurgy and Technology Practical Titanium Alloys;* Fujishiro, S., Eylon, D., Kishi, T., Eds.; TMS: Warrendale, PA, 1994; pp 3–8.
20. Nakayama, Y.; Yamamuro, T.; Kotoura, Y.; Oka, M. Proc. 3rd World Biomaterials Congr.; Koto, Japan, 1988; p 202.
21. Bhambri, S. K.; Shetty, R. H.; Gilbertson, L. N. In *Medical Applications of Titanium and Its Alloys*; Brown, S. A., Lemons, J. E., Eds.; ASTM STP 1272, ASTM: Philadelphia, PA, 1996; pp 88–95.
22. Ahmed, T.; Long, M.; Slivestri, J.; Ruiz, C.; Rack, H. J. In *Titanium '95: Science and Technology*; Blenkinsop, P. A., Evans, W. J., Flower, H. M., Eds.; Institute of Metals: London, 1996; vol. II, pp 1760–1767.
23. Kuroda, D.; Niinomi, M.; Morinaga, M.; Kato, Y.; Yashiro, T. *Mater. Sci. Eng. A* **1998,** *A243*, 244–249.
24. Ortiz, C. R.; Villa, R.; Gonzalez, G.; Trillo, E.; Stafford, S. W.; Murr, L. E. In *Structural Biomaterials for the 21st Century*; Niinomi, M., Okabe, T., Taleff, E. M., Lesure, D. R., Lippard, H. E., Eds.; TMS: Warrendale, PA, 2001; pp 35–42.
25. Uggowitzer, P. J.; Ba¨hre, W. F.; Speidel, M. O. *Adv. Powder Metall. Part Mater.* **1997,** *3*, 18.113–18.121.
26. Niinomi, M. *Function Mater.* **2000,** *20*, 36–44.
27. Gil, F. X.; Manero, J. M.; Planell, J. A. *J. Mater. Sci. Mater. Med.* **1996,** *7*, 403–406.
28. Nitta, K.; Watanabe, S.; Masahashi, N.; Hosoda, H.; Hanada, S. *Structural Biomaterials for the 21st Century*; Niinomi, M., Okabe, T., Taleff, E. M., Lesure, D. R., Lippard, H. E., Eds.; TMS: Warrendale, PA, 2001; pp 25–34.
29. Yoshitani, S.; Niinomi, M.; Kuroda, D.; Fukunaga, K.; Saito, T.; Furuta, T. *Proc. Powder Metallurgy World Congr. & Exh. 2000*; Kyoto, Japan, 2000.
30. Kobayashi, E.; Doi, H.; Takahashi, M.; Nakano, T.; Yoneyama, T.; Hamanaka, H. *J. Jpn. Soc. Biomater.* **1995,** *14*, 406–413.
31. Akahori, T.; Niinomi, M.; Isohama, R.; Suzuki, A. In *Structural Biomaterials for the 21st Century*; Niinomi, M., Okabe, T., Taleff, E.M., Lesure, D.R., Lippard, H.E., Eds.; TMS: Warrendale, PA, 2001; pp. 91–98.

32. Kato, K.; Izumi, T.; Simano, I.; Kikui, T.; Okada, H.; Nagayama, K. *J. Jpn. Soc. Dental Mater. Dev.* **1997,** 16 (30), 45.
33. Doi, H.; Yoneyama, T.; Kobayashi, E.; Hamanaka, H. *J. Jpn. Soc. Dental Mater. Dev.* **1998,** *17*, 247–252.
34. Kobayashi, E.; Doi, H.; Yoneyama, T.; Hamanaka, H.; Matsumoto, S.; Kudaka, K. *J. Jpn. Soc. Dental Mater. Dev.* **1995,** *4*, 321–328.
35. Yoshitani, S.; Niinomi, M.; Fukunaga, K.; Fukui, H.; Takeuchi, T.; Katsura, S. CAMP-ISIJ **2000,** *13*, 1345.
36. Berry, G.; Bolton, J. D.; Brown, J. B.; McQuaide, S. In *Cobalt Base Alloys for Biomedical Applications*; Disegi, J. A., Kennedy, R. L., Pilliar, R., Eds.; ASTM STP 1365, ASTM: Werst Conshohocken, PA, 1999; pp 11–31.
37. Pan, J.; Karlen, C.; Ulfvin, C. *Electrochem. Soc. Proc.* **1999,** 98 (17), 383–394.
38. Speidel, M. O.; Uggowitzer, P. J. High Manganese High Nitrogen Austenitic Stainless Steels, Proc. ASM Materials Week '92, Chicago, IL; ASM International Metals Park: OH, 1992; p 135.
39. Endo, K.;Abiko, Y.; Suzuki, M.; Oho, H.; Kaku, T. Zairyo-to Kankyo **1998,** *47*, 570–576.
40. Brown, R. S.; Gebeau, R. C. 6th World Biomaterials Congr. Trans; Kamuela, Hawaii, 2000; 828.
41. Iguchi, Y. *Bull. Iron Steel Inst. Jpn.* **1999,** *4*, 9–12.
42. Zenbo, N.; Esato, K. In *Stent-Grafting and Aortic Diseases*; Esato, K., Hoshino, S., Ishimaru, S., Matsunaga, N., Zenpo, N., Eds.; Ishiyaku Publishers: Tokyo, 1999; pp 10–15.
43. Templeton, A. C.; Wuelfing, W. P.; Murray, R. W. *Acc. Chem. Res.* **2000,** *33*, 27.
44. Parker, J. F.; Fields-Zinna, C. A.; Murray, R. W. *Acc. Chem. Res.* **2010,** *43*, 1289.
45. Yau, S. H.; Varnavski, O.; Goodson, T. *Acc. Chem. Res.* **2013,** *46*, 1506.
46. Schmid, G. *Chem. Soc. Rev.* **2008,** *37*, 1909.
47. Jin, R. *Nanoscale* **2010,** *2*, 343.
48. Diez, I.; Ras, R. H. A. *Nanoscale* **2011,** *3*, 1963.
49. Choi, S.; Dickson, R. M.; Yu, J. *Chem. Soc. Rev.* **2012,** *41*, 1867.
50. Li, H. W.; Yue, Y.; Liu, T. Y.; Li, D.; Wu, Y. *J. Phys. Chem. C* **2013,** *117*, 16159.
51. Kauffman, D. R.; Alfonso, D.; Matranga, C.; Qian, H.; Jin, R. *J. Phys. Chem. C* **2013,** *117*, 7914.
52. Mohanty, J. S.; Xavier, P. L.; Chaudhari, K.; Bootharaju, M. S.; Goswami, N.; Pal, S. K.; Pradeep, T. *Nanoscale* **2012,** *4*, 4255.
53. Goswami, N.; Giri, A.; Bootharaju, M. S.; Xavier, P. L.; Pradeep, T.; Pal, S. K. *Anal. Chem.* **2011,** *83*, 9676.
54. Jia, X.; Li, J.; Han, L.; Ren, J.; Yang, X.; Wang, E. *ACS Nano* **2012,** *6*, 3311.
55. Wei, W.; Lu, Y.; Chen, W.; Chen, S. *J. Am. Chem. Soc.* **2011,** *133*, 2060.
56. Tanaka, S. I.; Miyazaki, J.; Tiwari, D. K.; Jin, T.; Inouye, Y. *Angew. Chem. Int. Ed.* **2011,** *50*, 431.
57. Vajda, S.; Pellin, M. J.; Greeley, J. P.; Marshall, C. L.; Curtiss, L. A.; Ballentine, G. A.; Elam, J. W.; Stephanie, C. M.; Redfern, P. C.; Mehmood, F.; Zapol, P. *Nat. Mater.* **2009,** *8*, 213.
58. Tiwari, J. N.; Nath, K.; Kumar, S.; Tiwari, R. N.; Kemp, K. C.; Le, N. H.; Youn, D. H.; Lee, J. S.; Kim, K. S. *Nat. Commun.* **2013,** *4*, 2221.
59. Erdogan, H.; Metin, O.; Ozkar, S. *Phys. Chem. Chem. Phys.* **2009,** *11*, 10519.

60. Hyotanishi, M.; Isomura, Y.; Yamamoto, H.; Kawasaki, H.; Obora, Y. *Chem. Commun.* **2011,** *47*, 5750.
61. Brown, C. L.; Bushell, G.; Whitehouse, M. W.; Agrawal, D. S.; Tupe, S. G.; Paknikar, K. M.; Tiekink, E. R. T. *Gold Bull.* **2007,** *40*, 245.
62. Wen, F.; Dong, Y.; Feng, L.; Wang, S.; Zhang, S.; Zhang, X. *Anal. Chem.* **2011,** *83*, 1193.
63. Chen, X.; Zhou, Y.; Peng, X.; Yoon, J. *Chem. Soc. Rev.* **2010,** *39*, 2120.
64. Yuan, X.; Tay, Y.; Dou, X.; Luo, Z.; Leong, D. T.; Xie, J. *Anal. Chem.* **2012,** *85*, 1913.
65. Han, B.; Wang, E. *Biosens. Bioelectron.* **2011,** *26*, 2585.
66. Shang, L.; Dong, S. *Biosens. Bioelectron.* **2009,** *24*, 1569.
67. Huang, Z.; Pu, F.; Lin, Y.; Ren, J.; Qu, X. *Chem. Commun.* **2011,** *47*, 3487.
68. Liu, H.; Wu, X.; Zhang, X.; Burda, C.; Zhu, J. J. *J. Phys. Chem. C* **2011,** *116*, 2548.
69. Zhang, M.; Guo, S. M.; Li, Y. R.; Zuo, P.; Ye, B. C. *Chem. Commun.* **2012,** *48*, 5488.
70. Makarava, N.; Parfenov, A.; Baskakov, I. V. *Biophys. J.* **2005,** *89*, 572.
71. Yu, J.; Patel, S. A.; Dickson, R. M. *Angew. Chem. Int. Ed.* **2007,** *46*, 2028.
72. Yu, J.; Choi, S.; Richards, C. I.; Antoku, Y.; Dickson, R. M. *Photochem. Photobiol.* **2008,** *84*, 1435.
73. Lin, C. A. J.; Yang, T. Y.; Lee, C. H.; Huang, S. H.; Sperling, R. A.; Zanella, M.; Li, J. K.; Shen, J. L.; Wang, H. H.; Yeh, H. I.; Parak, W. J.; Chang, W. H. *ACS Nano* **2009,** *3*, 395.
74. Tao, Y.; Lin, Y.; Huang, Z.; Ren, J.; Qu, X. *Adv. Mater.* **2013,** *25*, 2594.
75. Wang, J.; Zhang, G.; Li, Q.; Jiang, H.; Liu, C.; Amatore, C.; Wang, X. *Sci. Rep.* **2013,** *3*, 1157.
76. Wang, Y.; Chen, J.; Irudayaraj, J. *ACS Nano* **2011,** *5*, 9718.
77. Sun, G.; Zhou, L.; Liu, Y.; Zhao, Z. *New J. Chem.* **2013,** *37*, 1028.
78. Liang, G.; Ye, D.; Zhang, X.; Dong, F.; Chen, H.; Zhang, S.; Li, J.; Shen, X.; Kong, J. *J. Mater. Chem. B* **2013**, *1*, 3545.
79. Liu, C. L.; Wu, H. T.; Hsiao, Y. H.; Lai, C. W.; Shih, C. W.; Peng, Y. K.; Tang, K. C.; Chang, H. W.; Chien, Y. C.; Hsiao, J. K.; Cheng, J. T.; Chou, P. T. *Angew. Chem. Int. Ed.* **2011,** *50*, 7056.
80. Simpson, C. A.; Huffman, B. J.; Gerdon, A. E.; Cliffel, D. E. *Chem. Res.Toxicol.* **2010,** *23*, 1608.

PART IV
Composite Materials

CHAPTER 12

Assessment of Technogenic Raw Materials for the Production of Glass Materials and Course on "Green" Growth

BERDZENISHVILI IRINE[1], SIRADZE MANANA[1], CHKHAIDZE EKATERINA[1], and ANTIA GIORGI[2]

[1]*Department of Chemical and Biological Technologies, Georgian Technical University, Tbilisi, Georgia*

[2]*Faculty of Medicine, Tbilisi State Medical University, Tbilisi, Georgia*

ABSTRACT

The prospects for the "green" growth of industries and the economy of Georgia are widely related to the need to take more effective measures to eliminate the negative consequences of the impact of technogenic raw materials on land resources. It is emphasized that technogenic loads have an extreme impact on the environment and a transition to a green economy is required for the survival and development of mankind. In this chapter, a methodological principle of PTEPD is proposed to assess the possibility of using technogenic raw materials from Georgia in the production of glass materials.

12.1 INTRODUCTION

The reserves of raw materials for the production of glass materials are limited; therefore, intensive investigations are being carried out to expand the raw material base of glassmaking and enamel production with a focus

Tailored Functional Materials for Clean and Sustainable Development. Tamara Tatrishvili, PhD, Neha Kanwar Rawat, PhD, Ann Rose Abraham, PhD & A. K. Haghi, PhD (Eds.)

on a new technological mode, on a wider use of natural and technogenic raw materials, and the concept of "green" growth. The concept of a green economy is based on the relationship of society with the environment.[1–3]

The main sources of large tonnage waste in Georgia are the developed mining and related processing industries. Georgia, despite its small territory, has valuable natural resources, such as large deposits of manganese, silver–lead, copper–barite and zinc ores, bentonite clays, and marble, etc.[3–7]

As a result of the extraction and processing of these resources, also as a result of the activities of the Rustavi metallurgical plant (Kvemo-Kartli region), and the Zestafoni ferroalloy plant (Imereti region), technogenic objects—accumulations of mineral substances on the surface of the earth—were formed. The level of the amount of waste equates them to technogenic deposits, which are mining dumps, tailings, and sludge storages. Therefore, the condition of individual districts and settlements in the regions of Georgia is assessed as unsatisfactory.

12.2 RESULTS AND DISCUSSION

Nowadays, considering the enormous risks of waste on the environment, the prevention of irreversible impact of waste components on land resources and renewal of consumer qualities of land as a factor of agricultural production becomes extremely important.[2,4,8]

In this chapter, the main attention is paid to the assessment of the mineral resource potential of industrial waste/technogenic objects belonging to the class of complex source materials. It should be noted that this should in no way belittle the importance of studying about rocks in order to use them in the production of various glass materials, especially in our reality.

Regardless of the nature of technogenic objects (overburden rocks, enrichment tailings, slag, sludge, etc.), the main criterion determining their suitability for glass and enamel production is the chemical composition.

Along with this, we note that the criterion of oxide composition has a limited scope as it was based on the definition of "glass" as an indefinite chemical compound of oxides capable of decomposing into specific compounds.[3,4]

The positive experience accumulated in recent years at the Georgian Technical University in the utilization of substandard raw materials in the production of glass materials allows us to conclude that the essence of the complexity of technogenic materials is not only that it can simultaneously inject several necessary oxides into glass composition but also that these oxides are initially presented in the form of "ready-made" compounds that have a peculiar effect on both the behavior of the charge during its heating and on the properties of the final product.[3,4]

A comparison of the indices of the difference between the maximum melting temperature of glass and the melting temperature of the most refractory component of the charge composition showed that the lower this index, the stronger the effect of structural fragments on the properties of glass materials. Based on the foregoing, we suppose that an adequate assessment of the suitability of each technogenic waste for glassmaking production requires the knowledge of both the oxide and mineralogical compositions of this raw material, as well as the course of long-term high-temperature heating (LHH) in the charge that occurs with its participation.

A methodological principle is proposed for predicting the phase compositions and structural phase transformations in the course of LHH of initial compositions, based on the complex use of petrochemical and thermodynamic calculation methods with simultaneous analysis of phase equilibrium diagrams of simple and complex systems (PTEPD).

This methodology was tested on the following objects: sludge from the enrichment of manganese ores of the Chiatura deposit. The chemical composition of the tested sludge is presented in Table 12.1.

TABLE 12.1 The Chemical Composition of Sludge.

Ingredient	Content, mass%
$\sum RO_2(SiO_2 + TiO_2 + MnO_2)$	62, 57
$\sum R_2O_3(Fe_2O_3 +_2 AlO_3)$	10, 84
$\sum RO(MnO + CaO + MgO + BaO)$	12, 38
$\sum R_2O(Na_2O + K_2O)$	2, 70
$\sum(P_2O_5 + SO_3)$	1, 11
$\sum(A, M)$	10, 4

The cycle of investigations was carried out in a wide range of conditions which include temperature, phase state, as well as during phase transformations of the components that make up the system. It is clear that the technogenic raw materials are pretreated in order to bring them to an acceptable state. The mineral composition of the sludge from the enrichment of manganese ores of the Chiatura deposit is presented in Figure 12.1.

The data on Figure 12.1 enable us to judge the mass percentage composed by each mineral of the sludge.

To predict the behavior of the sludge during LHH, the thermodynamic modeling method is used, which is based on Gibbs free energy (ΔG) minimization.[9] For any temperature, the change in Gibbs energy of a system is calculated using the first Ulikh approximation, in which

$$\Delta G_T^0 = \Delta H_{298}^0 - 298\Delta S_{298}^0.$$

where ΔH_{298}^0 and ΔS_{298}^0 are the standard enthalpy and entropy changes of the process.

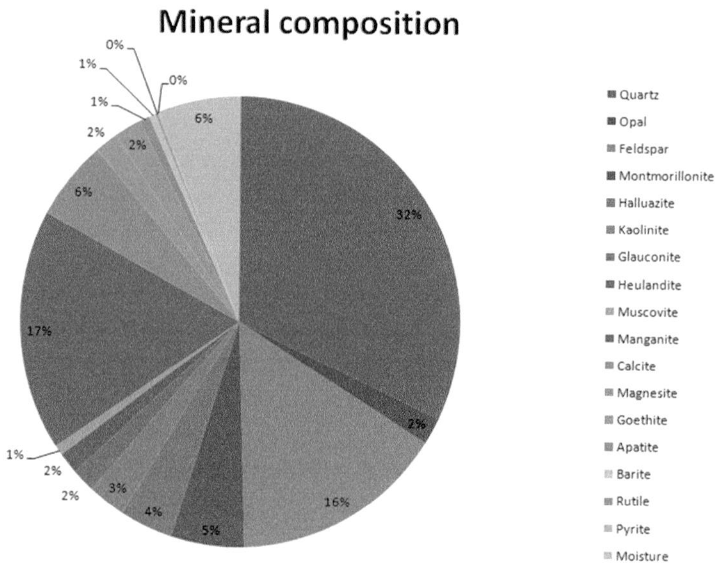

FIGURE 12.1 The mineral composition of sludge.

The thermodynamic data to computing DG were taken from Ref. [10,11]. Obviously, a process goes spontaneously in the direction that reduces the Gibbs energy of the system to a minimal value for these temperature and pressure. The results of thermodynamic analyses indicate that during the LHH a series of *interactions* between minerals, various *transitions of oxides, reactions between the phase assemblages, and* the *formations of eutectics at* $T \geq 1000°C$ are possible.

Phase composition of the technogenic raw material after heat treatment obtained according to the PTEPD principle shows that the system is represented by a liquid phase with crystalline inclusions distributed in it (34, 14%). Thermodynamic calculations allow us to predict that opal, anorthite, sillimanite, silicates, as well as manganese and iron oxides are presented in the largest amount in the liquid phase. These components will eventually also dissolve in the melt.

Thermodynamic analysis results reveal that to obtain a composite glass material from the charge mixture containing, along with other raw materials, the investigated sludge, the required temperature of synthesis is not less than 1300–1400°C.

Consequently, the main function of the PTEPD principle is to carry out a theoretical analysis of a wide variety of high-temperature processes occurring in charges with technogenic waste, without their experimental implementation. At the same time, a good correlation was observed between our calculations and experimental data.

12.3 CONCLUSIONS

The results of the investigations performed on the sludge can be used as estimates in order to improve the resource efficiency and reduce the negative impact of technogenic raw materials on the environment. All of the above suggests that only based on complex studies of each specific technogenic waste, is it possible to assess the fundamental possibility of using it as an alternative source of raw materials for the production of glass materials and heading for "green" growth.

KEYWORDS

- **technogenic raw materials**
- **production**
- **glass materials**
- **"green" growth**
- **principle of PTEPD**

REFERENCES

1. Zhironkin, S.; Cehlár, M. Green Economy and Sustainable Development: The Outlook. *Energies* 2022, *15*, 1167.
2. Tleubergen, M. A. *Economic Evaluation of Mineral Resources.* Kokshetau University: Kokshetau, 2004; p 382.
3. Berdzenishvili, I.; Siradze, M. Novel Nonfluor Corrosion Resistant Coatings Based on Material Resources of Georgia. *Int. J. Adv. Mater. Res.* 2018, *4* (3), 35–39.
4. Abzalov, A. A.; Aseeva, M. A.; Berdzenishvili, I. G.; et al. The Results of Studies of Natural and Exact Sciences: An Interdisciplinary Approach and a Super Additive Effect: Monograph; Erokhina, V. V., Terekhova, L. P., Podkopaeva, O. A., Ed.; Scientific and Research Center, Povolzhskaya Scientific Corporation: Samara, 2018; p 261.
5. Avaliani, G. A. *Manganese Deposits of Georgia*; M. Nauka, 1982; p 169.
6. https://www.advantour.com/rus/georgia/geography/resources.htm [Online] Natural resources of Georgia.
7. http://cyclowiki.org/wiki/%D0%93%D1%80%D1%83%D0%B7%D0%B8%D1%8F
8. Nageswara Rao, L. Environmental Impact of Uncontrolled Disposal of E-Wastes. *Int. J. ChemTech. Res.* 2014, *6* (2), 1343–1353.
9. *Basic Chemical Thermodynamics*, 5th Edition; Smith, E. B., 2004; p 166.
10. Yokokawa, H. Tables of Thermodynamic Properties of Inorganic Compounds. Journal of the National Chemical Laboratory for Industry; Tsukuba Ibaraki: Japan, 1988; vol 83, pp 27–118.
11. Knacke, O.; Kubaschewski, O. Hesselman, K. In *Thermochemical Properties of Inorganic Substances*, 2nd ed.; Springer-Verlag: Berlin, 1991; pp 1114–2412.

CHAPTER 13

Synthesis and Investigation of Environment-Friendly Silicon-Containing Carbohydrates Derivatives

N. N. SIDAMONIDZE, R. A. GAKHOKIDZE, R. O. VARDIASHVILI, and M. O. NUTSUBIDZE

Department of Chemistry, Ivane Javakhishvili State University, Tbilisi, Georgia

ABSTRACT

By hydrosilylation of 1-0-allyl-2,3,4-tri-O-acetyl-β-L-arabinopyranose (1) with 1,3-bis(dimethylsilyl)-2,2,4,4-tetramethylcyclodisilazane (2) in the presence of catalyst—benzoyl peroxide – $(C_6H_5COO)_2O$,, we obtained 1,3-di[3-(2,3,4-tri-O-acetyl-β-L-arabinopyranosyloxy)propyldimethylsilyl]-2,2,4,4-tetramethyl-cyclodisilazane (3). The reaction was carried out in dry chloroform at a molar ratio of 3: 1 at a temperature of 60–65°C. The reactions mainly occur according to Farmer's rule, although a small amount of the Markovnikov addition product is also formed. The structure of obtained compounds was established by physical–chemical methods of analysis.

13.1 INTRODUCTION

Silicon–nitrogen-containing organic compounds are used in many fields of activity, such as in medicine, agriculture, modern science-intensive technologies, and in the production of synthetic materials. They are used in

Tailored Functional Materials for Clean and Sustainable Development. Tamara Tatrishvili, PhD, Neha Kanwar Rawat, PhD, Ann Rose Abraham, PhD & A. K. Haghi, PhD (Eds.)

organic synthesis as silylating agents, synthons, and ligands in the preparation of complex compounds, and are used as vulcanizing agents for silicone rubbers. Silicon–nitrogen-containing organic compounds can be used in technological processes for the production of microelectronic products such as adhesives in microlithography processes, as well as precursors in chemical vapor deposition processes to obtain films and coatings of silicon carbonitride. Silicon carbonitride is a unique material that combines the properties of carbon nitride and silicon nitride. Silicon carbonitride films have high chemical and thermal stability, high hardness, good dielectric characteristics, optical transparency in the visible and infrared regions of the spectrum, and other functional properties. This determines the possibilities of their use for hardening the surface of materials, as well as in microelectronics as thin-layer structures for various functional purposes. Thus, the expansion of the synthetic potential and areas of application of silicon–nitrogen-containing compounds is an urgent task.

In order to obtain biologically active compounds, we have devised a new method for synthesizing glycosides containing silicon. Insertion of silicon atoms into known medicinal preparations can result in the essential change in the nature of the drug's action and sometimes can give them a number of new properties. They play an essential role in life processes, and many of them (especially organic) have a specific effect on living organisms.

Research in the recent years has shown that the synthesis of low-toxic compounds is becoming relevant in biological and pharmacological research. Therefore, it is of interest to use carbohydrates for the modification of some heterocyclic compounds, which can lead to a significant change in the nature of the action of drugs.[1–3]

13.2 RESULTS AND DISCUSSION

In some studies carried out earlier,[4–7] we have developed a new method for the synthesis of glycosides containing silicon. The introduction of silicon atoms into known drugs can lead to a significant change in the nature of the drug's action, and sometimes can give them a number of new properties.

By hydrosilylation 1-0-allyl-2,3,4-tri-O-acetyl-β-L-arabinopyranose (1) with 1,3-bis(dimethylsilyl)-2,2,4,4-tetramethylcyclodisilazane (2) in the presence of catalyst–$(C_6H_5COO)_2O$, we obtained 1,3-di[3-(2,3,4-tri-O-acetyl-β-L-arabinopyranosyloxy) propyldimethylsilyl]-2,2,4,4-tetramethylcyclodisilazane (3). The reaction was carried out in dry chloroform at a temperature of 60–65C.

The reaction mainly occurs according to Farmer's rule, although a small amount of Markovnikov addition product is also formed, that is, hydrogen joins a secondary carbon atom.

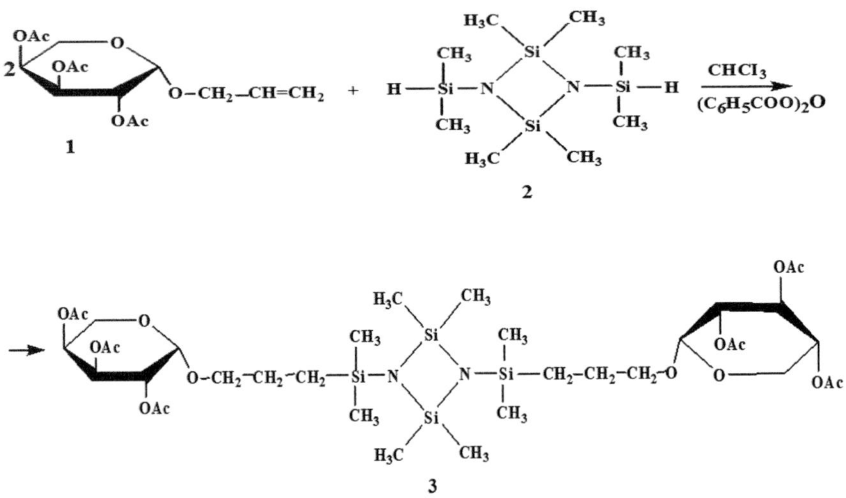

The reaction takes place by a radical mechanism. Benzoyl peroxide (initiator) decomposes under the influence of temperature and a free radical is formed, which in turn is easily decomposed into CO_2 and aromatic radicals

$$\begin{array}{c} C_6H_5-CO-O \\ | \\ C_6H_5-CO-O \end{array} \longrightarrow \begin{array}{c} C_6H_5-CO-O\bullet \\ \\ C_6H_5-CO-O\bullet \end{array} \longrightarrow \begin{array}{c} CO_2 + C_6H_5\bullet \\ CO_2 + C_6H_5\bullet \end{array}$$

The generated radical readily attacks the hydrogen atom of 1,3-bis(dimethylsilyl)-2,2,4,4-tetramethylcyclodisilazane, forming benzene and tetramethylcyclodisilazanes radical (4)

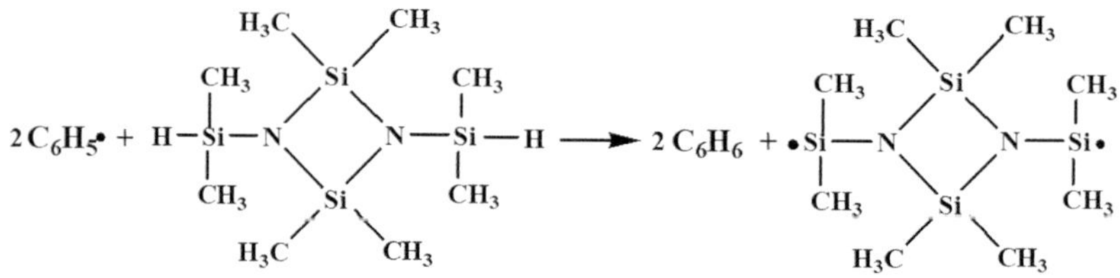

(1)

4

The resulting radical then acts on 1-0-allyl-2,3,4-tri-O-acetyl-β-L-arabinopyranose.

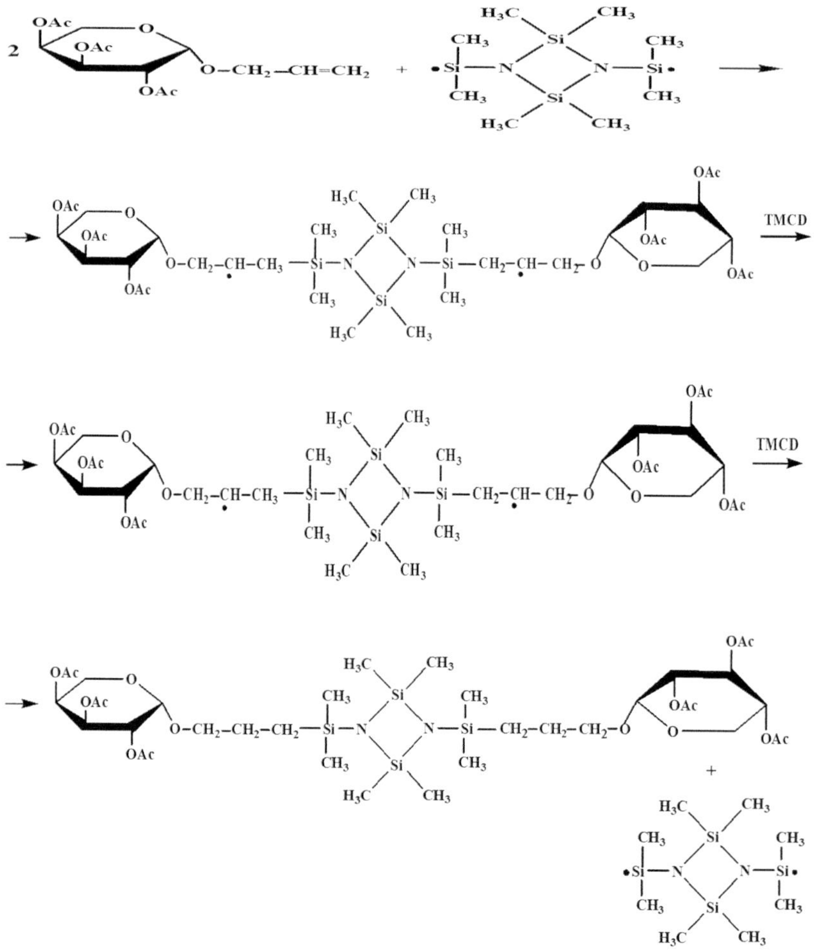

(2)

(2)

(3)

(4)

The first stage is the initializer, and the next steps go by the chain mechanism. The generated radical will still act on the allele, and so on.

13.3 EXPERIMENTAL PART

The IR spectra were obtained on a UR-20 in KBr disks. The ^{13}C NMR spectrum was taken on Bruker AM-300 spectrometer (75 MHz) in $CDCI_3$. The purity of the compounds obtained and the R_f values were determined on Silufol UV-254 plates. The optical rotation was measured on an SU-3 general purpose saccharimeter at 20 ± 2°C.

Synthesis of 1-O-allyl-2,3,4-tri-O-acetyl-β-L-arabinopyranose (1) is described in Ref [7].

1,3-di[3-(2,3,4-tri-O-acetyl-β-L-arabinopyranosyloxy) propyldimethylsilyl]-2,2,4,4-tetramethyl-cyclodisilazane (3). A mixture of 1,3-bis(dimethylsilyl)-2,2,4,4-tetramethylcyclodisi-lazane (2) (8.94 g, 0.01 mol) in dry $CHCI_3$ (50 mL) and benzoyl peroxide (0.5 g) was treated with a solution of 1-0-allyl-2,3,4-tri-O-acetyl-β-L-arabinopyranose (1) (6.32 g, 0.02 mol) in dry $CHCI_3$ (35 mL). The reaction was carried out under N_2 with constant stirring for 7–8 h (75–85°C).

The mixture was cooled and separated on a column (system $CH_3OH:CHCI_3$ 1:5, silica gel 50/100). The product was recovered by removing and evaporating a part of the solvent and then adding methanol in order to cause precipitation of the desired substance. The precipitate was washed several times in methanol at 60°C and then dried under vacuum to remove the remaining solvent. A chromatographically pure product is obtained (5.76 g, 64.5%). R_f 0.73 (system $C_6H_6:CHCI_3$), mp 151–153°C, $[\alpha]_D^{17}$ – 34.7° (c 0.52, $CHCI_3$). $C_{36}H_{66}O_{16}Si_4N_2$. IR spectrum, ν, cm^{-1}: 670 (Si-C); 1140, 1020 (C-O-C); 1715 (C=O); 930 (Si-N); 1460 (CH_3).

^{13}C NMR spectrum (δ, ppm): 170.00–174.66 (RO-C̲O-CH_3); 20.46 (RO-CO-C̲H_3), 72.3 (RO-C̲H_2-CH_2-SiR_3), 27.81–28.73 (RO-CH_2-C̲H_2-SiR_3), 21.6–22.65 (RO-CH_2-CH_2-C̲H_2-SiR_3),

8.15–13.35 (Si-CH_3); 91.90 and 100.88 ($C_{(1)}$ and $C_{(1')}$); 60.98 and 61.80 ($C_{(4)}$ and $C_{(4')}$); 34.80–43.51 ($C_{(5)}$ and $C_{(5')}$).

^1H NMR spectrum, δ, ppm (*J*, Hz): 4.30 (1H, d, $J_{1,2}$=8, H-1); 5.55 (1H, d, $J_{1,2}$=4, H-1'); 4.38 (1H, dd, $J_{2,1}$=8.1; $J_{2,3}$=9.4, H-2); 4.78 (1H, dd, $J_{2,1}$=4; $J_{2,3}$=10.6, H-2'); 5.67 (1H, dd, $J_{3,2}$=9.4; $J_{3,4}$=10, H-3); 5.60 (1H, dd, $J_{3,2}$=10.6; $J_{3,4}$=9.8, H-3'); 4.22–4.28 (dd, $J_{4,3}$=10; $J_{4,5}$=12.3, H-4); 5.29–5.30 (dd, $J_{4,3}$=9.8; $J_{4,5}$=9.9, H-4'); 4.80–4.10 and 4.11–4.14 (2H, d, H-5, and H-5',

$2CH_2$); 3.45–3.60 and 3.70–3.88 (2H, 2m, RO-CH_2 – CH_2 – CH_2 –SiR_3); 1.80–1.96 and 1.94–1.99 (2H, 2m, RO-CH_2 – CH_2 – CH_2 –SiR_3); 1.64–1.76 and 1.78–1.85 (2H, 2m, RO-CH_2 – CH_2 – CH_2 –SiR_3); 1.15–1.12 (24H, m, 8 Si- CH_3); 2.15–2.20 (21H, m, 7 CO-CH_3).

KEYWORDS

- **hydrosilylation**
- **benzoyl peroxide**
- **farmer reaction**
- **13C NMR spectrum**
- **1,3-bis(dimethylsilyl)-2,2,4,4-tetramethylcyclodisilazane**
- **β-L-arabinopyranose**
- **Markovnikov products**

REFERENCES

1. Zuev, V.; Kalinin, A. V. A Facile and Efficient Synthesis of Organocyclosiloxanes. *Phosphorus Sulfur Silicon Relat. Elem.* **2003,** *178* (6), 1289–1294.
2. Ionas, G.; Stadler, R.; Carbohydrate Modified Polysiloxanes. *Acta Polymer* **1994,** *45* (11), 14–20.
3. Kramarova, V. P.; Shipov, A. B.; Baukov, Yu. I. Synthesis of N-Silatranylmethyllactams. *J. General Chem.* **1992,** *62* (11), 2559–2567.
4. Sidamonidze, N. N.; VardiaShvili, R. O.; Isakadze, M. O. Hydrosilylation of Allyl Esters of Monosaccharides. *Chem. Heterocycl. Comp.* **2005,** *41* (12), 1534–1536.
5. Sidamonidze, N. N.; Vardiashvili, R. O.; Isakadze, M. O.; Djaniashvili, L. K. Boilogical Activity of some Aldose-Containing Compounds. *Pharm. Chem. J.* **2007,** *41* (3), 131–132.
6. Sidamonidze, N.**;** Vardiashvili, R.; Nutsubidze, M. Reaction Mechanism and Quantum-Chemical Computation of Hydrosilylation Reaction of Allyl Glycosides. *J. Nat. Sci. Sustain. Technol.* **2016,** *9* (3)**,** 679–684.
7. Sidamonidze, N. N.; Vardiashvili, R. O.; Nutsubidze, M. O. Reaction Hydrosilylsation of Allyl-2,3,5,6-di-O-Izopropylidene-D-Mannofuranoze with Methyl- and Phenylcyclodisilaza-zanes. In *Science and Technology of Polymers and Advanced Materials Applied Research Methods* ; Chapter 13**,** 2019; pp 174–180 .
8. Takano, T.; Nakatsubo, F. In *Carbohydrate Research*, 1999; vol. 203, p 341.

CHAPTER 14

Hot Shock Wave Consolidation and Syntheses of W-MG Composites

AKAKI PEIKRISHVILI[1], LASZLO KECSKES[2], ELGUJA CHAGELISHVILI[3], and BAGRAT GODIBADZE[1]

[1]*Ferdinad Tavadze Institute of Metallurgy and Materials Science, Tbilisi, Georgia*

[2]*John Hopkins University, Baltimore, MD, USA*

[3]*Institute of Mining and Technology, Tbilisi, Georgia*

ABSTRACT

Tungsten–metallic glass (W-MG) powder mixtures were consolidated and synthesized into cylindrical rods using a hot explosive consolidation (HEC) process. The W-MG composite consisted of W particles dispersed in an MG matrix. The consolidation temperature was varied from room temperature to 1000°C; the shock loading intensity was under 10 GPa.

The investigation showed that the combination of high temperatures and explosive compression was beneficial to the consolidation of the W-MG composites, resulting in near theoretical densities and high hardness values. The structure and property of the samples obtained depend on the intensity of shock loading and the phase content of precursors. Features of the structure/property relationship of the consolidated W-MG samples as a function of the loading conditions (experimental setup, intensity of loading, preloading temperature) are discussed.

Tailored Functional Materials for Clean and Sustainable Development. Tamara Tatrishvili, PhD, Neha Kanwar Rawat, PhD, Ann Rose Abraham, PhD & A. K. Haghi, PhD (Eds.)

14.1 INTRODUCTION

There is still a shortage of data concerning the study of explosive compacting of amorphous and ceramic powders. This is primarily caused by a number of technological difficulties associated with shock wave processing.[1–5] Difficulties are related to the impossibility of direct compaction and consolidation of fine powders. The problem is that during the compaction of a monophase or composite material powder, the short period of the loading cycle is not enough for plastic flow to occur in and outside of the consolidating particles, and as a result, no common boundaries can form under the shock wave front. Moreover, the stresses necessary to facilitate the plastic flow of the particles and for the formation of common boundaries between them exceed the ultimate strength of their steel containers. Typically, this happens where the consolidated powders are situated; therefore, after compressive loading, the sample container is always destroyed.

Application of a low-intensity shock wave, followed by sinter processing, to pure ceramic powders is not effective in most cases either.[1] The use of high sintering temperatures after consolidation leads to the annealing out of the as-processed crystalline structure of the shock-loaded particles. Thus, it is impossible to maintain the plastically deformed structure and other advantages of the passage of the shock wave through the compacted powders. This would be especially detrimental to and likely result in the disappearance of amorphous and nano-scale grain structures of certain precursor powders. Likewise, if ultimately the formation of the solid solutions within or between multiple bonding phases and additional sintering processes is desired, high temperatures (especially during sintering) are again not recommended. Lastly, the melting of the steel container may also take place which would interfere with obtaining an uncontaminated but well-consolidated composite material.

The positive role of the use of high temperatures during the consolidation processes of ceramic and refractory hard alloy powders has been well documented.[5–10] Application of high temperatures in hot explosive consolidation (HEC) allows the consolidation mechanism to work by plastic flow and mutual collision of the compacting particles, resulting in the melting of their surfaces and formation of common interfacial boundaries at essentially much lower loading intensities.[5,6] Variants of HEC using rapid heating cycles (15–45 s) allow not only to heat the consolidating particles to the necessary temperature but also allows us to limit and minimize the oxidation at temperature as high as 1000°C.

Metallic glasses (MGs) are an interesting class of materials. In their vitreous state, their structure is amorphous; however, upon heating, they devitrify into a crystalline solid. Conventional processing of pure and MG-based composite materials typically does not retain the amorphous structure. Because MGs have extremely high ultimate strengths, we wanted to create a system wherein both components making up the composite body would respond to the shock front in a similar manner. For the second component, we chose W. The objectives of this study were thus threefold. First, we wanted to demonstrate that using HEC, the initial W-MG precursor powders could be consolidated to near theoretical densities and appropriate microstructures. Second, we wanted to show that under certain heating and loading conditions, the initial amorphous and nano-crystalline structure of the MG could be maintained. Third, we wanted to investigate the shock-induced structural changes in both the W and MG powders after HEC. The results of this effort are described.

14.2 EXPERIMENTAL PROCEDURES AND MATERIALS

In this work, we used a fairly large-sized, mostly spherical 5–15 micrometer MG powder, obtained from Dr. Daniel Brannagan while he was still at Idaho National Engineering and Environmental Laboratory, Idaho Falls, ID, USA. The powder is a multicomponent Fe-based MG (FeMG) with a relatively high crystallization temperature of 580°C, a barely detectable glass transition temperature of 555°C, and a very narrow under-cooled region of 25°C. Though X-ray amorphous form has relatively poor thermal characteristics, it was assumed that it will most likely devitrify (even at low temperatures) during the explosive consolidation experiments. The W powder was obtained from Osram-Sylvania, M-10 grade, nominally had 1-micrometer average particle size. The W powder, obtained from Kennametal Corporation, also had 1-micrometer nominal average particle size.

In order to consolidate the W-FeMG powder mixtures to near theoretical densities in hot conditions, the two-stage cylindrical setup of loading described in Ref. [11] was used.

Only Ammonite, NH4NO3+20%TNT, with a detonation velocity of 3.1 km/s, Ammonite diluted with NH_4NO_3, used as a passive were used as explosive materials. The corresponding intensity of loading on the wall of the steel container was under 10 GPa.

14.3 RESULTS AND DISCUSSION

Due to the relatively low initial packing density of the precursor powders, the typical single-stage HEC was not successful. Therefore, consolidation of the W-FeMG powders to high densities, while maintaining the initial amorphous and/or nano-scale features of the FeMG matrix required a two-stage process. A first stage was necessary to increase the density of the starting powders. This was accomplished with a preliminary shock wave consolidation step at room temperature, using the same cylindrical loading geometry. The second stage, as described above, was the HEC of the preliminarily consolidated powders up to 1000°C. The positive role of the preliminary densification before the shock wave consolidation and fabrication of high-density samples from AlN has been in detail described in Ref. [3].

In brief, the application of a preliminary densification and compression step to the initial precursors at room temperature increases the relative stiffness of the powder mixture whereby a stiffer body allows for the build-up of higher pressures under the shock wave front during HEC. Additionally, the process activates individual particle surfaces for localized sintering. Lastly, preconsolidation helps to maintain the correct cylindrical geometry within the consolidated samples after HEC.

The experimental effort showed that the application of the high velocity explosive (Hexogen) for the preliminary densification step always led to the formation of a Mach's Stem or Zone in the central part of the cylindrical sample. Recompressions of such samples at high temperatures were carried out without success and the consolidated cylindrical rods always contained a hole at their center.

Figure 14.1 illustrates the central holes that formed in the hot consolidated W-FeMG rods, using the first-stage pre-densification procedure with Hexogen, detonation velocity under 6.0 km/s).

Application of the low-intensity explosives of ammonite and the mixture of ammonite with saltpeter (NH_4NO_3) prevented the formation of the hole in the central part of consolidated rod, and it thus also helped to maintain the correct cylindrical geometry of the samples.

Reloading of the preliminary consolidated samples in hot conditions with intensity of loading under 10 GPa allowed to consolidate samples near to theoretical densities; however, with different properties depending on the compaction temperature.

It must be strongly emphasized that depending on the loading conditions and temperature of consolidation, definite changes in the structure of the

Fe-based MG matrix as well as in the separate W grains were observed. The investigation showed that the consolidation W–FeMG compositions up to about 500°C allowed the retention of the amorphous structure of the FeMG particles. However, further increase of the temperature led to recrystalization processes within the MG particles. Consolidation at 800°C resulted in a completely nanocrystalline structure, corresponding to the devitrification of the FeMG matrix Ref. [12].

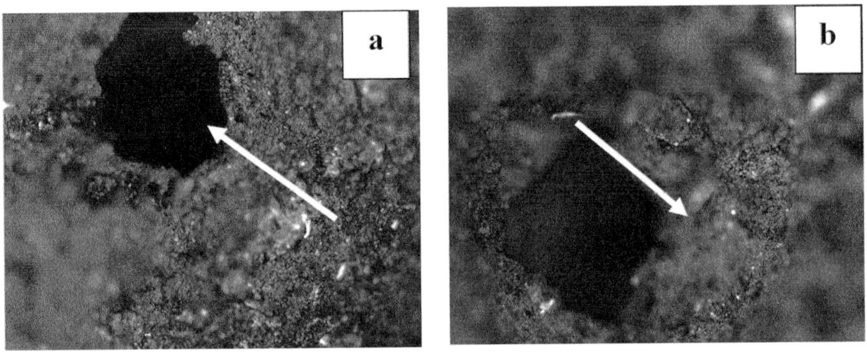

FIGURE 14.1 Macrostructure of two-stage consolidated (first stage P = 20 GPa; second stage P = 10 GPa) of W-20%FeMG composite powders; (a) consolidation temperature 380°C and (b)consolidation temperature ~600°C. The Mach's Stems or Zones (hole) in the central part of consolidated rods can be clearly observed. The formation of the void at the central part of the rods may be explained as stemming from the influence of high pressures (20 GPa) during the first pre-compacting stage shifting the amorphous FeMG binder phase to the central part of the rods. Because of the much lower melting temperature of the FeMG component, compared to W during compaction the amorphous FeMG is likely melted and vaporized.

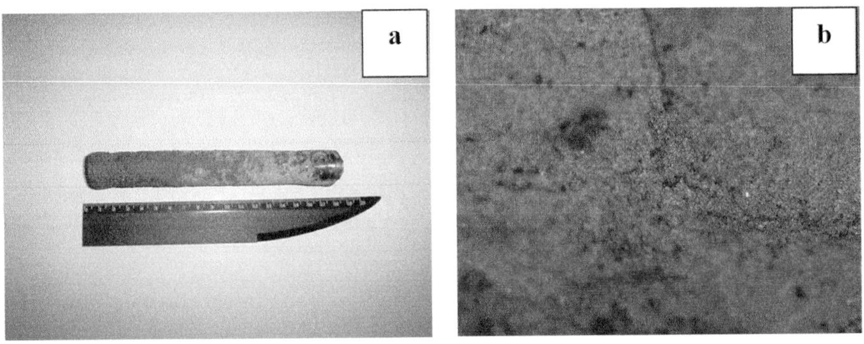

FIGURE 14.2 Examples of the consolidation of the W-MG powder mixture by HEC. The view of the consolidated rod after two-stage hot consolidation at 800°C is shown in (a). The loading intensity of preliminary shock wave densification was under 10 GPa. The view of the single-stage preconsolidated W-MG powder blends at T=20°C is shown in (b).

Polished cross-sections of the samples shown in Figure 14.3 represent the typical morphology of the hot explosively consolidated W-FeMG composites. As seen from the surface structures, the use of elevated temperature has had a positive influence on the consolidation process. The crack in room temperature consolidation (black arrow) could be noticed. Despite the considerable particle size mismatch between the W and FeMG particles, hot consolidation using a second-stage allows the fabrication of highly dense samples without cracks and low-volume fraction of porosity. The pulled-out FeMG particles (see blue arrow) are caused by the relatively weak bonding between the two components.

X-ray diffraction measurements of the consolidated W-FeMG composite samples (Fig. 14.4) show that depending on the consolidation temperature, it is possible to regulate the post-consolidation structure of the FeMG matrix. As seen from the individual scans in Figure 14.5, it is possible to maintain the amorphous structure of the matrix below 500°C. Devitrification, that is, loss of the amorphous structure is more pronounced at higher temperatures; peaks associated with the devitrified structure appear to grow. Besides the regularly spaced primary W peaks, there are no secondary peaks, indicating that the interaction between the two components is minimal or not detected by X-ray diffraction.

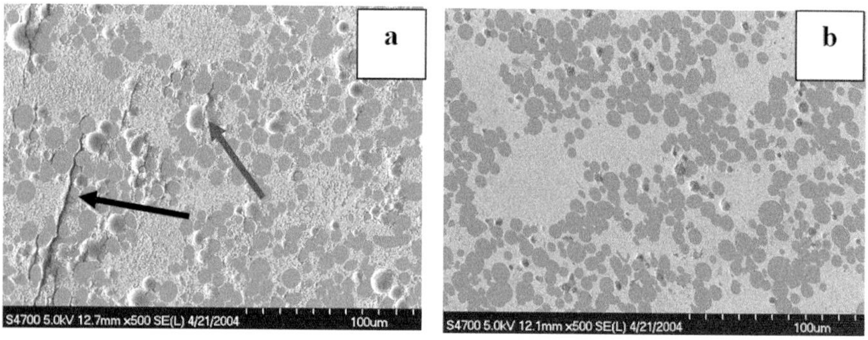

FIGURE 14.3 The surface morphology is seen in polished cross-sections of two-stage explosively consolidated W-FeMG powders. Compression intensity is at 10 GPa; (a) consolidation at room temperature, 20°C; (b) consolidation at 580°C.

The presence of the broad "background" feature in the 20°C scan, corresponding to the amorphous FeMG could be noted. Similar to that found in the microscopic investigation, consolidation at higher temperatures leads to the formation of nanocrystalline structures in the FeMG. This is demonstrated by the appearance of the series of smaller peaks at and above 580°C. Further increase of the consolidation temperature up to 1000°C leads to the growth of nanocrystallites and the formation ultradispersed structures in the matrix.

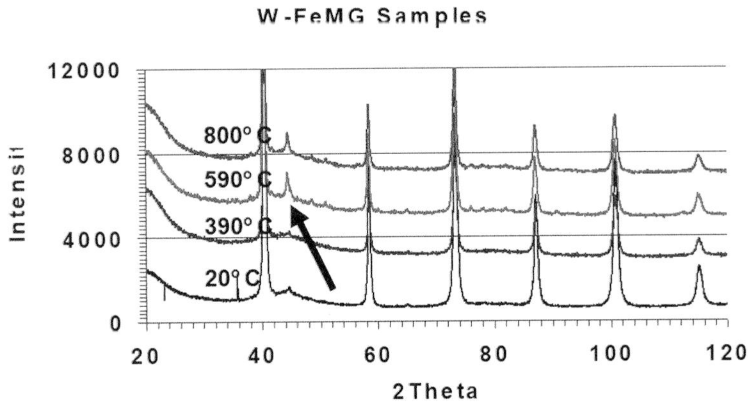

FIGURE 14.4 X-ray diffraction scans of the W-FeMG composite samples after HEC as a function of temperature; loading intensity is 10 GPa.

The X-ray diffraction results were further confirmed by microstructural investigations illustrated in Figure 14.5. As seen from the sample microstructures, the increase of the consolidation temperature above 500°C leads to the onset of recrystallization within the FeMG matrix particles. It should be noted that the formation of nano-scale features (Fig. 14.5b). The backscattered electron micrograph reveals the partitioning of the FeMG into different intermetallic species. In contrast, the microstructure of the FeMG matrix during consolidation below the 500°C threshold always stays amorphous (Fig. 14.5a).

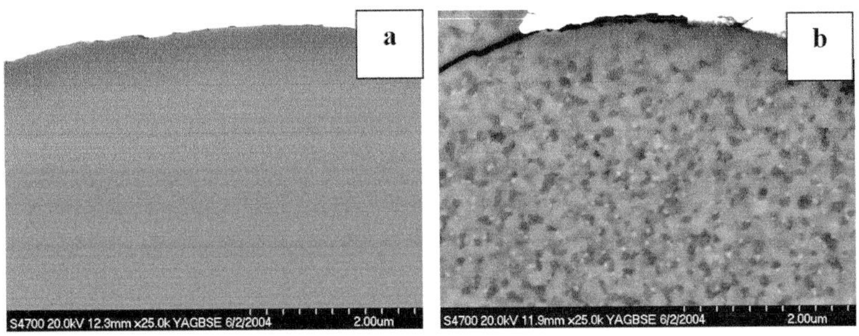

FIGURE 14.5 The microstructure of FeMG matrix after hot explosive consolidation of W-FeMG precursor powders at different consolidation temperatures at a loading intensity of 10 GPa; (a) consolidation at 380°C and (b) consolidation at 580°C.

14.4 CONCLUSIONS

We have demonstrated that using HEC, mixtures of W-FeMG can be consolidated with moderate success. The compact use of a two-stage consolidation process was free of extensive cracking. Results also show that below 500°C, the amorphous structure of the MG matrix is retained. Above 500°C, the matrix devitrifies giving rise to the formation of nano-crystalline structures. At the same time, little or no change took place in the W particles.

KEYWORDS

- **MG powder**
- **cylindrical rods**
- **Osram-Sylvania**
- **FeMG matrix**
- **cylindrical sample**
- **W-FeMG composites**
- **surface morphology**

REFERENCES

1. Prummer, R. *Explosivverdichtung Pulvriger Substanzen*; Springer-Verlag: Berlin Germany, 1987.
2. Kiiski, A. A.; Ruuskanen, P. R.; Deribas, A. A.; Shtertser, A. A. Explosive Compaction of Copper and Graphite Powder Mixtures. In *Metallurgical and Materials Applications of Shock-Wave and High-Strain Rate Phenomena*; Murr, L. E., Staudhammer, K. P., Meyers, M. A., Eds.; Elsevier: Amsterdam, Netherlands, 1995; pp 109–115.
3. Andreev, B. D.; Lukash, V. A.; Voloshin, M. N.; Markov, A. I.; Sozin, Yu. I.; Dub, S. N. Certain Properties of Aluminum Nitride Compacts Produced by Shock-Wave Loading . *Poroshkovaya Metall.* **1991,** *10*, 27–32.
4. Hoenig, C. L.; Yust, C. S. Explosive Compaction of AlN, Amorphous Si_3N_4, Boron, and Al_2O_3 Ceramics. *Bull. Amer. Cer. Soc.* **1981,** *60* (11), 1175–1224.
5. Mote, J. D.; Fitzpatrick, J. J. Investigation of a Method to Consolidate Hard Materials in a Tough Matrix. In *Emergent Process Methods for High Technology Ceramics*; Davis, R. F., Palmour, H., Porter, R. L., Eds.; Plenum Press: New York, NY, 1984; pp 695–710.
6. Bhalla, A. K. Hot Explosive Compaction of Metal Powders. *Trans. Powder Metall. Assoc. India* **1980,** *7* (9), 1–8.
7. Peikrishvili, A., Chikhradze, N. Explosive Working of Some Metals and Alloys at High Temperatures. In *Metallurgical Application of Shock Wave and High Strain Rate*

Phenomena; Murr, L. E., Staudhammer, K. P., Meyers, M. A., Eds.; Marcel Dekker Inc.: New York, NY, 1986; pp 905–915.

8. Japaridze, L.; Peikrishvili, A.; Chikhradze, N.; Gotsiridze, G. Importance of Preheating at Dynamic Consolidation of Some Hard Materials. In *Shock-Wave and High-Strain Rate Phenomena in Materials*; Meyers, M. A., Murr, L. E., Staudhammer, K. P., Eds.; Marcel Dekker, Inc.: New York, NY, 1992; pp 463–472.
9. Peikrishvili, A.; Japaridze, L.; Chikhradze, N.; Chagelishvili, E. Possibilities of Obtaining Combined Samples From Tungsten-Based Alloys by High-Temperature Shock Wave Treatment. In *Metallurgical and Materials Applications of Shock-Wave and High-Strain Rate Phenomena*; Murr, L. E., Staudhammer, K. P., Meyers, M. A., Eds.; Elsevier: Amsterdam, Netherlands, 1995; pp 99–108.
10. Simonsen, I.; Horie, Y.; Akashi, T.; Sawaoka, A. B. Diamond Formation in Aluminum Compressed with Ni-Graphite Under Shock Loading. *J. Mat. Sci.* **1992,** *27* (7), 1735–1740.
11. Kesckes, L. J.; Szewczyk, S. T.; Peikrishvili, A. B.; Chikhradze, N. M. In *Exothermic Reactions In Hot Explosively Compacted Al-Ni Composites*, Proceeding of PM^{2}TEC Orlando, Florida, June 19, 2002.
12. Kesckes, L.; Peikrishvili, A.; Chikhradze, N. *Hot Explosive Consolidation of Tungsten-Metallic Glass Matrix Composite Materials*, Proceeding of Fifth International Symposium on Impact Engineering University of Cambridge, Churchill College Cambridge, UK, July 11–15, 2004.

CHAPTER 15

Some Aspects Concerning Conformation of Polypeptide Chains in Biosystems

G. A. KORABLEV[1], V. I. KODOLOV[2], and G. E. ZAIKOV[3]

[1]*Department of Physics at Izhevsk State Agricultural Academy, Izhevsk State Agricultural Academy, Izhevsk, Udmurt Republic, Russia*

[2]*Department of Chemistry and Chemical Engineering of Kalashnikov Izhevsk State Technical University, Basic Research-Educational Center of Chemical Physics and Mesoscopy, UdSC, UrD, RAS, Russia*

[3]*N.M. Emmanuel Institute of Biochemical Physics, RAS, Russia*

ABSTRACT

The notion of spatial-energy parameter (P-parameter), which is a complex characteristic of important atomic values, is used to evaluate energy criteria of stable biostructure formation. The rationality of applying such methodology when investigating the formation process of some nanosystems and the conformation of polypeptide chains and fragments of DNA molecules is demonstrated. Polypeptide is a polymer of amino acids. The principles of biosystem formation and stabilization have a definite analogy with the conditions of wave processes. The results obtained do not contradict the experimental data.

15.1 INTRODUCTION

To obtain the dependence between energy parameters of free atoms and dynamics of structural formations in simple and complex systems is one

Tailored Functional Materials for Clean and Sustainable Development. Tamara Tatrishvili, PhD, Neha Kanwar Rawat, PhD, Ann Rose Abraham, PhD & A. K. Haghi, PhD (Eds.)

of the strategic tasks in physical chemistry. Classical physics and quantum mechanics widely use Coulomb interactions and their varieties for this.

Thus, in Ref. [6], van der Waals orientation and charge–dipole interactions are referred to electron–conformation interactions in biosystems and as a particular case–exchange-resonance transfer of energy. But biological and many cluster systems are electroneutral in structural basis and non-Coulomb equilibrium-exchange spatial-energy interactions, that is, non-charge electrostatic processes, are mainly important for them.

The structural interactions of summed electron densities of valence orbitals of corresponding conformation centers take place—processes of equilibrium flow of electron densities due to overlapping of their wave functions.

Heisenberg and Dirac[10] proposed the exchange Hamiltonian derived from the assumption of direct overlapping of wave functions of interacting centers. In this model, electrostatic interactions are modeled by effective exchange Hamiltonian acting in the space of spin functions.

In particular, such an approach is applied to the analysis of structural interactions in cluster systems.[7] It is demonstrated in Anderson's works[9] that in compounds of transition elements when the distance between paramagnetic ions considerably exceeds the total of their covalent radii, "super-exchange" processes of overlapping cation orbitals take place through the anion between them.

In this chapter, similar equilibrium-exchange processes are evaluated through the notion of spatial-energy parameter–P-parameter.

15.1.1 INITIAL CRITERIA

In the systems in which the interaction proceeds along the potential gradient (positive work), the resultant potential energy is found based on the principle of adding reciprocals of corresponding energies of subsystems.[2]

Thus, the energy of atom valence orbitals (responsible for interatomic interactions) can be calculated by the principle of adding reciprocals of some initial energy components based on the following equations:

$$\frac{1}{q^2/r_i}+\frac{1}{W_i n_i}=\frac{1}{P_E} \text{ or } \frac{1}{P_0}=\frac{1}{q^2}+\frac{1}{(Wrn)_i}; \; P_E=P_0/r_i \qquad (15.1), (15.2), (15.3)$$

Here W_i – electron orbital energy[14]; r_i – orbital radius of *i*–orbital [13]; $q=Z^*/n^*$,[11,12] n_i – number of electrons of the given orbital, Z^* and n^* – nucleus effective charge and effective main quantum number, r – bond-dimensional characteristics.

P_0 is called a spatial-energy parameter (SEP), and P_E – effective P–parameter (effective SEP). Effective SEP has a physical sense of some averaged energy of valence electrons in the atom and is measured in energy units, for example, electron volts (eV).

Calculation results by the equations (15.1–15.3) are given in Table 15.1.

TABLE 15.1 P-Parameters of Atoms Calculated Via the Electron Bond Energy.

Atom	Valence electrons	W (eV)	r_i (Å)	$q^2{}_0$ (eVÅ)	P_0 (eVÅ)	R (Å)	P_0/R (eV)
H	$1S^1$	13.595	0.5292	14.394	4.7969	0.5292	9.0644
						0.375	12.792
						0.28	17.132
C	$2P^1$	11.792	0.596	35.395	5.8680	0.77	7.6208
						0.67	8.7582
						0.60	9.780
	$2P^2$	11.792	0.596	35.395	10.061	0.77	13.066
						0.67	15.016
						0.60	16.769
	$2P^3{}_r$				13.213	0.77	17.160
	$2S^1$	19.201	0.620	37.240	9.0209	0.77	11.715
	$2S^2$				14.524	0.77	18.862
	$2S^1 + 2P^3{}_r$				22.234	0.77	28.875
	$2S^1 + 2P^1{}_r$				13.425	0.77	17.435
	$2S^2 + 2P^2$				24.585	0.77	31.929
					24.585	0.67	36.694
						0.60	40.975
N	$2P^1$	15.445	0.4875	52.912	6.5916	0.70	9.4166
						0.55	11.985
	$2P^2$				11.723	0.70	16.747
						0.63	18.608
	$2P^3$				15.830	0.70	22.614
						0.55	28.782
	$2S^2$	25.724	0.521	53.283	17.833	0.70	25.476
	$2S^2 + 2P^3$				33.663	0.70	48.090

TABLE 15.1 *(Continued)*

Atom	Valence electrons	W (eV)	r_i (Å)	q^2_0 (eVÅ)	P_0 (eVÅ)	R (Å)	P_0/R (eV)
O	$2P^1$	17.195	0.4135	71.383	6.4663	0.66	9.7979
	$2P^1$					0.55	11.757
	$2P^2$	17.195	0.4135	71.383	11.858	0.66	17.967
						0.59	20.048
	$2P^4$	17.195	0.4135	71.383	20.338	0.66	30.815
						0.59	34.471
	$2S^2$	33.859	0.450	72.620	21.466	0.66	32.524
	$2S^2 + 2P^4$				41.804	0.66	63.339
						0.59	70.854

In the systems in which the interactions proceed against the potential gradient (negative performance), the algebraic addition of their masses as well as the corresponding energies of subsystems is performed.[2] Thus, for the interaction of similarly charged (homogeneous) subsystems, the principle of algebraic addition of their P-parameters is performed:

$$\sum_{i=1}^{m} P_0 = P_0^{'} + P_0^{''} + \ldots + P_0^{m} \tag{15.4}$$

$$\Sigma P_E = \frac{\Sigma P_0}{R} \tag{15.5}$$

where R – dimensional characteristic of atom (or chemical bond).

Modifying the rules of adding the reciprocals of the energy magnitudes of sub-systems as applicable to complex structures, we can obtain the equations to calculate a P_C-parameter of a complex structure:

$$\frac{1}{P_C} = \left(\frac{1}{NP_E}\right)_1 + \left(\frac{1}{NP_E}\right)_2 + \ldots \tag{15.6}$$

where N_1 and N_2 – number of homogeneous atoms in subsystems.

The calculation results in this equation for some atoms and radicals of biosystems are given in Table 15.2.

The value of the relative difference of P-parameters of interacting atoms-components—the structural interaction coefficient α is used as the main numerical characteristic of structural interactions in condensed media:

$$\alpha = \frac{P_1 - P_2}{(P_1 + P_2)/2} 100\% \tag{15.7}$$

Applying the reliable experimental data, we obtain the nomogram of structural interaction degree dependence (ρ) on coefficient α, the same for a wide range of structures (Fig. 15.1). This approach gives the possibility to evaluate the degree and direction of the structural interactions of phase formation, isomorphism, and solubility processes in multiple systems, including molecular ones.

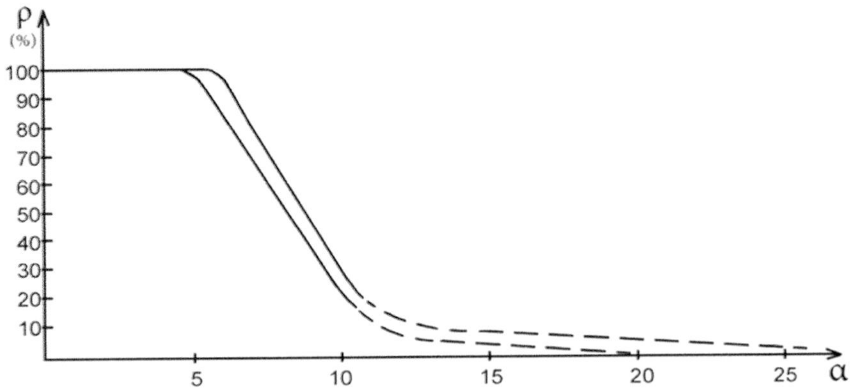

FIGURE 15.1 Nomogram of structural interaction degree dependence (ρ) on coefficient α.

TABLE 15.2 Structural P_C-Parameters Calculated Via Electron Bond Energy.

Radicals, molecule fragments	P'_i **(eV)**	P''_i **(eV)**	P_C **(eV)**	**Orbitals**
OH	9.7979	9.0644	4.7084	O ($2P^1$)
	30.815	17.132	11.011	O ($2P^4$)
	17.967	17.132	8.7710	O ($2P^2$)
H_2O	2·9.0644	17.967	9.0237	O ($2P^2$)
	2·17.132	17.967	11.786	O ($2P^2$)

TABLE 15.2 *(Continued)*

Radicals, molecule fragments	P'_i (eV)	P''_i (eV)	P_C (eV)	Orbitals
CH_2	17.160	2·9.0644	8.8156	C $(2S^1 2P^3_i)$
	31.929	2·17.132	16.528	C $(2S^2 2P^2)$
	36.694	2·9.0644	12.134	C $(2S^1 2P^3_r)$
CH_3	31.929	3·17.132	19.694	C $(2S^2 2P^2)$
	15.016	3·9.0644	9.6740	C $(2P^2)$
CH	36.694	17.132	11.679	C $(2S^2 2P^2)$
	17.435	17.132	8.6423	C $(2S^2 2P^2)$
NH	16.747	17.132	8.4687	N$(2P^2)$
	19.538	17.132	9.1281	N$(2P^2)$
	48.090	17.132	12.632	N$(2S^2 2P^3)$
NH_2	19.538	2·9.0644	9.4036	N$(2P^2)$
	16.747	2·17.132	12.631	N$(2P^2)$
	28.782	2·17.132	18.450	N$(2P^3)$
C_2H_5	2·31.929	5·17.132	36.585	C $(2S^2 2P^2)$
NO	19.538	17.967	9.3598	N$(2P^2)$
	28.782	20.048	11.817	N$(2P^3)$
CH_2	31.929	2·9.0644	11.563	C $(2S^2 2P^2)$
CH_3	16.769	3·17.132	12.640	C $(2P^2)$
CH_3	17.160	3·17.132	12.865	C $(2P^3_r)$
CO–OH	8.4405	8.7710	4.3013	C $(2P^2)$
CO	31.929	20.048	12.315	C $(2S^2 2P^2)$
C=O	15.016	20.048	8.4405	C $(2P^2)$
C=O	31.929	34.471	16.576	O $(2P^4)$
CO=O	36.694	34.471	17.775	O $(2P^4)$
C–CH_3	31.929	19.694	12.181	C $(2S^2 2P^2)$
C–CH_3	17.435	19.694	9.2479	C $(2S^1 2P^1)$
C–NH_2	31.929	18.450	11.693	C $(2S^2 2P^2)$
C–NH_2	17.435	18.450	8.8844	C $(2S^1 2P^1)$
C–OH	8.7572	8.7710	4.3821	

15.1.2 FORMATION OF CARBON NANOSTRUCTURES

After different allotropic modifications of carbon nanostructures (fullerenes, tubules) have been discovered, a lot of research papers dedicated to the investigations of such materials, for instance, were published, determined by the perspectives of their vast application in different fields of material science.

The main conditions of stability of these formulated structures based on modeling the compositions of over 30 carbon clusters are given[8]:

1. Stable carbon clusters look like polyhedrons where each carbon atom is three-coordinated.
2. More stable carbopolyhedrons containing only 5- and 6-term cycles.
3. 5-term cycles in polyhedrons – isolated.

Let us demonstrate some possible explanations of such experimental data based on the application of spatial-energy concepts. The approximate equality of effective energies of interacting subsystems is the main condition for the formation of stable structure in this model[4] based on the following equation:

$$\left(\frac{P_0}{\kappa R}\right)_1 \approx \left(\frac{P_0}{\kappa R}\right)_2 ; P_1 \approx P_2 \qquad (15.8)$$

where K is the coordination number, R is the bond-dimensional characteristic.

At the same time, the phase-formation stability criterion (coefficient α) is the relative difference of parameters P_1 and P_2 that is calculated following the equation (15.1) and is αST < (20–25)% (according to the nomogram).

During the interactions of similar orbitals of homogeneous atoms, $P'_0 = P''_0$, we have

$$K_1R_1 \approx K_2R_2 \qquad (15.8a)$$

Let us consider these initial notions as applicable to certain allotropic carbon modifications:

1. Diamond: modification of structure where K_1=4, K_2=4; $P'_0 = P''_0$, R_1=R_2, P_1=P_2 and α=0. This is absolute bond stability.
2. Non-diamond carbon modification for which P'0 = P"0, K_1=1; R_1=0,77Å; K_2=4; $R_2^{4+} = 0,2$Å, α=3,82%. Absolute stability is due to ionic–covalent bonds.
3. Graphite: P'0 = P"0, K_1=K_2=3, R_1=R_2, α=0. This has absolute bond stability.
4. Chains of hydrocarbon atoms consisting of a series of homogeneous fragments with similar values of P-parameters.

5. Cyclic organic compounds as a basic variant of carbon nanostructures. Apparently, not only does inner-atom hybridization of valence orbitals of carbon atom takes place in cyclic structures but also total hybridization of all cycle atoms.

Not only the distance between the nearest similar atoms by bond length (d) is the basic dimensional characteristic but also the distance to the geometric center of cycle interacting atoms (D) as the geometric center of total electron density of all hybridized cycle atoms.

Then the basic stabilization equation for each cycle atom will take into account the average energy of hybridized cycle atoms:

$$\left(\frac{\Sigma P_0}{\kappa d}\right)_i^{'} \gg \left(\frac{\Sigma P_0}{\kappa Д}\right)_i^{''} \quad (15.9)$$

$$P^{'} \approx P^{''} \quad (15.9a)$$

where $\Sigma P_0 = P_0 N$; N is the number of homogeneous atoms, P_0 is the parameter of one cycle atom, K is the coordination number relative to the geometric center of cycle atoms. Since in these cases, K'=K" and N'=N", K=N, the simple correlation for paired bond appears as the following:

$$\frac{P_0^{'}}{d} \approx \frac{P_0^{''}}{Д}; P_E^{'} \approx P_E^{''}; \quad (15.10)$$

During the interactions of similar orbitals of homogeneous atoms, $P_0^{'} \approx P_0^{''}$, and then:

$$d \approx D \quad (15.11)$$

Equation (15.10) reflects a simple regularity of stabilization of cyclic structures:

The main condition of formation and stability of structures is an approximate equality of effective interaction energies of atoms along all directions of the interatomic bond.

The corresponding geometric comparison of cyclic structures consisting of 3, 4, 5, and 6 atoms results in the conclusion that only in 6-term cycle (hexagon), the bond length (d) equals the length of the geometric center of atoms (D): d=D.

Such calculation of α following the equation (11.7), gives for hexagon $\alpha=0$ and absolute bond stability. And for pentagon $d\approx1.17D$ and the value of $\alpha=16\%$, that is, this is the relative stability of the structure being formed. For the other cases, $\alpha>25\%$, therefore, many biosystems such as, for example, nucleic acids, contain only 5- and 6-term cyclic fragments.

The calculations and comparisons performed based on spatial-energy concepts allow explaining some formation peculiarities of hexagonal structures in biosystems.

15.1.3 FORMATION OF POLYPEPTIDE CHAIN

The main components of organic compounds constituting 98% of cell element composition include carbon, oxygen, hydrogen, and nitrogen. The polypeptide bond formed by COOH and NH_2 groups of amino acid CONH is the binding base of protein biopolymers of a cell. At the same time, carbon atoms are more frequently found in polypeptide chain nodes, and sometimes in nitrogen atoms.

Fragments of polypeptide chain nodes are formed of CH, OH, CO, NH, NH_2, CO-OH atom groups, and some radicals. In accordance with the conformation energy criteria of such a chain, the approximate equality of their P_E-parameters needs to be followed, both separately for all fragments and chain atomic nodes. The calculations of possible variants are given in Table 15.2. Its analysis results in the conclusion of the existence possibility of three series of such relations. Their summarized data are given in Table 15.3. Such cyclicity of functional correlations can be evaluated from the point of quantum-wave properties of P-parameter.[5]

The interference minimum, oscillation weakening (in antiphase) takes place if the difference in wave move (Δ) equals the odd number of semiwaves:

$$\Delta = (2n+1)\frac{\lambda}{2} = \lambda(n+\frac{1}{2}), \text{ where } n = 0, 1, 2, 3, \ldots \qquad (15.12)$$

It means that the minimum of interactions take place if P-parameters of interacting structures are also "in antiphase" —either oppositely charged systems or heterogeneous atoms are interacting (e.g., during the formation of valence-active radicals CH, CH_2, CH_3, NO_2 ..., etc.).

In this case, P-parameters are summed up based on the principle of adding the reciprocals of P-parameters – equations (15.1–15.3).

The difference in wave move (Δ) for P-parameters can be evaluated via their relative value ($\gamma = \frac{P_2}{P_1}$) or via the relative difference of P-parameters (coefficient α), which gives an odd number at a minimum of interactions:

$$\gamma = \frac{P_2}{P_1} = \left(n + \frac{1}{2}\right) = \frac{3}{2}; \frac{5}{2} \dots \text{ At n=0 (basic state) } \frac{P_2}{P_1} = \frac{1}{2} \qquad (15.12a)$$

Interference maximum, oscillation enhancing (in phase) takes place if the difference in wave move equals an even number of semiwaves:

$$\Delta = 2n\frac{\lambda}{2} = \lambda n \quad \text{or} \quad \Delta = \lambda(n+1) \qquad (15.13)$$

As applicable to the P-parameters, the maximum enhancement of interaction in the phase corresponds to the interactions of similarly charged systems or systems homogeneous by their properties and functions (e.g., between the fragments or blocks of complex organic structures, such as CH_2 and NNO_2).

Then

$$\gamma = \frac{P_2}{P_1} = (n+1) \qquad (15.13a)$$

By this model, the maximum of interactions corresponds to the principle of algebraic addition of P-parameters—equations (15.4, 15.5). When n=0 (basic state), we have $P_2 = P_1$, or: the maximum structure interaction occurs if their P-parameters are equal. This postulate and equation (15.13a) are used as basic conditions for the formation of stable structures.[4]

Hydrogen atom, element No. 1 with orbital $1S^1$ defines the main energy criteria of structural interactions (their "ancestor"). Table 15.1 shows its three P_E-parameters corresponding to three different characteristics of the atom.

$R_1 = 0.5292\ A^0$ – orbital radius (quantum-mechanical characteristic) gives the initial main value of P_E-parameter equaled to 9.0644 eV;

$R_2 = 0.375\ A^0$ – distance equaled to half of the bond energy in H_2 molecule. But if a hydrogen atom is bound with other atoms, its covalent radius is $\approx 0.28\ A^0$.

In accordance with the equation (13a), $P_2 = P_1\ (n+1)$; therefore, $P_1 \approx 9.0644$ eV, $P_2 \approx 18.129$ eV.

TABLE 15.3 Bio-Structural Spatial-Energy Parameters (eV).

Series	H	C	N	O	CH	CO	NH	NH_2	OH	C–NH_2	C–CH_3	$\langle P_E \rangle$	$\langle \alpha \rangle$
I	9.0644	8.7582	9.4166	9.7979	8.6423	8.4405	8.4687	9.4036	8.7710	8.8844	9.2479	8.9905	0.82
	($1S^1$)	($2P^1$)	($2P^1$)	($2P^1$)	($2S^2 2P^2$–$1S^1$)	($2P^2$–$2P^2$)	($2P^2$–$1S^1$)			$2S^1 2P^1_r$–	$2S^1 2P^1_r$–		
		9.780					9.1281			($2P^3$–$1S^1$)	($2S^2$ $2P^2$–$1S^1$)		
		($2P^1$)					($2P^2$–$1S^1$)						
II	12.792	13.066	11.985	11.757	11.679	12.315	12.632	12.404	11.011	11.693	12.181	12.138	5.25
	($1S^1$)	($2P^2$)	($2P^1$)	($2P^1$)	($2S^2 2P^2$–$1S^1$)	($2S^2 2P^2$–$2P^2$)	($2S^2 2P^3$–$1S^1$)			$2S^2 2P^2$–($2P^3$ –$1S^1$)	$2S^2 2P^2$– ($2S^2 2P^2$–$1S^1$)		
		11.715			12.081								
		($1S^1$)			($2S^2 2P^2$–$1S^1$)								
III	17.132	16.769	16.747	17.967	Blocks	16.576	Blocks	18.450	Blocks	Blocks	Blocks	17.104	0.16–4.92
	($1S^1$)	($2P^2$)	($2P^2$)	($2P^2$)	C and H	($2S^2 2P^2$–$2P^4$)	N and H		CO and OH	C and NH_2	C and NH_2		
		17.435											
		($2S^1 2P^1$)											

Note: Names of interacting orbitals are indicated in brackets.

These are the values of possible energy criteria of stable (stationary) structures. The dimensional characteristic 0.375 A^0 does not satisfy them, therefore, there is a transition onto the covalence radius ≈ 0.28 A^0, which provides the value of P-parameter approximately equaled to P_2.

Three series with approximately equal values of P-parameters of atoms or radicals at α < 7.5% are given in Table 15.3.

First series for P_E= 9.0644 eV – main, initial, where H, C, O, N atoms have P_E-parameters only of the first electron, and interactions proceed in the phase.

Second series for P''_E = 12.792 eV is the non-rational, pathological as it more corresponds to the interactions in anti-phase: by equation (12a) P''_E = 13.596 eV.

Third series for P''''_E = 17.132 eV – stationary as the interactions are in the phase: by equation (12a) P''''_E = 18.129 eV (α = 5.5%).

With specific local energy actions (electromagnetic fields, radiation, etc.), the structural formation processes can grow along the pathological series II, which can result in the destruction of the main conformation chain.[5]

Therefore, during the transplantation and use of stem cells, the condition of approximate equality of P-parameters of the corresponding structures should be observed (not by the series II).

15.1.4 ON CONFORMATION OF FRAGMENTS OF DNA MOLECULES

Pyrimidines—cytosine (C) and thymine (T), as well as purines—adenine (A) and guanine (G) are canonic foundations in DNA molecules. They comprise nitrogen and carbon atoms, as well as molecular groups CH, OH, NH, NH_2, and CH_3. The main property of pyrimidines and purines—nitrogen atoms can attach protons.[1] The values of P-parameters of all these fragment components are given in Tables 15.1–15.3.

Such data are also systemized in operation series, with numerical values to be quantized from the initial value of hydrogen atom (9.0644 eV), that is, approximately two times more. Obviously, the series of structures with the parameter value two times less from the initial one (about 4.533 eV) are also possible. If we take 9.0644 eV as the unit of energy content (X=1), then in the first series X=1 with deviations of coefficient α to ±0.82% for practically all structures. And for radical OH, the value of X can be both 0.5 and 1. In Table 15.4, we can see the numerical composition of purines and pyrimidines, as well as the values of their energy contents. Out of the two values of X for OH group, one (X=1) corresponds to the possibility of DNA fragment formation. Another one (X=0.5), obtained taking into account the interactions of the most valent-active orbitals of hydrogen and oxygen atoms, apparently characterizes the possibility of structural interactions in the formations alrcady formcd that is considered in Table 15.4.

Despite a rather simplified approach in this model, the results obtained match the experimental data. Thus, from Table 15.4, it follows that X_A+X_T equals the sum of X_C+X_G, which indicates the vertical stability of DNA branches. It is known that the horizontal stability of such a structure is defined, in particular, by the fact that molecules A и T are connected with two hydrogen bonds and G and C—with three.

TABLE 15.4 Numerical and Energy Composition of DNA Fragments.

	X	A	T	C	G	X_A+X_T	X_C+X_G
C	1	3	3	2	4		
N	1	3	2	2	3		
CH	1	2	1	2	1		
NH	1	1	–	–	1		
NH_2	1	1	–	1	1		
CH_3	1	–	1	–	–		
OH	0.5	–	2	1	1		
Σ <x>		10	8	7.5	10.5	18	18

Such correlation defines the equality of their energy contents. From Table 15.4, we have $X_T + 2 = X_A$ (two hydrogen bonds are added), and $X_C + 3 = X_G$ (three hydrogen bonds are added).

According to Chargaff's rule,[1] for DNA molecule, the number of fragments A equals T and the number of fragments C equals G, which also corresponds to the equality principle of summed-up P-parameters for all fragments of the DNA molecule.

15.2 CONCLUSIONS

The application of spatial-energy parameter methodology allows in evaluating the possibility of conformation processes in different biosystems based on the energy characteristics of free atoms.

KEYWORDS

- **spatial-energy parameter**
- **nanosystems**
- **conformation**
- **biosystems**
- **DNA molecules**

REFERENCES

1. Volkenshtein M. V. Biophysics; M. Nauka, 1988; p 592.
2. Korablev G. A. Spatial-Energy Principles of Complex Structures Formation; Brill Academic Publishers and VSP: Netherlands, 2005; p 426 (Monograph).
3. Korablev G. A.; Zaikov G.E. *J. Appl. Polymer Sci.* **2006,** *101* (3), 2101–2107.
4. Korablev, G. A.; Zaikov, G. E. Formation of Carbon Nanostructures and Spatial-Energy Criterion of Stabilization; Mechanics of Composite Materials and Structures: RAS – IPM, 2009; vol 15 (1), pp 106–118.
5. Korablev, G. A.; Zaikov, G. E. Bio-Structural Energy Criteria of Functional States in Normal and Pathological Conditions. *HEI News* 2012, vol 1 (2), pp 118–124.
6. Rubin, A. B. Biophysics, Book 1. In *Theoretical Niophysics*; M. Vysshaya shkola, 1987; p 319.
7. Sviridov, D. T.; Sviridova, R. K.; Smirnov, Yu. F. Optical Spectra of Ions of Transition Metals in Crystals; M. Nauka, 1976; p 256.
8. Sokolov V. I.; Stankevich I. V. Fullerenes – New Allotropic Forms of Carbon: Structure, Electron Composition and Chemical Properties; Success in Chemistry, 1993; vol 62 (5), pp 455–473.
9. Anderson, P. W. Magnetism; Acad. Press, 1963; vol 1, p 25.
10. Dirac, P. A.; Quantum Mechanics, Oxford University Press: London, 1935.
11. Clementi, E.; Raimondi, D. L. Atomic Screening Constants from S.C.F. Functions. *J. Chem. Phys.* **1963,** *38* (11), 2686–2689.
12. Clementi, E.; Raimondi, D. L. *J. Chem. Phys.* **1967,** *47* (4), 1300–1307.
13. Waber, J. T.; Cromer, D. T. *J. Chem. Phys.* **1965,** *42* (12), 4116–4123.
14. Blokhintsev, D. I. Basics of Quantum Mechanics; M. Vysshaya shkola, 1961; p 512.
15. Fischer, C. F. *Atomic Data* **1972,** *4*, 301–399.

CHAPTER 16

Polymer-Based Fibers: The Role of Solution Rheology in Fibers Production

ANDREEA IRINA BARZIC

Laboratory of Physical Chemistry of Polymers, "Petru Poni" Institute of Macromolecular Chemistry, Iasi, Romania

ABSTRACT

The production of polymer fibers is of great importance since it has implications in many areas, like life sciences, sensors, cosmetics, and other industries. The chapter depicts some introductive aspects of solution properties' role in attaining polymer fibers. The main categories of polymers used for this goal are enumerated. Also, the principal techniques reported to render polymer fibers are shortly classified and presented. The shear flow behavior implications in fiber formation (avoiding defects) are described based on the most recent reports on this subject. The rheological data are evaluated regarding polymer structure characteristics and interactions from the system. Also, the influence of working concentration domain, solvent features, and shear response of the polymer solution on the resulting fiber morphology is evaluated and the corresponding applications of polymer fibers are shortly mentioned.

16.1 INTRODUCTION

Shaping of polymer materials into films, fibers or porous architectures has drawn the interest of the research communities from around the world for

Tailored Functional Materials for Clean and Sustainable Development. Tamara Tatrishvili, PhD, Neha Kanwar Rawat, PhD, Ann Rose Abraham, PhD & A. K. Haghi, PhD (Eds.)

many years.[1] For fine processing, the polymer must be found in a fluid state and thereby a deep knowledge on the solution or melt properties is mandatory.[2,3] Much research has been devoted for elucidation of the behavior of the macromolecular compounds in solution.[4] The properties of such systems are affected by chemical composition, molecular weight (MW), solvent thermodynamic characteristics, specific interactions, temperature, pressure, and so on.[5–10] The conformation of polymer chains in the solvent environment and their arrangement is dictated by the working concentration domain. In dilute domain, the coils are surrounded by solvent molecules and do not have the chance to interact with each other. Further incorporation of polymer amounts in a good solvent is shortening the distances among the coils, and in this way they begin to interact with each other.[11] These changes are reflected in the physical properties of the solutions, namely in its viscosity. For less concentrated polymer solutions, viscometry is a suitable technique of analysis,[12,13] whereas for more concentrated systems it is recommendable to employ rheology as a characterization tool.[14–16]

A special attention is granted to the production of polymer filaments or fibers owing to their applicability in various domains, such as biomedical engineering,[17] photonic and light-emitting sources,[18] textile industry,[19] hollow fibrous membranes,[20] and shape-memory devices.[21] There is a variety of polymer structures that are proved to be suitable for attaining fibers. Their features (morphology, length, diameter, size uniformity, and mechanical resistance) are especially influenced by their compound chemical properties, the preparation technique, and the processing conditions.[22–29] The control of the fiber characteristics is considered paramount in many applications since it can conduct the improvement of the material's performance. Moreover, the reinforcement of the polymer with particular inorganic particles, followed by the production of fibers starting from the corresponding composite is offering additional properties to the system. In this way, the performance of the hybrid material is extended as a function of the properties of the inserted fillers in the matrix. Fibers based on polymer composites have found practical importance in several areas, and recently the scientists are aiming towards "greener" materials that contain natural components and have less negative impact on the environment.[30–32]

This chapter is focused mainly on polymer fibers fabricated from solution phase. Therefore, the manuscript begins by presenting certain basic aspects regarding polymer fibers, particularly which types of macromolecules lead to fibers and what are the principal methods employed in fiber production. Another aspect that is addressed is related to the role of the polymer solution properties in the fiber accomplishment. For this purpose, some relevant rheological studies are reviewed for some representative polymer systems. In addition to this, the fiber properties are presented in relation to their morphological aspects. The chapter briefly mentions the practical importancc of thc polymcr fibers in many fields.

16.2 TYPES OF POLYMERS USED IN FIBER PRODUCTION

In the field of polymer fibers, literature describes a considerable variety of polymer structures that are able to be processed as fibers. Mainly, given the origin of the polymer used in fiber production, it is possible to mention two large categories, namely:

- Natural macromolecular compounds: cellulose,[33] silk,[34] chitosan,[35] collagen,[36] and starch[37];
- Synthetic macromolecular compounds: polyimides,[38] polystyrene,[39] poly (methyl methacrylate),[40] polysulfone,[41] polyethylene,[42] polyurethane,[43] and polymer blends.[44–48]

Certain polymers are naturally structured under the form of fibers. For instance, the source of fibers in cellulose (including its derivatives), cotton, sisal, or flax, is represented by plants. In addition to this, silk is a protein yielded by the cocoons of silkworms. Another protein in fibrous structure is wool collected from particular animals, like sheep, goats, alpacas, and sometimes angora rabbits. On the other hand, there are synthetic polymers which can be processed as fibrillated products. Fibers form when the macromolecules become arranged in the same direction so that the chains may be packed together closely. The macromolecules do not require fully straight disposition to render fibers, but a higher level of straightness facilitates better packing and tougher fibers properties.

Aside from the chemical features of the polymer, the properties of the fibers are also depending on the physical properties (thermal behavior and rheological properties) and the method used for processing. In the following section of the chapter, the main techniques for fiber forming are described.

16.3 TECHNIQUES FOR POLYMERS FIBER PRODUCTION

Generally, processing of polymer fluids into fibers requires the use of either melts or solutions. Polymer micro/nano-fibers can be accomplished from natural or synthetic compounds by means of typical procedures, which could be largely imparted in:

- Top-down approaches: bulk material is transformed into fibers via special treatments;
- Bottom-up approaches: requires assembling the substance at the molecular level to attain the desired fibers.

The main techniques currently reported to be suitable for manufacturing of polymer micro/nano-fibers are illustrated in Figure 16.1.

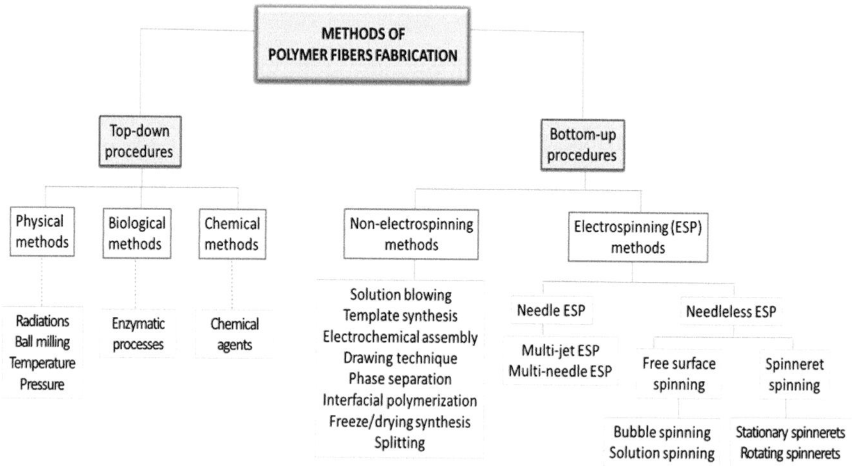

FIGURE 16.1 The schematic representation of the most utilized procedures of polymer micro/nano-fibers fabrication.

The top-down procedures are based on slicing or cutting of a bulk material by attrition or milling. Therefore, such methods involve the polymer treatment by utilizing one of the following routes:

- Physical[49,50] treatments: demand use of high energy radiations, ball milling or a combination of elevated temperature and high pressure, to inflict polymer melting, abrasion, evaporation, or condensation to generate fibers;
- Biological[51] treatments: enzymatic reactions might also determine formation of polymer fibers;

- Chemical[52–54] treatments: utilization of oxidization agents, cross-linking compounds or other types of substances is useful for chemical adaptation of polymers to render the targeted fibers.

The bottom-up procedures are generating the assembling of molecules through one of the following routes[55,56]:

- Non-electrospinning methods:
  - Solution blowing[57]: needs the use of two concurrent concentric fluid streams: a polymer mixed with a volatile solvent and a pressurized gas, which moves around the solution, generating fibers, which can be casted along the gas flow;
  - Template synthesis[58]: requires a metal oxide or polymer template having cylindrical pores of uniform diameter to make fibers within the pores of the template membrane;
  - Electrochemical assembly[59]: is limited to electrically conductive polymers (polythiophene, polypyrrole, and polyaniline) which are capable of organizing under electrical forces;
  - Drawing technique[60]: is divided in three stages: (1) rolling or stacking polymer films into a multilayer preform system, (2) consolidation by subjecting the system to temperatures beyond the glass transition temperature (Tg), and (3) drawing the preform into fibers using a drawing tower (at $T > Tg$) and the fibers keep the structured profile of a preform, having complex microstructures;
  - Phase separation[61,62]: can be triggered by temperature or non-solvent, and might determine condensation and render polymer fibers;
  - Interfacial polymerization[63]: is step-growth polymerization, where the reaction takes place at the separation zone between the immiscible phases. This renders a material which is constrained to the interface and gains distinct topologies including fibers;
  - Freeze/drying synthesis[64]: process is mainly employed to remove residual solvent from a material to produce fibers;
  - Splitting[65]: begins with microcracks in the polymer and might spread along the macrofiber producing new ones;
- Electrospinning (ESP) methods[55,56,66]: represent a set of fabrication procedures for achieving polymer fibers at reduced costs and fine control of the fibers features. In the case of ESP, the polymer is brought in a fluid state and subsequently inserted in the syringe tube for ESP.

The positive part of the DC power supply is linked to the hollow needle, while the negative one is jointed with the metal collector. At a certain level of the electric field, the repulsive electrostatic interactions are exceeding the surface tension, so that the charged jet of polymer is brought out from the Taylor cone. The discharged sample jet is subjected to elongation process, which enables the material in the jet to be longer and to diminish the fiber diameter. The solvent is released from the polymer during this procedure and the evaporation is affected by the tip-collector distance and the working temperature. A constant environmental regime is beneficial for acquiring good quality polymer fibers. They can be largely imparted in the next subcategories[66]:

- Needle ESP: this sort of approach includes (1) multijets ESP that uses several jets which are employed for fiber fabrication and (2) multi-needle ESP implies employment of a number of needles as spinnerets, which comprise one or different sorts of polymer solutions. Owing to multijets, uniform web of fibers might be hard to attain given the repletion action of the jets;
- Needleless ESP: is an innovative alternative of ESP which makes fibers on a substantial scale. Several planes are modeled concomitantly from the needleless fiber generator (spinneret) excluding the implications of capillary effect that is often linked to the needle ESP. Two main sorts of approaches are remarked here: (1) free-surface spinning (like bubble spinning and solution spinning) and (2) spinneret spinning (like stationary spinnerets or rotating spinnerets).

16.4 INVOLVEMENT OF POLYMER SOLUTION PROPERTIES IN FIBER PRODUCTION

The ESP technique can lead to products with tunable features via a number of variables. A major importance should be ascribed to the flow behavior of the polymer fluid (solution or melt) used for fiber fabrication. The specific factors that dictate the solution properties might be divided into the following categories:

- The sample characteristics: polymer molecular weight, electrical features (dipole moment and dielectric constant), deformation response of the sample (viscosity at variable shear rates and visco-elasticity), and surface tension;

- Flow rate, imposed electric field intensity, ESP device design (needle type, needle-collector distance, etc.), and ambient conditions, related to temperature and humidity.[67]

Given the exposed arguments, the rheological analysis of the solution behavior of the polymer under precise conditions is mandatory. Viscosity is a relevant parameter that depicts the resistance to deformation at an imposed rate. It reveals information on the frictional forces that are appearing among the adjacent layers within the liquid sample, which are in relative movement. The flow curves attained from rheological experiments are illustrating the manner in which the viscosity of the polymer fluid depends on the shear rate, temperature, and even pressure. Aside from the molecular weight of the polymer, the concentration of the solution is also affecting the shape of the flow curves. Literature on this topic[38,68–71] presents some detailed studies regarding the correlation between the polymer solution rheological behavior and the morphology of the produced fibers by electrospinning. For instance, it is demonstrated that in order to accomplish fibers with the desired morphology, it is essential to work in a typical concentration regime. The latter is ranging from a polymer structure to another as a function of its chemical and physical features. The viscosity of polymer is enhanced as the dimensions of the macromolecules and their amount in the system is augmented.[72] A general sorting of the polymer solutions, which accounts the polymer concentration (noted here c), is based on the following four types[73]:

- The dilute domain;
- The unentangled semidilute domain;
- The entangled semidilute domain;
- The concentrated domain.

At a concentration in the dilute range, the macromolecular coils are very far from each other, so they cannot reach contact. As the concentration becomes higher, the distances among the macromolecular coils are being reduced and they can touch their neighboring ones. At a certain point the chains might overlap (c^* here is denoted as being the critical chain overlap concentration). Further increase in the concentration ($c > c^*$) designates the semidilute regime that can be imparted as a function of the possibility of chains to entangle themselves. The limit among the unentangled and the entangled semidilute zones is known as the critical entanglement concentration (c_e). The c_e describes the point where the macromolecular coil interpenetration starts. The explanation why coils entanglement is not happening, until c_e is reached compared to c^*, resides in the fact that a considerable quantity of chain overlapping is demanded by the macromolecules to form entanglements. The unentangled semidilute zone is ascertained to be $c^* < c < c_e$, hence in this interval the chains are not fully superimposing to generate entanglements. The entangled semidilute regime is noted at concentrations beyond c_e and in this range the polymer chains entangle.[71,74] An augmentation of the molecular weight indicates that the mean chain length per molecule is larger, which provides a bigger amount of entanglements per volume. The aforementioned aspects are graphically sketched in Figure 16.2, where the slope of the specific viscosity (denoted η_{sp}) against concentration ranges as indicated in literature[11] for the linear polymers.

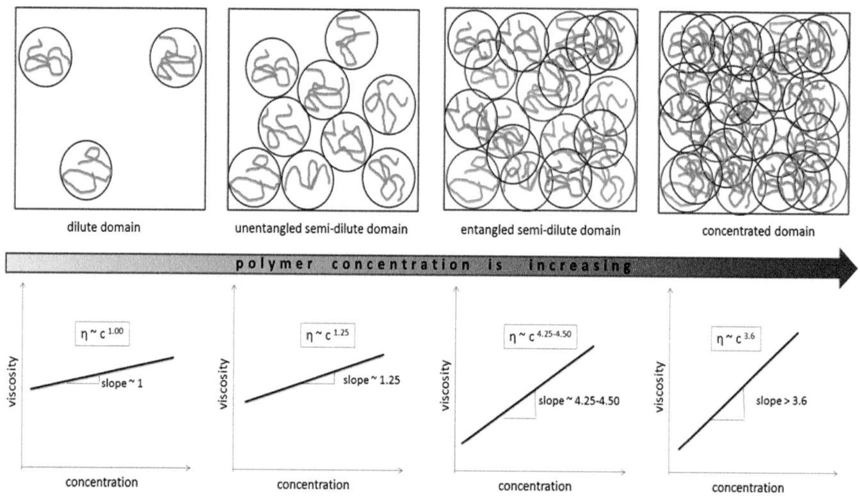

FIGURE 16.2 Schematic illustration of the four types of concentration domains and the variations in viscosity with polymer concentration in solution for each corresponding regime.

Colby and coworkers[75] have evidenced the implications of the concentration together with thermodynamic interactions on the balance between the viscous and elastic components of polybutadiene solutions in a theta solvent and a good solvent. In dilute domain the viscosity can be properly described by the Huggins formula, but intrinsic viscosities and Huggins coefficients are differing in the used solvents. In the entangled semidilute interval and the concentrated one, the free volume adjusted viscosity does not differ as a function of the solvent environment, and could be depicted by a power law of concentration with exponent of about 4.7 in semidilute range and becomes 3.6 at larger concentrations. The storage modulus is not much influenced by the solvent nature, while variation of the entanglement molecular weight and characteristic molecular weight for the first appearance of the chains interpenetration is slightly distinct. Based on this study, a couple of years later, Colby et al.[71] attempted to elucidate the rheology effects on the electrospun fiber occurrence. To this scope, they used polymers with linear or branched structures derived from polyethylene terephthalate. The entanglement concentration was determined for all systems by monitoring the viscosity as a function of polymer concentration. The changes in the graph slope were useful for delimiting the regimes where the macromolecular coils are unentangled and interpenetrated. The concentration domains for the examined polymers were evidenced by means of viscosity tests performed with a Rheometer. The experiments revealed a Newtonian plateau for the solutions in the imposed shear range, but it was possible to remark that pseudoplasticity was seen at larger shear rates. The polymer contribution to the viscosity in the absence of shearing was emphasized by means of the specific viscosity. The fluctuations of the η_{sp} values, as the concentration increases, were evaluated for distinguishing among the onset of the principal four concentration regimes. The theoretical predictions[11] are supported by the data extracted from rheological tests for linear polymers, while for the branched structures a poor dependence of η_{sp} on concentration was noticed in the semidilute entangled interval. The solutions of the linear and branched macromolecular compounds were subjected to ESP. Variable concentrations were employed to elucidate the implications of concentration domain and molecular topology on the resulted fiber morphology. The viscosity (at null shear rate) was extracted from the flow curves and correlated with the average fiber diameter and morphology. For the structures with molecular weights exceeding the molecular weight of chain interpenetration, it was found that c_e represents the minimum concentration necessary for ESP of fibers with beaded defects. Furthermore, it was showed that the achievement of uniform, bead-free fibers by ESP imposed to use concentrations of at least 2–2.5 times c_e. When the concentration was reported to c_e, the impact of chain size and topology on the ESP process was less relevant, and the fiber diameter was generally proportional with the normalized concentration having an exponent of 2.6. Similar observations were reported by Chisca et al.[38] for a thermostable polyimide. Rheological studies indicated a shear thinning at low shear rates followed by a Newtonian plateau. The slope of the pseudoplastic range decreased upon addition of a higher amount of polymer in the solution. This interval of flow was depicted by the power law, from which the flow index and consistency parameters were estimated. It was remarked that the consistency index changes suddenly after increasing the concentration (around 15–20%) due to chains interpenetration in the solvent environment. The viscoelastic behavior changed with concentration showing an enhancement of the storage modulus, since the solid-like character of the sample is favored at larger concentrations. The relative percent of energy loss in the polyimide fluid during shearing was monitored by tan δ values, entailing more interactions among the macromolecules that diminish the space between them, making the system more elastic (meaning lower tan δ values). Thus, the appearance of chain entanglements avoided the loss of energy and kept the mechanical features in solution adequate for fiber fabrication. Examination of the polyimide fiber morphology[38] in the working concentration intervals was proved to follow a general mechanism of fiber formation, which could be assumed universal for a variety of polymers.[76–79] According to literature,[41,68,70,73,76–79] an augmentation of the polymer amount in the system generates a specific progression in regard to the produced polymer fiber morphology, namely:

- Irregular/regular beads only;
- Beads with incipient fibers;
- Beaded fibers;
- Fibers without defects.

Based on the aforementioned context, the evolution of the polymer fibers morphology as a function of the employed concentration regime is schematically drawn in Figure 16.3.

In the case of the reported aromatic polyimide solutions,[38] the scanning electron microscopy recordings proved that fibers lacking defects are accomplished at high concentrations (i.e., $c \geq 2\ c_e$), as reported for other polymers.[76–79]

McKee et al.[80] have investigated the role of self-complementary hydrogen bonding in poly(alkyl methacrylate)s/solvent systems in fiber formation by ESP method. The analyzed structures exhibited side chains of carboxylic acid or self-complementary multiple (SCM) hydrogen bonding. The cosolvent amount was ranged to tune the dielectric constant of the ESP solvent and in this way the intermolecular interactions could be screened to emphasize the action of hydrogen bonding on fiber morphology. For homopolymers, the fiber diameter fitted the theoretical predictions, while for the copolymers containing sequences of methacrylic acid, the size of the fibers was lower than foreseen. This aspect might be the result of dissociation of the carboxylic acid group that enhanced the fluid conductivity. The samples displaying

SCM hydrogen bonding led to considerable associations that diminish the dielectric constant. This was conducted to higher viscosity and smaller c_e magnitude. Furthermore, strong hydrogen bonding between the structures with SCM groups in solvents of reduced polarity enhanced the apparent MW and determined bigger electrospun fibers in comparison to the theoretical predictions.

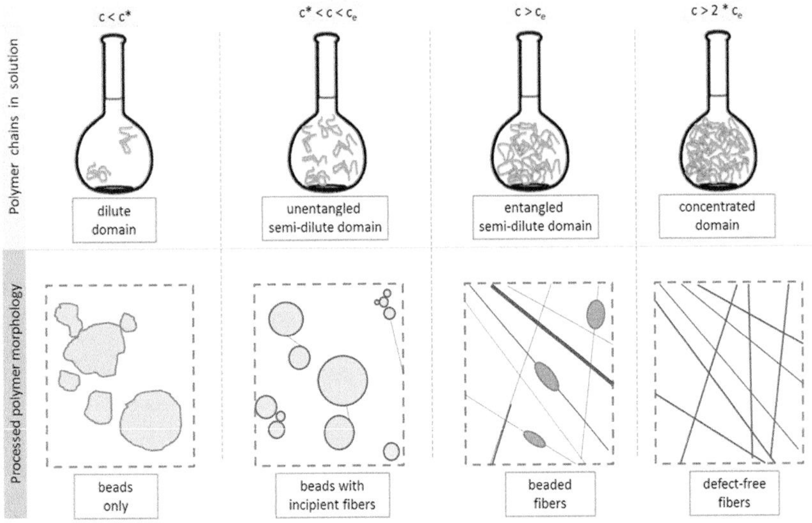

FIGURE 16.3 The graphical representation of the role of the solution concentration regime in the morphology of the processed polymer structure.

Shenoy and collaborators[69] have attempted to develop a semiempirical approach to elucidate the transition from electrospraying (ESPR) to ESP by accounting on the good solvent and nonspecific polymer–polymer interaction limit. Macromolecular coils overlapping might be responsible for the physical interlocking of polymer chains and is necessary to clarify how many entanglements could be enough for stabilizing the fiber formation. In a polymer solution, the amount of chain interpenetrations is influenced by concentration (or volume fraction) and MW. The entanglement number in a solution could be estimated by the ratio of the MW of the polymer and its solution entanglement molecular weight. Therefore, for moderately concentrated or concentrated solutions, the entanglement number in solution is described by eq 16.1[69]:

$$(n_e)_{soln} = \frac{MW}{(M_e)_{soln}} = \frac{(MW\ \phi_p)}{M_e} \tag{16.1}$$

where $(n_e)_{soln}$ is the entanglement number in a solution, ϕ_p is the volume fraction of polymer, and M_e is the entanglement molecular weight.

The zero shear viscosity dependence on MW involves an upturn (MW^1 to $MW^{3.4}$) that happens when $(n_e)_{soln} \sim 2$. So, the number of entanglements per chain might be provided for $(n_e)_{soln} \sim 1$ because a single entanglement implies two chains. As a result, $(n_e)_{soln} = 2$, two entanglements are formed, but only one interpenetration per chain. Therefore, the evolution of the fiber morphology during ESPR and ESP could be explained. These processes are similar but the main difference resides in the fact that ESPR produces micro-beads, while ESP leads to fibers. The main reason is that the macromolecular coils' overlap is minimal for ESPR solutions.[69]

Polymer entanglements inside the beads confine their subdivision. A suitable mechanism for this might consider that upon solvent evaporation, the next competing effects are taking place[69]:

- Polymer concentration is enhanced and interpenetration is commencing impeding the bead to additionally divide;
- Surface charge is augmented exceeding the bead surface tension, offering a driving force that favors subdivision;
- Heat transfer during the fast removal of the solvent could also hinder the droplet dimension, namely as the bead is cooled, solvent removal is sluggish, skin occurrence stabilizes the beads so that the surface charge is stopped form increasing.

The concentrations regions, where chains are able to overlap a mixture of fibers and beads, can be noted since insufficient coil interpenetrations occur and the jet is not entirely stable. It can be assumed that at larger concentrations (beyond c*), the enhanced coil entanglements could, for the moment, ensure the stabilization of the ESP jet by preventing its breakup. Also, "spinnable" fluids must present elastic properties. A supplementary consequence of solvent removal resides in the cooling of the jet, which ease the skin appearance and hence the fiber stabilization. Considering these aspects, Shenoy et al.[69] showed that incomplete fiber occurrence is well foreseen for several polymer/solvent systems at one entanglement per chain. Fully stable fibers are fabricated at minimum (or higher) 2.5 entanglements per chain.

Uyar and coworkers[39] reported the effect generated by the solvent conductivity on the morphological aspects of polystyrene (PS) fibers. They have utilized various grades of dimethylformamide (DMF) as a solvent taken from several suppliers and noted that it displays different conductivities. Accordingly, the PS solutions in the purchased DMF samples present distinct conductivities at identical polymer concentrations. Even the rheological curves do not show significant changes in the viscosity of these systems; the changes in conductivity determine the modification of PS fibers morphology from beaded to defect-free uniform fibers during processing in fixed conditions. They attained bead-free PS fibers by using PS solutions in the interval of 20–30% (w/v) as a function of the solvent grade. In other words, it was remarked that polymer solutions having greater conductivity rendered bead-free fibers at smaller PS concentrations. This proves that the conductivity of the employed solvent is a paramount factor in making bead-free uniform polymer fibers.

Härdelin et al.[73] have prepared fibers based on solutions made from a natural polymer (acid-treated cellulose) and ionic liquids. All systems are characterized by a shear thinning behavior, which is accentuated by increasing polymer concentration. The presence of the clearly defined zero shear viscosity (η_0) and since the Cox–Merz rule is applicable, while the viscosity scales with a concentration in agreement to the theory for entangled semidilute solutions, led to the conclusion that the cellulose systems are dissolved on the molecular level. In this case, the macromolecules are overlapping and interactions among the coils are ruled by topological constraints. In the working concentration range, the relation of Onogi et al.[81] was used:

$$\eta_0 = Kc^{\alpha} MW^{\beta} \tag{16.2}$$

where K, α, and β are specific constants to the studied system.

This power law correlation is suitable for a number of concentrated polymer solutions implying flexible chains.

The pseudoplastic zones commence at higher shear rates as the acid treatment time is prolonged and MW is reduced. At considerably small shear rates, a deviation in the shape of the graphs is noted that might be ascribed to the fact that the experiments were recorded at shear rates that are below the device reliability. The magnitude of η_0 at distinct cellulose concentrations reveals that all studied systems could be integrated in the entangled semidilute domain (sufficient overlap for chain interpenetration). Interestingly, in the studied regimes, it was found that neither cellulose amount nor MW demonstrated to be decisive for if a solution can be electrospun into fibers or not. It is rather the viscosity of the solution that is decisive for ESP performance, even if the polymer fluid contains considerable entanglements.

Wang et al..[53] used crosslinkers to improve ESP properties of low molecular weight polymers. The proposed approach resides in formation of a supramolecular structure through hydrogen bonding or coordination interactions in solution. The addition to poly(4-vinylpyridine) (P4VP) in the system containing DMF, 4,4′-biphenol (BiOH), and nitromethane, a poor solvent to the polymer, enhances the interactions among BiOH and P4VP. The presence of CH_3NO_4 reduces the concentrations required to attain fibers and facilitates formation of fibers lacking beads. The rheological analyses prove the occurrence of hydrogen bonding that renders a physical network. BiOH was eliminated via sublimation and deterioration of the fibers was not observed. The crosslinking efficiency was enhanced by substituting BiOH component with 4-hydroxy-4′-biphenylcarboxylic acid (HBCA), or by inserting 1% $ZnCl_2$ (relative to pyridine). In the first case, HBCA develops stronger interactions with P4VP, whereas using $ZnCl_2$ determines metal coordination interactions with the polymer. Interestingly, addition of HBCA favors ESP of very low molecular weight P4VP (far from M_e) and the obtained fibers have no beads. The work of Wang et al.[53] reveals that supramolecular crosslinking is a successful method to achieve fibers (having nanosized diameter) of polymers with known limited ESP.

Kong and coworkers[70] also reported a methodology of making fibers from starch, which relates rheological behavior of the polymer dispersions and ESP. The values of η_0 were extrapolated from flow curves and used to estimate the specific viscosity for establishing the boundary between the concentration regimes containing or lacking interpenetrated chains. The systems of amylose starch demanded 1.2–2.7 times the entanglement amount for proper ESP. Aside from the polymer concentration, its macromolecular shape and shear viscosity were essential for examining the ESP performance. It was demonstrated that the amount of amylose in the system influences the interpenetration concentrations, so this leads to distinct flow behavior and implicit ESP. At lower percent of amylose, the prepared dispersions led to very poor fibers.

Poudel and collaborators[79] have tested the ESP characteristics of some biocompatible, biodegradable, and natural polymers. For instance, they accomplished fibers (with submicron dimensions) from multifunctional blend. This was obtained based on an exopolysaccharide and hydroxypropyl-β-cyclodextrin (CD), having good hosting ability. The approach involved utilization of a nontoxic solvent, namely water. The exopolysaccharide solutions at concentrations under 5% (w/v) present properties similar to viscoelastic liquids. At this point, the shear moduli are overlapping indicating solid-like features. The bigger gap $G' > G''$ recorded at small frequencies could be indicative that the natural polymer chains are entangling. In any case, there is still a visible dependence of loss modulus on the frequency regarding storage, one revealing a shear response affected by the viscous behavior at 8–15% (w/v). At concentrations in the interval of 20–25% (w/v), the gap between rheological moduli is lowered at lower frequencies, reflecting the occurrence of a weak entangled network. Incorporation of CD strengthened the viscoelastic features of the exopolysaccharide determining larger shear moduli in solutions containing 10–20% (w/v) of natural polymer. The effect is highly predominant with the insertion of 30% w/v CD. ESP experiments showed that the pure polymer demanded three times c_e to make bead-free fibers, while the presence of CD reduced the required amount of entanglements for proper ESP, since it acts as a crosslinker.

Owing to their versatility and advantages, polymers with fibrous morphology have brought advances in numerous applicative areas,[18,30,37,42,45,56,66,82,83] such as:

- Medicine: drug delivery systems, scaffolds employed for tissue regeneration, membranes for filtration of biological fluids, and wound dressing;
- Cosmetics: mask for skin care (cleansing, therapy, and healing);
- Sensors: biochemical devices, thermal devices, piezoelectric devices, and fluorescence-based devices;
- Environment: water depollution and gas filtration;
- Electronics: electrostatic dissipation, liquid crystal displays, and electromagnetic interference shielding;
- Energy: photovoltaic panels;
- Catalysis: functional catalysts;
- Military: clothing having antibiochemical properties;
- Airspace: ultralightweight components for spacecraft.

16.5 CONCLUSIONS

The research subdomain of polymer fibers is of great interest for the current technologies since it introduces considerable advantages, such as lightweight, large surface area, mechanical resistance and so on. In order to fabricate fibers with the targeted morphology, it is important to carefully select a polymer with special chemical features and high molecular weight. The conformational characteristics and the functional groups could be beneficial for establishing specific interactions in the system that might stabilize the formed fibers. Also, the solution properties are indicative of the peculiarities of the polymer used for ESP. In particular, rheological analyses allow extraction of valuable data regarding the viscosity and viscoelasticity of the investigated fluids. Viscosity in the absence of shearing is utilized for evaluation of the specific viscosity whose dependence on concentration helps to elucidate the limit of the regime where chain interpenetration might occur or not. Oscillatory experiments provide analysis of the balance between the viscous and elastic components of the polymer system that could reveal the proper composition for appearance of entangled network, which is useful for defect-free fiber production. Therefore, it seems that ESP experiments must be correlated with rheological studies to determine the optimal processing window that leads to the desired products. Polymer fibers are involved in several fields of applications, such as biomedicine, electronics, solar cells, environment decontamination, aircraft components, military clothing, catalysis, sensors, and cosmetics.

ACKNOWLEDGMENT

This work is dedicated in the memory of Dr. Silvia Ioan, a remarkable senior scientist from "Petru Poni" Institute of Macromolecular Chemistry of Romanian Academy, who devoted her life with admirable passion and dedication towards mentoring young scientists in the field of polymer solutions.

KEYWORDS

- **polymers**
- **fibers**
- **rheology**
- **fabrication techniques**
- **applications**

REFERENCES

1. Agassant, J. F.; Avenas, P.; Carreau, P. J.; Vergnes, B.; Vincent, M. *Polymer Processing. Principles and Modelling*, 2nd ed.; Hanser: USA, 2017.

2. Teraoka, I. *Polymer Solutions: An Introduction to Physical Properties*, Wiley, USA, 2002.
3. Cogswell, F. N. *Polymer Melt Rheology: A Guide for Industrial Practice*, Woodhead Publishing Limited: UK, 2003.
4. Fujita, H. *Polymer Solutions*, Elsevier: USA, 2012.
5. Kulicke, W. D.; Clasen, C. *Viscosimetry of Polymers and Polyelectrolytes*, Springer: Germany, 2004.
6. Simionescu, C. I.; Simionescu, B. C.; Ioan, S. Solution Properties of Ultrahigh Molecular Weight Polymers, 1 Methyl Methacrylate-styrene 50:50 Random Copolymers. *Makromol. Chem.* **1983**, *184*, 829–835.
7. Simionescu, B. C.; Ioan, C.; Ioan, S.; Simionescu, C. I. Specific Interactions in Dilute Polymer Solutions. *Marcromol. Symp.* **1995**, *98*, 1045–I068.
8. Grigorescu, G.; Ioan, S.; Simionescu, B. C. Short- and Long-range Interaction Parameters for Polydimethylsiloxanes with Different Terminal Groups. *Marcromol. Symp.* **1999**, *138*, 197–202.
9. Ioan, S.; Aberle, C.; Grigorescu, G.; Stanciu, A.; Aberle, T. Compositional Parameters of Segmented Poly (Estersiloxane)-urethanes and Limitations in the Study of Dilute Solution Properties. *Int. J. Polym. Anal. Charact.* **2002**, *7*, 210–226.
10. Cosutchi, A. I.; Hulubei, C.; Ioan, S. Rheological Study of Some Epiclon-based Polyimides. *J. Macromol. Sci. B* **2007**, *46*, 1003–1012.
11. De Gennes, P. G. *Scaling Concepts in Polymer Physics*, Cornell University Press, Ithaca, 1979.
12. Albu, R. M.; Avram, E.; Musteata, V. E. Homocianu, M.; Ioan, S. Opto-electrical Properties of Some Quaternized Polysulfones. *High Perform. Polym.* **2011**, *23*, 85–96.
13. Ioan, S.; Lupu, M.; Macocinschi, D. Viscometric Study of Poly (Ester Urethane) Solutions: 2. Influence of Hard-segment Composition on the Unperturbed Dimensions. *High Perform. Polym.* **2003**, *15*, 319–328.
14. Barzic, A. I.; Albu, R. M.; Gradinaru L. M.; Buruiana, L. I. New Insights on Solvent Implications in Flow Behavior and Interfacial Interactions of Hydroxypropylmethyl Cellulose with Cells/Bacteria. *e-Polymers* **2018**, *18*, 135–142.
15. Barzic, A. I.; Soroceanu, M.; Albu, R. M.; Ioanid, E. G.; Sacarescu, L.; Harabagiu, V. Correlation between Shear-flow Rheology and Solution Spreading during Spin Coating of Polysilane Solutions. *Macromol. Res.* **2019**, *27*, 1210–1220.
16. Albu, R. M.; Hulubei, C.; Stoica, I.; Barzic, A. I. Semi-alicyclic Polyimides as Potential Membrane Oxygenators: Rheological Implications on Film Processing, Morphology and Blood Compatibility, *eXPRESS Polym. Lett.* **2019**, *13*, 349–364.
17. Nicolae, A.; Grumezescu, A. M. Polymer Fibers in Biomedical Engineering. In *Materials for Biomedical Engineering. Biopolymer Fibers*, Grumezescu, V.; Grumezescu, A. M., Eds., Elsevier: Amsterdam, 2019, 1–20.
18. Asmatulu, R.; Khan, W. S. Electrospun Nanofibers for Photonics and Electronics Applications. In *Synthesis and Applications of Electrospun Nanofibers*, Asmatulu, R.; Khan, W. S.; Eds.; Elsevier: Amsterdam, 2019, 239–256.
19. Taşdemir, M.; Akalin, M.; Koçak, D.; Usta, İ.; Merdan, N. Investigation of Properties of Polymer/Textile Fiber Composites. *Int. J. Polym. Mater.* **2010**, *59*, 200–214.
20. Pagliero, M.; Khayet, M.; García-Payo, C.; García-Fernández, L. Hollow Fibre Polymeric Membranes for Desalination by Membrane Distillation Technology: A

Review of Different Morphological Structures and Key Strategic Improvements. *Desalination* **2021**, *516*, 115235.
21. Xia, Y.; Zhang, F.; Wang, L.; Liu, Y.; Leng, J. Electrospun Shape-memory Polymer Fibers and their Applications. In *Electrospun Polymers and Composites. Ultrafine Materials, High Performance Fibres and Wearables*, Dong, Y.; Baji, A.; Ramakrishna, S.; Eds., Woodhead Publishing Series in Composites Science and Engineering: USA, 2021, 567–596.
22. Korycka, P.; Mirek, A.; Kramek-Romanowska, K.; Grzeczkowicz, M.; Lewińska, D. Effect of Electrospinning Process Variables on the Size of Polymer Fibers and Bead-on-string Structures Established with a 23 Factorial Design. *Beilstein J. Nanotechnol.* 2018, *9*, 2466–2478.
23. Liu, H.; Gough, C. R.; Deng, Q.; Gu, Z.; Wang, F.; Hu, X. Recent Advances in Electrospun Sustainable Composites for Biomedical, Environmental, Energy, and Packaging Applications. *Int. J. Molec. Sci.* **2020**, *21*, 4019.
24. Fu, S.; Lauke, B.; Mai, Y.; Major Factors Affecting the Performance of Short Fibre-reinforced Polymers. In *Science and Engineering of Short Fibre-reinforced Polymer Composites*, Elsevier: USA, 2019, 33–66.
25. Ziyadi, H.; Baghali, M.; Bagherianfar, M.; Mehrali, F.; Faridi-Majidi, R. An Investigation of Factors Affecting the Electrospinning of Poly (Vinyl Alcohol)/Kefiran Composite Nanofibers. *Adv. Compos. Hybrid Mater.* **2021**, *4*, 768–779.
26. El-hadi, A. M.; Al-Jabri, F. Y. Influence of Electrospinning Parameters on Fiber Diameter and Mechanical Properties of Poly (3-hydroxybutyrate) (PHB) and Polyanilines (PANI) Blends. *Polymers* **2016**, *8*, 97.
27. Pant, H. R.; Nam, K.-T.; Oh, H.-J.; Panthi, G.; Kim, H.-D.; Kim, B.; Kim, H. Y. Effect of Polymer Molecular Weight on the Fiber Morphology of Electrospun Mats. *J. Coll. Interf. Sci.* **2011**, *364*, 107–111.
28. Afifi, A. M.; Yamane, H.; Kimura, Y. Effect of Polymer Molecular Weight on the Electrospinning of Polylactides in Entangled and Aligned Fiber Forms. *Fiber* **2010**, *66*, 35–42.
29. Ko, J.; Jun, S.; Lee, J. K.; Lee, P. C.; Jun, M. B. G. Effects of Molecular Weight and Temperature on Fiber Diameter of Poly (ε-caprolactone) Melt Electrospun Fiber. *J. Korean Soc. Manuf. Technol. Eng.* **2015**, *24*, 160–165.
30. Mohammed, L.; Ansari, M. N. M.; Pua, G.; Jawaid, M.; Islam, M. S. A Review on Natural Fiber Reinforced Polymer Composite and Its Applications. *Int. J. Polym. Sci.* 2015, *2015*, 1–15.
31. Das, P. P. Chaudhary, V. Moving Towards the Era of Bio-fibre based Polymer Composites. *Cleaner Eng. Technol.* **2021**, *4*, 100182.
32. More, A. P. Flax Fiber–based Polymer Composites: A Review. *Adv. Compos. Hybrid Mater.* **2022**, *5*, 1–20.
33. Zhang, Y.; Zhang, C.; Wang, Y. Recent Progress in Cellulose-based Electrospun Nanofibers as Multifunctional Materials. *Nanoscale Adv.* **2021**, *3*, 6040–6047.
34. Mohammadzadehmoghadam, S.; Dong, Y. Electrospinning of Silk Fibroin-based Nanofibers and their Applications in Tissue Engineering. In *Electrospun Polymers and Composites. Ultrafine Materials, High Performance Fibres and Wearables*, Dong, Y.; Baji, A.; Ramakrishna, S.; Eds., Woodhead Publishing Series in Composites Science and Engineering: USA, 2021, 111–146.

35. Homayoni, H.; Ravandi, S. A. H.; Valizadeh, M. Electrospinning of Chitosan Nanofibers: Processing Optimization. *Carbohydr. Polym.* **2009**, *77*, 656–661.
36. Lu, W. P.; Guo, Y. Electrospinning of Collagen and its Derivatives for Biomedical Applications. In *Novel Aspects of Nanofibers*, Lin, T.; Ed.; InTech: Croatia, 2018, 39–65.
37. Hemamalini, T.; Giri Dev, V. R. Comprehensive Review on Electrospinning of Starch Polymer for Biomedical Applications. *Int. J. Biol. Macromol.* **2018**, *106*, 712–718.
38. Chisca, S.; Barzic, A. I.; Sava, I.; Olaru, N.; Bruma, M. Morphological and Rheological Insights on Polyimide Chain Entanglements for Electrospinning Produced Fibers. *J. Phys. Chem. B.* **2012**, *116*, 9082–9088.
39. Uyar, T.; Besenbacher, F.; Electrospinning of Uniform Polystyrene Fibers: The Effect of Solvent Conductivity. *Polymer* **2008**, *49*, 5336–5343.
40. Mohammad Khanlou, H.; Chin Ang, B.; Talebian, S.; Muhammad Afifi, A.; Andriyana, A.; Electrospinning of Polymethyl Methacrylate Nanofibers: Optimization of Processing Parameters using the Taguchi Design of Experiments. *Text. Res. J.* **2014**, *85*, 356–368.
41. Rasouli, M.; Pirsalami, S.; Zebarjad, S. M.; Optimizing the Electrospinning Conditions of Polysulfone Membranes for Water Microfiltration Applications. *Polym. Int.* **2019**, *68*, 1610–1617.
42. Ma, J.; Zhang, Q.; Mayo, A.; Ni, Z.; Yi, H.; Chen, Y.; et al. Thermal Conductivity of Electrospun Polyethylene Nanofibers. *Nanoscale* **2015**, *7*, 16899–16908.
43. Akduman, C.; Akçakoca Kumbasar, E. P. Electrospun Polyurethane Nanofibers. In *Aspects of Polyurethanes*, Yilmaz, F.; Ed., Intech: Croatia, 2017, 17–52.
44. Yoon, J. W.; Park, Y.; Kim, J.; Park, C. H. Multi-jet Electrospinning of Polystyrene/Polyamide 6 Blend: Thermal and Mechanical Properties. *Fash. Text.* **2017**, *4*, 9.
45. Tipduangta, P.; Belton, P.; Fábián, L.; Wang, L. Y.; Tang, H. Eddleston, M.; Qi, S.; Electrospun Polymer Blend Nanofibers for Tunable Drug Delivery: The Role of Transformative Phase Separation on Controlling the Release Rate. *Mol. Pharm.* **2015**, *13*, 25–39.
46. Kurusu, R. S.; Demarquette, N. R. Surface Properties Evolution in Electrospun Polymer Blends by Segregation of Hydrophilic or Amphiphilic Molecules. *Eur. Polym. J.* **2017**, *89*, 129–137.
47. Martinova, L.; Lubasova, D.; Reasons for using Polymer Blends in the Electrospinning Process. *AIP Conf. Proc.* **2012**, *1502*, 115.
48. Hanumantharao, S. N.; Rao, S. Multi-functional Electrospun Nanofibers from Polymer Blends for Scaffold Tissue Engineering. *Fibers* **2019**, *7*, 66.
49. Suzuki, A.; Sasaki, C.; Asada, C.; Nakamura, Y. Production of Cellulose Nanofibers from Aspen and Bode Chopsticks using a High Temperature and High Pressure Steam Treatment Combined with Milling. *Carbohydr. Polym.* **2018**, *194*, 303–310.
50. Piras, C. C.; Fernández-Prieto, S.; De Borggraeve, W. M. Ball Milling: A Green Technology for the Preparation and Functionalisation of Nanocellulose Derivatives. *Nanoscale Adv.* **2019**, *1*, 937–947.
51. Bondancia, T. J.; Corrêa, L. J.; Cruz, A. J. G.; Badino, A. C.; Mattoso, L. H. C.; Marconcini, J. M.; Farinas, C. S. Enzymatic Production of Cellulose Nanofibers and Sugars in a Stirred-tank Reactor: Determination of Impeller Speed, Power Consumption, and Rheological Behavior. *Cellulose* **2018**, *25*, 4499–4511.
52. Khiari, R.; Rol, F.; Brochier Salon, M. C.; Bras, J.; Belgacem, M. N. Efficiency of Cellulose Carbonates to Produce Cellulose Nanofibers. *ACS Sustain. Chem. Eng.* **2019**, *7*, 8155–8167.

53. Wang, X.; Pellerin, C.; Bazuin, C. G. Enhancing the Electrospinnability of Low Molecular Weight Polymers using Small Effective Cross-linkers. *Macromolecules* **2016**, *49*, 891–899.
54. Isogai, A.; Saito, T.; Fukuzumi, H. TEMPO-oxidized Cellulose Nanofibers. *Nanoscale* **2011**, *3*, 71–85.
55. Gugulothu, D.; Barhoum, A.; Nerella, R.; Ajmer, R.; Bechelany, M. Fabrication of Nanofibers: Electrospinning and Non-electrospinning Techniques. In *Handbook of Nanofibers*, Barhoum A.; Bechelany, M.; Hamdy Makhlouf, A. S.; Eds., Springer: Switzerland, 2019, 45–78.
56. Dos Santos, D. M.; Correa, D. S.; Medeiros, E. S.; Oliveira, J. E.; Mattoso, L. H. C.; Advances in Functional Polymer Nanofibers: From Spinning Fabrication Techniques to Recent Biomedical Applications. *ACS Appl. Mater. Interf.* **2020**, *12*, 45673–45701.
57. Medeiros, E. S.; Glenn, G. M.; Klamczynski, A. P.; Orts, W. J.; Mattoso, L. H. C. Solution Blow Spinning: A New Method to Produce Micro- and Nanofibers from Polymer Solutions. *J. Appl. Polym. Sci.* **2009**, *113*, 2322–2330.
58. Meyer, B.; Croce, F. Materials | Nanofibers. In *Encyclopedia of Electrochemical Power Sources*, Garche, J.; Ed., Elsevier: USA, 2009, 607–612.
59. Li, S.; Macosko, C. W.; White, H. S. Electrochemical Processing of Conducting Polymer Fibers. *Science* **1993**, *259*(5097), 957–960.
60. Gorgutsa, S.; Skorobogatiy, M.; Soft Capacitance Fibers for Touch-sensitive Smart Textiles. In *Multidisciplinary Know-How for Smart-Textiles Developers*, Kirstein, T.; Ed., Woodhead Publishing Limited: USA, **2013**, 154–188.
61. Zhao, J.; Han, W.; Chen, H.; Tu, M.; Zeng, R.; Shi, Y.; Cha, Z.; Zhou, C. Preparation, Structure and Crystallinity of Chitosan Nano-fibers by a Solid–liquid Phase Separation Technique. *Carbohydr. Polym.* **2011**, *83*, 1541–1546.
62. Nayani, K.; Katepalli, H.; Sharma, C. S.; Sharma, A.; Patil, S.; Venkataraghavan, R. Electrospinning Combined with Nonsolvent-induced Phase Separation to Fabricate Highly Porous and Hollow Submicrometer Polymer Fibers. *Ind. Eng. Chem. Res.* **2011**, *51*, 1761–1766.
63. Mohammadifakhr, M.; Trzaskus, K.; Kemperman, A. J. B.; Roesink, H. D. W.; de Grooth, J. Increasing the Success Rate of Interfacial Polymerization on Hollow Fibers by the Single-step Addition of an Intermediate Layer. *Desalination* **2020**, *491*, 114581.
64. Iqbal, H.; Mahar, F. K.; Razzaq, A.; Kamal, R.; Khan, N. U.; Ullah, K.; Iqbal, S. Green Synthesis of Cefadroxil Loaded Chitosan/PVA Nanofibers by Freeze Drying. *Mater. Res. Express* **2019**, *6*, 125094.
65. Mikheyev, P. V.; Berlin, A. A. Effect of Splitting of Polymer Fibers on the Strength of Unidirectional Composites. *Mech. Compos. Mater.* **2019**, *55*, 267–274.
66. Munir, M. W.; Ali, U. Classification of Electrospinning Methods. In *Nanorods and Nanocomposites*, Ghamsari, M. S.; Dhara, S.; Eds., InTech: Croatia, 2020, 1–19.
67. Fong, H.; Reneker, D. H. Electrospinning and the Formation of Nanofibers. In *Structure Formation in Polymeric Fibers*, Salem, D. R.; Ed., Hanser: Munich, 2001, 225–246.
68. Gupta, P.; Elkins, C.; Long, T. E.; Wilkes, G. L. Electrospinning of Linear Homopolymers of Poly (Methyl Methacrylate): Exploring Relationships between Fiber Formation, Viscosity, Molecular Weight and Concentration in a Good Solvent. *Polymer* **2005**, *46*, 4799–4810.
69. Shenoy, S. L.; Bates, W. D.; Frisch, H. L.; Wnek, G. E. Role of Chain Entanglements on Fiber Formation during Electrospinning of Polymer Solutions: Good Solvent, Non-specific Polymer–polymer Interaction Limit. *Polymer* **2005**, *46*, 3372–3384.

70. Kong, L.; Ziegler, G. R. Role of Molecular Entanglements in Starch Fiber Formation by Electrospinning. *Biomacromolecules* **2012**, *13*, 2247–2253.
71. McKee, M. G.; Wilkes, G. L.; Colby, R. H.; Long, T. E. Correlations of Solution Rheology with Electrospun Fiber Formation of Linear and Branched Polyesters. *Macromolecules* **2004**, *37*, 1760–1767.
72. Ferry, J. D. *Viscoelasticity Properties of Polymers*, Wiley-Interscience: New York, 1980.
73. Härdelin, L.; Perzon, E.; Hagström, B.; Walkenström, P.; Gatenholm, P. Influence of Molecular Weight and Rheological Behavior on Electrospinning Cellulose Nanofibers from Ionic Liquids. *J. Appl. Polym. Sci.* **2013**, *130*, 2303–2310.
74. Colby, R. H. Structure and Linear Viscoelasticity of Flexible Polymer Solutions: Comparison of Polyelectrolyte and Neutral Polymer Solutions. *Rheol. Acta.* **2010**, *49*, 425–442.
75. Colby, R. H.; Fetters, L. J.; Funk, W. G.; Graessley, W. W. Effects of Concentration and Thermodynamic Interaction on the Viscoelastic Properties of Polymer Solutions. *Macromolecules* **1991**, *24*, 3873–3882.
76. Felice, B.; Prabhakaran, M. P.; Zamani, M.; Rodríguez, A. P.; Ramakrishna, S. Electrosprayed Poly (Vinyl Alcohol) Particles: Preparation and Evaluation of their Drug Release Profile. *Polym. Int.* **2015**, *64*, 1722–1732.
77. Chang, H.-Y.; Chang, C.-C.; Cheng, L.-P. Preparation of Hydrophobic Nanofibers by Electrospinning of PMMA Dissolved in 2-propanol and Water. *MATEC Web Conf.* **2019**, *264*, 03004.
78. Lee, K. H.; Kim, H. Y.; Bang, H. J.; Jung, Y. H.; Lee, S. G. The Change of Bead Morphology Formed on Electrospun Polystyrene Fibers. *Polymer* **2003**, *44*, 4029–4034.
79. Poudel, D.; Swilley-Sanchez, S.; O'keefe, S.; Matson, J.; Long, T. Fernández-Fraguas, C.; Novel Electrospun Pullulan Fibers Incorporating Hydroxypropyl-β-cyclodextrin: Morphology and Relation with Rheological Properties. *Polymers* **2020**, *12*, 2558.
80. McKee, M. G.; Elkins, C. L.; Long, T. E. Influence of Self-complementary Hydrogen Bonding on Solution Rheology/Electrospinning Relationships. *Polymer* **2004**, *45*, 8705–8715.
81. Onogi, S.; Masuda, T.; Miyanaga, N.; Kimura, Y. Dependence of Viscosity of Concentrated Polymer Solutions upon Molecular Weight and Concentration. *J. Polym. Sci. Part A-2: Polym. Phys.* **1967**, *5*, 899–913.
82. Hu, J.-J.; Liu, C.-C.; Lin, C.-H.; Tuan-Mu, H.-Y. Synthesis, Characterization, and Electrospinning of a Functionalizable, Polycaprolactone-based Polyurethane for Soft Tissue Engineering. *Polymers* **2021**, *13*, 1527.
83. Luzio, A.; Canesi, E.; Bertarelli, C.; Caironi, M. Electrospun Polymer Fibers for Electronic Applications. *Materials* **2014**, *7*, 906–947.

PART V
Energy Materials

CHAPTER 17

Smart Nanomaterials in Energy Applications

N. G. DIVYA[1], MOHAMMOD SAJU[1], ANN ROSHNA[2], and RIJU K. THOMAS[1]

[1]*Bharata Mata College, Thrikkakara, Edappally, Kochi, Kerala, India*

[2]*Cochin University of Science and Technology, South Kalamassery, Kochi, Kerala, India*

ABSTRACT

A tremendous progress has been made in the development of energy devices in recent decades. Nanotechnology boosted functional nanomaterials and made substantial improvements in the performance of energy devices. In particular, the development of smart nanomaterials has been emerged as a promising solution to tackle several drawbacks of energy devices, and the recently developed multifunctional, smart nanomaterials have been highly pursued to accelerate the development of smart energy devices. Energy applications, including energy production, conversion, efficient storage, and transportation, are dependent upon the properties of nanomaterials such as chemical, electrical, mechanical, electrocatalytic, and thermal. In this chapter, the development of multifunctional smart nanomaterials in energy devices is discussed from the perspective of energy applications. In addition, the key issues of energy devices, and major challenges and advantages caused by the use of nanomaterials are also included with possible remedies to tackle the challenges.

Tailored Functional Materials for Clean and Sustainable Development. Tamara Tatrishvili, PhD, Neha Kanwar Rawat, PhD, Ann Rose Abraham, PhD & A. K. Haghi, PhD (Eds.)

17.1 INTRODUCTION

Nanoscience and nanotechnology are interdisciplinary fields that bring together chemists, physicists, material scientists, and engineers, and are recently growing fast and explosively to meet the potential challenges of energy applications. Energy has been the foundation of the sustainable development of our society. Thus, the dramatic energy requirement should be as unrestricted and fulfilled as possible, although that causes conflicting feedback loops on our ecosystem because the world energy consumption mainly relies on the combustion of nonrenewable energy sources such as fossil fuels. By considering the environmental impacts of nonrenewable sources, renewable energy sources are widely being investigated to replace nonrenewable sources. Appropriate steps are required to diminish the negative environmental impact of energy consumption involving energy-saving, generation, conversion, storage, etc.

Nanotechnology has opened up new frontiers by providing novel materials with unique properties for energy applications. Researchers have spent several years developing synthetic strategies to produce smart nanomaterials with desirable properties to meet the various requirements of energy applications. As a result, nanomaterials such as nanoparticles, nanofibers, nanocrystalline materials, thin-film nanocoating, etc. have emerged as promising candidates to tackle several drawbacks of conventional energy applications, and tremendous progress has already been made recently for the efficient use of smart and functional nanomaterials in various energy applications.

The energy applications involve various processes such as energy generation, conversion, transportation, and storage. Unique properties of nanomaterials such as good electrical and thermal conductivity, large surface area, thermal and chemical stability, etc. make nanomaterials indispensable in energy devices. The unusual quantum confinement effect at nanoscale benefits the electron transport kinetics and band structure engineering in nanomaterials, which brings up an excellent performance from energy devices made up of nanomaterials.[5,7]

Nanomaterials provide a tremendous potential to enhance energy efficiency across all the sectors of the industry and have found significant applications in supercapacitors, secondary batteries, photovoltaic cells, hydrogen production, hydrogen storage, etc. Moreover, the properties of nanomaterials, which heavily depend upon the number of atoms and atomic arrangement of constituent atoms and molecules of the nanomaterials, density of exposed atoms or ions on the surface of nanomaterials, etc., are significantly modified with the recent advancement of nanotechnology.

Multifunctional nanocomposite materials are more promising to meet the material requirements of most of the energy applications. Integrating different functions in a single, hybrid functional nanomaterial or nanocomposite leads to the development of one material with multiple responsibilities for energy applications. Advanced nanomaterials with intriguing physical, electrical, mechanical, electronic, and optical properties are significant in designing and developing sustainable energy devices such as photovoltaic cells, batteries, supercapacitors, etc., which would potentially produce a great impact on various kinds of ever-growing applications in energy sectors, and help to reduce the dependence on fossil fuels for day-to-day energy requirements which will also have the advantage of substantial reduction of greenhouse gas emissions.

17.2 NANOMATERIALS IN BATTERIES

17.2.1 INTRODUCTION TO BATTERIES

Energy storage using battery and supercapacitor is an essential element in the present scenario which is closely coupled with energy production, transportation, and usage. Over the past few decades, batteries have become important and popular energy storage devices in many areas of technology such as transportation, medical devices, portable electronics, and the efficient storage of electricity produced from various renewable sources like wind energy, solar energy, tidal energy, etc.

A battery consists of one or more voltaic cells that can convert chemical energy into electrical energy. Each voltaic cell consists of two half cells connected in series by an electrolyte having anions and cations. The two major parts of a battery include a positive electrode called an anode, and a negative electrode called a cathode, apart from the electrolyte with good ionic conductivity. The schematic representation of a battery is shown in Figure 17.1. The anode or the negative electrode is in one half of the cell and the cathode or positive electrode is in the other half of the cell. In the redox process (i.e., oxidation–reduction reaction) that powers the battery, the cations are reduced at the cathode electrode, while the anions are oxidized at the other electrode called the anode. The anode and cathode electrodes do not come into contact during the electrochemical oxidation–reduction process, but they are electrically connected through the electrolyte.

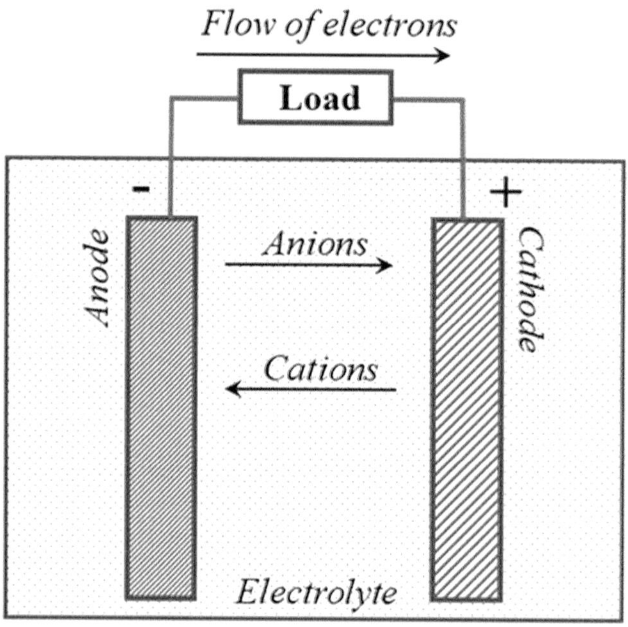

FIGURE 17.1 Schematic representation of a battery and its positive and negative electrodes.

Batteries require electrode materials with good chemical and thermal stability, excellent electrochemical reversible reactivity, low toxicity, and sustainability. The electrolyte must have good ionic conductivity for ionic transportation through it during the lithiation process. Recently investigated special features such as flexibility of batteries also require high-performance, flexible electrodes under bending. Thermoelectric, piezoelectric, and photovoltaic nanomaterials have made considerable contributions in this regard. The recent advances in crystal engineering of nanomaterials at the nanoscale have tremendous potential to propel further the prospectus of tuning the hybrid properties of nanomaterials for battery applications, which would lead to the development of promising and novel, next-generation nanomaterials for more promising energy applications.

The development of high energy density and low-cost lithium-based rechargeable batteries is highly desirable nowadays, and several research works are still going on in this direction to propel its performance further. A large number of studies have been devoted to developing high-performance anode nanomaterials. The possibility of using solid electrodes in batteries is also under investigation.

17.2.2 NANOMATERIALS AS HIGH-PERFORMANCE BATTERY ANODE

Nanotechnology has provided solutions to many technical issues in achieving high-performance anode materials. An anode is one of the most important parts of a battery apart from the cathode and the electrolyte. An anode is the negative electrode of a rechargeable battery. However, the important performance indexes such as energy density, power density, volumetric energy density, specific capacity, coulombic efficiency, and cycle life are determined by the combined effect of the physicochemical properties of all the components of a battery. The majority of these performance indexes of the anode material are determined by the material properties such as crystal structure, physical properties, mechanical stability, electronic conductivity, ionic diffusivity, chemical properties, and so on.[9] Among the different materials investigated over the past decades as the anodes of Li–ion, Na–ion, and Li–S battery systems, carbon-based nanomaterials, and their hybrids have been more explored and successfully implemented in commercial batteries for its applications in portable electronic devices, electric vehicles, etc.[10]

17.2.3 WHY NANOMATERIALS ARE USED IN BATTERIES

In recent decades, with the advancement of nanotechnology, tremendous progress has been made in the development of energy-efficient lithium-based rechargeable batteries. The development of novel electrode materials and storage mechanisms of rechargeable batteries has substantially improved the performance of rechargeable batteries. Nanomaterial-based electrode design has emerged as an efficient method to tackle several drawbacks of materials used in conventional batteries.

Some of the major drawbacks of conventional batteries include poor electronic and ionic transport during an electrochemical process, change in the volume of an electrode during the lithium intercalation process, the solid electrode–liquid electrolyte interphase formation, etc.[1] During the lithiation/delithiation process, anode materials encounter volume expansion or reduction of about 260%, which can cause irreversible damages to the anode such as pulverization of the active materials after long cycles, poor contact with the current collector, etc. that may lead to capacity fade and very poor cycle life of the batteries during practical applications.[10] Some of these drawbacks are effectively tackled by employing nanomaterials for the development of various components of rechargeable batteries because nanoparticles could endure stress and prevent the active materials from pulverization. Due to the high surface area, surface-to-volume ratio, and porous structure, functional nanomaterials with different morphologies, such as nanotubes, nanowires, etc., can incorporate volume expansion without much strain to preserve the structural integrity of the electrodes. The structural integrity and mechanical stability of batteries can be further boosted by empowering nanomaterials for developing different components of the batteries.

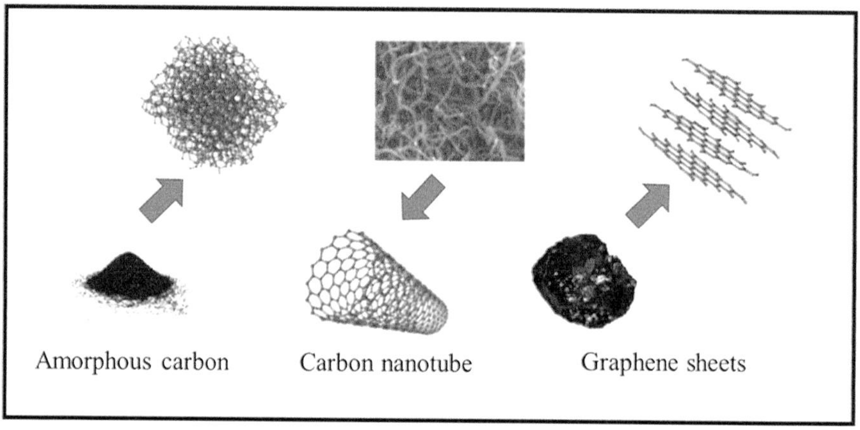

FIGURE 17.2 Schematic representation of allotropes of carbon such as amorphous carbon, carbon nanotubes (1D), and graphene sheets (2D).
Source: Reproduced with permission from Villarreal et al., 2017[4] ©2022. Elsevier.

Nanostructured material can also help to prevent the agglomeration of active nanomaterial, which may reduce the deterioration of its performance. The nanostructured materials can also provide flexibility and robustness to the batteries and help the development of batteries for modern and flexible electronic devices.[3] Moreover, batteries should be small in size and light in weight to find applications in lightweight and compact electronic devices; and it must also provide a constant and stable output voltage for device applications. In this regard, nanomaterials can contribute a lot for developing lightweight and small batteries with better stability, structural integrity, and performance efficiencies.

The power efficiency of a battery and the time required to recharge the battery can be reduced by using smart nanomaterials. These benefits are achieved by coating the surface of an electrode with nanomaterials, which increases the surface area of the electrode during the lithiation process. The nanomaterials also help to increase the shelf life of a battery. The nanomaterials help to prevent the low-level discharge that occurs in a conventional battery, which dramatically increases the battery's shelf life.[11]

Low-dimensional nanostructures, such as 1D (one-dimensional) nanotubes, 2D (two-dimensional) nanosheets, etc., possess several favorable properties for battery applications, such as good transport kinetics, and mechanical and chemical stabilities. The 1D nanostructures with a high aspect ratio include nanotubes, nanowires, nanorods, etc. The 2D nanostructures are thin sheet-like structures with thickness in the nanometer scale range that can provide strain relaxation generated from the volume change during the lithiation process, which also offers flexibility to the anode. This also provides good cycle life to the batteries as required in most practical applications. The schematics of various 1D, 2D, and 3D (three-dimensional) nanostructures are as shown in Figure 17.2. Moreover, the 2D nanomaterials also provide selective exposure of electrochemically active crystallographic facets that boost the storage capacity of the anode. The 3D nanostructures consist of a combination of 1D and/or 2D structures that can offer all the major advantages of 1D and 2D nanomaterials, along with some added advantages of 3D materials for electrochemical battery applications. The 3D nanostructures may be porous, hollow, core–shell type, or hybrids. Due to the high porosity of 3D nanostructures, the anode provides a good specific capacity and better cycle life. The hollow nanostructures with very thin shells provide better diffusion kinetics and structural stability against volume change during the lithium intercalation process. This is because the hollow structure can accommodate more volume expansion, which also prevents it from breaking the structure that can improve the electrochemical performance of an active electrode material.[10]

17.2.4 CHALLENGES OF NANOMATERIALS FOR LITHIUM-BASED RECHARGEABLE BATTERIES

The conventional intercalated electrode materials of a battery with high capacity undergo large volume changes during the electrochemical lithiation process, which induces dramatic structural and volume changes that may cause subsequent mechanical degradation of both the electrodes and active materials, and thus results in short cycle life of the device.[3] The use of multifunctional hybrid nanomaterials for battery applications also has some significant practical challenges. However, the limitations of nanotechnology from the perspective of industrial production include changes in the electrical, physical, chemical, and mechanical properties of nanomaterials as a function of their size and morphology, which generally determine the functional properties of nanomaterials.[2]

Generally, the difficulty in large-scale production of nanomaterials with uniform size and morphology is the property variations of nanomaterials for energy applications. The density of exposed atoms on the surface of the nanomaterial is also a major factor that can alter the fundamental properties of nanomaterials. These changes in material properties of the nanomaterials used in designing the batteries affect their functional properties and subsequently deteriorate the performance of the battery such as cycle life, shelf life, stable output voltage, etc. during their practical applications.

A promising strategy to mitigate several drawbacks of nanomaterials in battery applications is the hybridization of nanoparticles with carbon-based 1D, 2D, or 3D nanomaterials to develop hybrids, so that the carbon nanomaterials efficiently prevent the agglomeration, buffer the stress associated with the volume expansion and thus efficiently enhance the electroconductibility contributed by the conductive carbon nanostructures.

17.3 NANOMATERIALS FOR ENERGY STORAGE APPLICATIONS IN SUPERCAPACITORS

17.3.1 INTRODUCTION TO SUPERCAPACITORS

Supercapacitors can be effectively used for diverse applications such as hybrid vehicles, mobile phones, communication devices, military warheads, and uninterruptible power supplies. Supercapacitors consist of two electrodes, a negative electrode and a positive electrode, insulated with a membrane separator soaked in an electrolyte (Fig. 17.3).[8] The membrane separator favors the mobility of the ions, however it prevents the electric contact between the electrodes. The materials chosen as electrodes of the supercapacitor heavily determine its properties. As an example, carbon-based nanomaterials such as carbon nanotubes (CNTs), graphene sheets, etc. have been playing a major role in the development of supercapacitors. Further, the electrostatic charge transfers give high recyclability to these devices.

Supercapacitors are under investigation since 1853 with the study of Helmholtz, who revealed that the electrical charges occupy not only the surface of a conductor but also at the interface between the electrode and the electrolyte. Then, almost 100 years later, several research investigations and patents were produced in this field as the consequence of research studies related to graphitic carbon-based fuel cells and electrodes' materials.

Recently, supercapacitors have attained great research attention as efficient devices for electrochemical energy storage applications. They have high power and high energy density, and are capable of fast charging and discharging with long cycle life.[6] High capacitances of supercapacitors are due to the high specific surface areas of porous and conductive electrode nanomaterials. Supercapacitors bridge the gap between traditional capacitors and fuel cells. The conventional capacitors show high power density, while fuel cells are high-energy storage systems. In other words, supercapacitors have more capacitances than conventional capacitors due to the large surface areas of constituent electrodes. It can also produce greater power compared to batteries. For a supercapacitor, the charge accumulation involves only the surfaces but not the bulk volume whereas the reverse occurs in batteries.

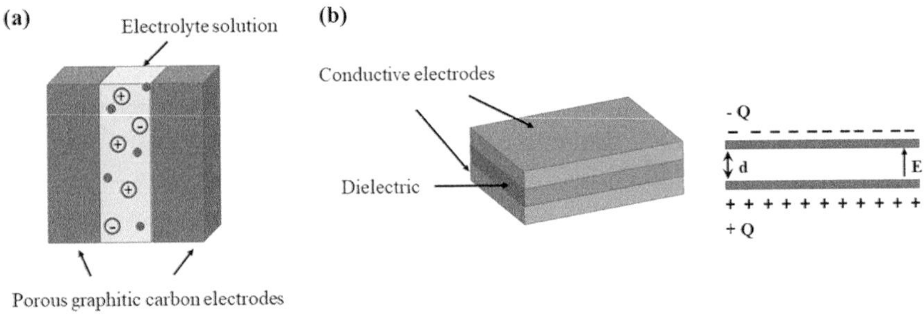

FIGURE 17.3 Schematic representation of (a) supercapacitor based on porous graphitic carbon electrodes and liquid electrolyte, and (b) an electrostatic capacitor.

In supercapacitors, the electrical energy is delivered by the movement of ions at the electrode/electrolyte interface, associated with the flow of electrons through the external circuit. Moreover, in supercapacitors, there is no volumetric change of electrochemical energy in electrodes during charging and discharging. However, a high power density, a high energy density, fast charging and discharging, and a long cycle life are possible to attain by using supercapacitors. The underlying mechanism and design of the supercapacitors determine the performance of the supercapacitor, which in turn depends upon the materials used. Several types of supercapacitors such as stretchable supercapacitor, editable supercapacitor, microsupercapacitors, etc. are developed and they find applications in displays, sensors, portable electronics, wearable electronics, flexible electronics, and so on.[6]

17.3.2 BENEFITS OF NANOMATERIALS FOR SUPERCAPACITOR APPLICATIONS

Recently, nanomaterials, especially carbon nanomaterials such as CNTs, graphene, and graphite have been widely studied as effective electrodes of supercapacitors due to their unique properties, such as large surface area, excellent electrical conductivity, and mechanical stabilities. Carbon-based nanomaterials and their composites mediate volume change during the cycling process, which can also help to restrain the agglomeration of metal oxide/hydroxides with high surface energy.[8]

The electrode materials of a supercapacitor must have a high specific surface area for which the use of porous nanomaterials are promising candidates due to their large surface area that improves the electrochemical reaction kinetics and the interaction between electrolyte and electrodes through ions or electrons. The 1D nanomaterials such as nanotubes and nanorods exhibit high surface area compared to 2D and 3D nanomaterials such as graphene, graphite, etc. that may offer high specific capacitance but with poor cyclic stability compared to the 1D nanomaterial. The 1D nanomaterials may also undergo agglomeration, which may block ionic transportation. However, the 3D nanomaterials offer more uninterrupted porous channels that provide good cyclic performance and specific capacitance. Some highly conducting metal oxide nanomaterials as the electrodes of a supercapacitor also help to minimize the charge transfer resistance of the electrodes, and improve the specific capacitance, energy density, and cyclic performance of the supercapacitor. The nanoarrays developed on substrates with high porosity exhibit good stability, both structural and mechanical stability, with enhanced electrochemical properties including cycle life.

17.3.3 DRAWBACKS OF NANOMATERIALS FOR SUPERCAPACITOR APPLICATIONS

Despite several advantages of supercapacitors, the high cost and low energy density prevent the complete replacement of supercapacitors with batteries. These drawbacks should be addressed without compromising their stability and performance. Further, despite several benefits of supercapacitors, performance efficiencies such as a sustainable source of energy, extensive commercialization, etc. have yet to be realized due to their higher cost to performance ratio and low energy density. Thus, tremendous research efforts must be devoted to developing low-cost nanomaterial-based electrodes with better performance.

Self-discharging is an inherent drawback of all the supercapacitors as in the case of any kind of energy device, which is mainly associated with an uncontrollable Faradaic process due to the redox reaction of impurities present in the electrode nanomaterial. This should be addressed by minimizing the probable impurities present in the nanomaterials used for developing the electrodes of the supercapacitors.

Some metal oxide nanomaterials have attained huge research interest owing to their promising structural properties and electrochemical performances, but with several drawbacks as usual, such as poor electrochemical and cyclic stability, low conductivity, damages of electrodes, etc. The formation of mixed transition metal oxides or composites of nanomaterials with conducting materials such as CNTs, graphene, etc., or other conducting agents such as silver and platinum for the development of electrode materials helps to improve the electrical conductivity.

Although great research efforts have been devoted to addressing several drawbacks of supercapacitors, such as obtaining high specific capacitance, stable performance, good cyclic reactions, etc., but are yet to be realized for commercialization and practical applications of supercapacitors.[6]

17.4 HYDROGEN: THE NEED OF THE HOUR

Hydrogen fuel is a renewable energy source, which is an ideal replacement for fossil-based and other types of nonrenewables. Because of its nonpolluting nature and abundance in its availability, this fuel is becoming a hot subject in the energy industry. Human energy consumption increases daily resulting in huge energy demand. The conventional nonrenewable fuels meet more than 80% of the current energy demand; however, due to the better efficiency of hydrogen fuel, it is identified as an ideal contender for fossil fuels.[18]

On one hand, technological advances are bringing up more hybrid production methods such as biomass conversion, solar photochemical–photobiological water decomposition, and water electrolysis. Most of these methods are eco-friendly and can be achieved at a higher efficiency.[18] On the other hand, the economic challenges still need to be resolved, especially the high cost of hydrogen production.[18] These insufficient hydrogen storage methods challenge the production of hydrogen on a large scale and its efficient management. The gravimetric energy density and available energy per unit mass are very high for hydrogen compared to fossil fuels. However, the low volumetric energy of hydrogen is a crisis for its transport and efficient utilization.[19]

17.4.1 DIFFERENT METHODS OF HYDROGEN STORAGE

A common technique among hydrogen storage facilities involves the necessity of height–weight tanks capable of withstanding pressure up to 350–700 bars.[20] These tanks contain a lot of hardware and chemical reactants. Another notable factor is the risk associated with its safety. The highly combustible nature of hydrogen can cause a large explosion, even due to a tiny spark. So, proper safety mechanisms should be met to monitor and avoid leakage during hydrogen storage and transportation.

Compressed gaseous hydrogen and liquid hydrogen are widely explored in current physical storage technologies.[20] It is an alarming fact that these techniques result in relatively low hydrogen capacity and contain hydrogenation and dehydrogenation reactions and some complicated purification processes.[18] Solid-state storage options are more promising for large-scale on-board and off-board storage applications than the conventional storage systems. Nanomaterials use all these solid-state technologies, especially nanoporous carbon materials, metal–organic frameworks, covalent organic frameworks, porous aromatic frameworks, nanoporous organic polymers, and nanoscale hydrides. The storage performances in solid hydrogen storage systems can be affected by three absorption processes: physisorption, chemisorption, and Kubas interaction between hydrogen and nanomaterials.[17]

In physisorption, van der Waals forces cause molecular hydrogen to be absorbed onto the surface of nanomaterials. However, the nonpolar hydrogen molecules are released from the matrix at high temperatures. Chemisorption occurs when the atomic hydrogen bonds react chemically with the nanomaterials and are released during chemical reactions. The third absorption type is Kubas interaction, where hydrogen is attracted to the material. This is either due to the charge transfer or by the polarization induced by the metal in the nanomaterials.[18]

17.5 NANOMATERIALS IN PHOTOVOLTAIC CELLS

17.5.1 INTRODUCTION TO PHOTOVOLTAIC CELLS

Photovoltaic cells (PVCs), or more commonly known as solar cells are very important in today's world for sustainable energy production. A photovoltaic cell converts light energy into electrical energy using the photovoltaic effect. A p–n junction diode when illuminated with sufficient radiation incident upon it produces an electric current. The efficiency of a solar cell depends upon many factors including the type of material used, the amount of radiation falling on the module, etc. The most commonly used semiconductor material for the production of solar cells is silicon due to its higher efficiency, lower cost, and longevity. However, using nanotechnology for the manufacture of solar cells, we can create more efficient and cost-effective solar cells.

In bulk materials, the surface area is much smaller compared to nanoparticles, which may effectively increase the light per unit surface area falling on the surface of the nanomaterial. Another thing to consider is the size of the particle and emergent quantum effects. Nanocrystal diameters are smaller than the Bohr radius and hence the radiation is no longer a continuous spectrum, rather the energy levels become discrete and quantum confinement effects take place. The size of the crystal starts to affect the color of radiation absorbed with smaller crystals absorbing shorter wavelengths and larger crystals absorbing higher wavelengths.

Researchers started to convert bulk silicon into discrete nanosized particles to make nanosized PVCs. Once nanoparticles of the desired size are achieved, they disperse them into isopropyl alcohol and are dropped onto the substrate of the solar cell. A nanothin film solar cell is achieved as and when the alcohol is evaporated.

As expected, the solar cells show fluorescence in different regions for different sizes. The enhanced power of 60% was observed in the UV range of the spectrum for a film of 1 nm, but an enhancement of only 3% was observed in the visible range. On the other hand, a film of 3 nm-sized particles, red particles showed an enhancement of about 67% in the UV region of the spectrum and 10% in the visible range. We can further increase the efficiency if we can tune the region of a wavelength absorbed. For this, nanorods can be used. In this chapter, we will try to explore how nanotechnology is used to further boost the performance of solar cells.[12]

17.5.2 NANOMATERIALS IN PHOTOVOLTAIC CELLS

17.5.2.1 NANOTUBES

Carbon nanotubes are a molecular scale of tubes of carbon that is made up of rolled-up graphene sheets. They have a very high aspect ratio, very good conductance of heat and electricity, and are chemically stable.[13] These tubes can be either semiconductors or metallic in nature based on how they are rolled up. By coating these nanotubes with p-type and n-type semiconductors, we can use them for photovoltaic applications.

To generate energy, a solar cell must have a transparent electrode that is also highly conductive. CNTs act as a hole extraction layer or a transparent conductive layer. They can form heterojunctions with conducting polymers and thus can separate carriers. They also can transfer electrons or holes efficiently.[14]

Even though single-walled carbon nanotubes (SWCNTs) were ideal as thin-film photovoltaics because of their ability to absorb wavelengths over a wide range from visible to near IR and possess fast-moving charge carriers, they suffer from poor power conversion efficiencies.

In a recent study, a new type of solar cell was designed containing polychiral SWCNTs and fullerenes that maximize the amount of photocurrent produced by absorbing a broader range of wavelengths. These cells could absorb the near-infrared portion of the spectrum, which is currently inaccessible to many other leading thin-film photovoltaic technologies. They are made up of a one atom thick sheet of carbon rolled up into a tube of diameter of about 1 nm; they are polychiral and hence can absorb across a wider portion of the solar spectrum.[15,16] By controlling the interface between the active and photovoltaic layer, the underlying hole transport layer photovoltage produced can be maximized. This interface layer allows the efficient recombination of generated electrons and holes.

17.5.2.2 QUANTUM DOTS

Quantum dots (QDs) are another way of improving the efficiency of solar cells. Quantum dots are semiconducting crystals of nanometer dimensions and they have a tunable bandgap. They behave like a special class of semiconductor materials composed of elements of periodic groups II–VI,

III–V, or IV–VI. The larger the quantum dot, the longer the wavelength of the spectrum it emits. A solar cell with a greater bandgap of semiconductor absorbs more energetic photons causing the greater output voltage. On the other hand, a cell with the lower bandgap captures more photons including those in the red end of the solar spectrum, resulting in a higher output current and lower output voltage.[17]

In a traditional solar cell, light absorbed by a semiconductor is used for the production of an exciton or electron–hole pair. These excitons are separated by the internal potential produced due to the p–n junction diodes. The flow of electrons and holes as a result creates an electric current. Conventionally, this internal potential is created by doping the semiconductors involved with one part doped with electron donors or n-types and the other part doped to act as electron acceptors (p-type). This gives rise to a p–n junction.[17] When photons with energy greater than the bandgap of the semiconductor are incident on it, electron–hole production occurs. So, if the energy of the photon is less than that of the bandgap, then electron–hole production does not happen. However, those with higher energy can quickly thermalize to the band edges. Both of these reduce the efficiency with the former reducing current and the latter reducing voltage. This limits the ability of a solar cell massively and an efficiency of 33% cannot be exceeded even if silicon semiconductor is used with a bandgap of 1.1 eV, which is close to an ideal bandgap of 1.34 eV. This limit is known as the Shockley–Queisser limit.[17] This limit can be surpassed if somehow it is possible to stack cells with different band-gaps vertically. This is known as a multijunction or tandem approach. Theoretically, an infinity layer cell has an efficiency of 86%. This is where QDs turn up as most effective. By using QDs of different sizes which have tunable bandgaps, multijunction semiconductor solar cells can be created.

Researchers in recent years have also focused on the advancement of photovoltaic conversion efficiency in third-generation PVCs using QDs. By forming the QDs into an order of 3D array, there will be strong electronic coupling between QDs and hence excitons or electron–hole pairs will have a longer life for collecting and transporting "hot carriers" enabling them to generate electricity at a higher voltage. When the energy of the absorbed photon is considerably larger than the semiconductor bandgap, multiple excitations are possible as different QDs have different bandgaps. This leads to better power generation efficiency.[17]

17.5.3 ADVANTAGES OF USING NANOTECHNOLOGY IN PHOTOVOLTAIC CELLS

Using nanotechnology, the production of efficient photovoltaic cells is more cost-effective and hence can be used on an industrial scale to help the environment. Most of the methods of PVCs using nanotechnology are environment-friendly in themselves. As we can use nanotechnology with polymers to create efficient PVCs, covering the roof of a house with inexpensive plastic PVCs can generate enough power to power the entire house. This in turn can reduce the dependence on fossil fuels and other conventional energy production methods.

i. Nanotechnology in solar cells would have military implications. Using nanotechnology can scale down the size of solar cells, thereby making them easy to set up or carry. The US army uses nanotechnology to help design a better way to power their soldiers' electrical devices. With the help of nanotechnology, the weight of batteries a soldier has to carry can be reduced from 140 pounds to approximately 60 pounds. This can increase the mobility of soldiers.
ii. Inexpensive solar cells can help provide electricity for rural areas or third-world countries. In the areas where electrical connections are impossible or hard to set up or areas are highly spaced-out, high-efficiency solar cells with the help of QDs can be very practical. It can also be used for many practical applications like cooking, hot water, medical devices, etc. other than providing electrical connection.
iii. With improved efficiency, solar-powered vehicles with decent mileage and power are possible which can then, in turn, reduce the use of fossil fuels. Not only solar cells can be made compact that can help in scaling down the size of solar vehicles, but they can also be made more efficient with the help of nanotechnology resulting in effective solar-powered transportation.[17]

17.6 SUMMARY AND FUTURE PROSPECTS

Several hybrid nanomaterials, nanoallotropes, nanopolymers, etc. are the subject of recent intensive scientific research to develop more energy-efficient electronic devices such as batteries, supercapacitors, photovoltaic cells, etc. for practical applications because several applications require in-depth knowledge about the structure and properties of these materials.

Therefore, much research effort has been devoted to this direction over the past decades. However, several materials aspects should be exhaustively elucidated to reach the commercial-scale production of many energy devices using nanomaterials.

The use of low-cost, natural materials or waste materials should also be highly investigated on priority to develop nontoxic components of various energy devices that can be considered as a highly promising alternative for economic benefits and practical applications with a minimum toxicity index. Research is also going on to develop efficient green synthesis routes to enable the utilization of the complete potential of nanomaterials for the benefit of society with minimum environmental risks.

KEYWORDS

- **nanomaterials in batteries**
- **nanomaterials in supercapacitors**
- **nanomaterials in photovoltaic cells**
- **battery**
- **carbon nanotubes**
- **energy devices**
- **graphene**
- **photovoltaic cell**
- **supercapacitor**

REFERENCES

1. Sun, Y.; Liu, N.; Cui, Y. Promises and Challenges of Nanomaterials for Lithium-based Rechargeable Batteries. *Nat. Energy* **2016,** *1*, 16071.
2. Huilin, W.; Xitong, L.; Jiutian, W.; Shengjian, J.; Dongfeng, X. Multifunctional Inorganic Nanomaterials for Energy Applications. *Nanoscale* **2020,** *12*, 14.
3. Jun-Fan, D.; Rui, X.; Chong, Y.; Ye, X.; Lei, X.; Hong-Jie, P.; Ho, S. P.; Ji, L.; Jia-Qi, H. Review on Nanomaterials for Next-generation Batteries with Lithium Metal Anodes. *Nanoselect* **2020,** *1* (1), 94.
4. Villarreal, C. C.; Pham, T.; Ramnani, P. Mulchandani, A. Carbon Allotropes as Sensors for Environmental Monitoring. *Curr. Opin. Electrochem.* **2017,** *3*, 106.
5. Law, M.; Greene, L. E.; Johnson, J. C.; Saykally, R.; Yang, P. Nanowire Dye-sensitized Solar Cells. *Nat. Mater.* **2005,** *4* (6), 455.

6. Sun, J.; Cui, B.; Chu, F.; Yun, C.; He, M.; Li, L.; Song, Y. Printable Nanomaterials for the Fabrication of High-Performance Supercapacitors. *Nanomaterials* **2018,** *8*, 528.
7. Bao, W.; Wan, J.; Han, X.; Cai, X.; Zhu, H.; Kim, D.; Ma, D.; Xu, Y.; Munday, J. N.; Drew, H. D.; Fuhrer, M. S.; Hu, L. Approaching the Limits of Transparency and Conductivity in Graphitic Materials through Lithium Intercalation. *Nat. Commun.* **2014,** *5*, 4224.
8. Verma, S.; Arya, S.; Gupta, V.; Mahajan, S.; Furukawa, H.; Khosla, A. Performance Analysis, Challenges and Future Perspectives of Nickel-based Nanostructured Electrodes for Electrochemical Supercapacitors. *J. Mater. Res.* **2021,** *11*, 564.
9. Wang, Y.; Xie, K.; Guo, X.; Zhou, W.; Song, G.; Cheng, S. Mesoporous Silica Nanoparticles as High-performance Anode materials for Lithium-ion Batteries. *New J. Chem.* **2016,** *40*, 8202.
10. Yang, Z.; Wu, H.; Zheng, Z.; Cheng, Y.; Li, P.; Zhang, Q.; Wang, M. Tin Nanoparticles Encapsulated Carbon Nanoboxes as High-Performance Anode for Lithium-Ion Batteries. *Front. Chem.* **2018,** *6*, 1.
11. Majdi, H. S.; Latipov, Z. A.; Borisov, V.; Yuryevna, N. O.; Kadhim, M. M.; Suksatan, W.; Khlewee, I. H.; Kianfar, E. Nano, and Battery Anode: A Review. *Nanoscale Res. Lett.* **2021,** 16, 177.
12. Sethi, V. K.; Pandey, M.; Shukla, P. Use of Nanotechnology in Solar PV Cell. *Int. J. Chem. Eng.* **2011,** *2* (2), 76–80.
13. Maruyama, T. *Hand-book of Carbon-based Nanomaterials*. Elsevier **2021,** 299–319.
14. Alturaif, H. A.; Othman, Z. A.; Shapter, J. G.; Wabaidur, S. M. Use of Carbon Nanotubes (CNTs) with Polymers in Solar Cells. *Molecules* **2014,** *19* (11), 17329–17344.
15. Dumé, I. Making Better Solar Cells with Polychiral Carbon Nanotubes. *Phys. World* **2014,** *15*, 1–10.
16. Manna, T. K.; Mahajan, S. M. Nanotechnology in the Development of Photovoltaic Cells. *IEEE* **2007,** 1–10.
17. Ehrler, B.; Alarcón-Lladó, E.; Tabernig, W. S.; Veeken, T.; Garnett, E. C.; Polman, A. Photovoltaics Reaching for the Shockley–Queisser Limit. *ACS Energy Lett.* **2020,** *5* (9), 3029–303.
18. Zheng, J.; Wang, C. G.; Zhou, H.; Ye, E.; Xu, J.; Zibiao, L., Xian, J. L. Current Research Trends and Perspectives on Solid-State Nanomaterials in Hydrogen Storage. *Research* **2021,** 39, 2021.
19. Singh, R.; Altaee, A.; Gautam, S. Nanomaterials in the advancement of hydrogen energy storage. *Heliyon* **2020,** *6* (7), 2405–8440.
20. Eberle, U.; Felderhoff, M.; Schueth, F. Chemical and physical solutions for hydrogen storage. *Angew. Chem.* **2009,** 48, 6608–6630.

CHAPTER 18

Nanofibers for Energy Application

SAURABH PANDEY[1], ABHISHEK ANAND[1], AMRITANSHU SHUKLA[1,2], and ATUL SHARMA[1]

[1]*Non-Conventional Energy Laboratory, Rajiv Gandhi Institute of Petroleum Technology, Jais, Amethi, India*

[2]*Department of Physics, Lucknow University, Lucknow, India*

ABSTRACT

Everyone knows the concerns related to the scarcity of fossil fuels and their bad effects on our environment. The best way to overcome these effects and drawbacks is to have clean and efficient energy sources and energy storage devices. They are spreading their feet very fast nowadays in which nanofiber is playing an important role and is being used on a large scale. The reason behind their widespread use is their porosity and high surface area. Currently, electrospinning is an efficient and versatile method used extensively to fabricate nanofibers. In this chapter, the authors will mainly discuss the application of nanofibers for energy storage and generation like triboelectric nanogenerators (TENG), lithium-ion batteries, supercapacitors, dye-sensitized solar cells, and phase change materials (PCMs). The structural properties of nanofibers are also briefly and systematically discussed.

18.1 INTRODUCTION

The world of nanofibers contains a wide range of fascinating fibers which have amazing physical and chemical properties in which high surface-to-volume ratio, high porosity, high absorbency, and nanoscale interstitial

Tailored Functional Materials for Clean and Sustainable Development. Tamara Tatrishvili, PhD, Neha Kanwar Rawat, PhD, Ann Rose Abraham, PhD & A. K. Haghi, PhD (Eds.)

space interconnectivity are the distinguished qualities. The greater surface area to volume ratio and porosity of nanofiber makes them tempting for composites and filtration application.[1,2] These materials include quantum dots, one-dimensional nanowires, two-dimensional nanosheets, nanorods, and nanotubes.[3,4] At present, nanofibers are prepared from some classified materials, that is, natural polymers, polymer blends, carbon-based nanomaterials, synthetic polymers, semiconductor nanomaterials, composite nanomaterials, and ceramic precursors in multiple frameworks like ribbon-shaped, porous, beaded, and core-shell fibers are one of them.[5,6] Keeping in view the rapid progress in the synthesis and characterization of nanofibers over the last few years, amazing efforts have been made to realize the potential of nanofibers in a variety of functional applications including energy generation and storage, water and environmental treatment, healthcare, and biomedical engineering.[7] Recent studies on the energy application of nanofiber show that the electrochemical properties and cycling performance of nanofiber-based energy harvesting devices have increased but still, nanofiber has not been fully exposed in this area. Authors are looking at the current use of nanofibers for energy applications, mainly electrospun nanofibers.[8]

Nanofiber fabrication should be done either by physical or chemical fabrication method as shown in Figure 18.1. Among all these currently available strategies, electrospinning.[9] is an established and widely used inexpensive and efficient technique for producing one-dimensional nanofibers, with its roots in the 1934 technology of making textile yarns from Formhals.[10] Although electrospinning did not become a popular technique until the 1990s, when many research groups, especially Renekar[11] found that nanofibers could be made from any organic polymer solution with this technique, after which this technique was further developed.

Discussion should be done earlier that there are many techniques of nanofiber fabrication in today's time, among them, electrospinning is the cheapest and most flexible way to make nanofiber. In this technique, the high voltage source is used, that is, several kilovolts of DC voltages are required to create a high electric field[12] so that nanofibers of the range from nanometers to micrometers can be made, and electrospinning apparatus mainly consists of three things.[13]

1. A grounded conductive collector
2. A syringe with a metallic needle with a thick-headed tip, and
3. High-voltage power supply

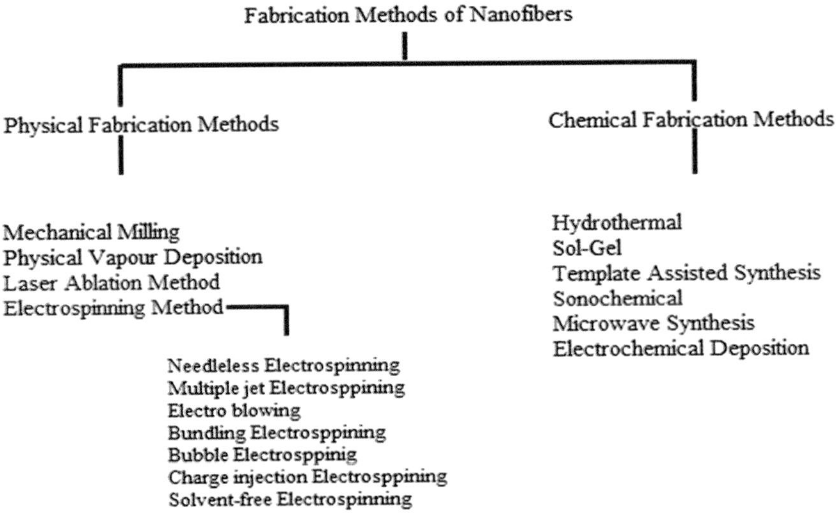

FIGURE 18.1 Fabrication methods of nanofibers.

The polymer solution is filled inside the syringe, a metallic needle is attached to the syringe (or spinneret) and grounded conductive collector on which fibers are collected as shown in Figure 18.2.[16] Because of the high-voltage electrostatic field, electrostatic forces overcome the surface tension on the droplet on the needle which takes the shape of a cone, and then droplet on top of the needle comes out as a fiber jet then further splits into thinner fiber depending on the nature of the field and then it is collected on collector.[14] A collector may be of a metal frame type or rotating drum or rotating mandrel type.[15]

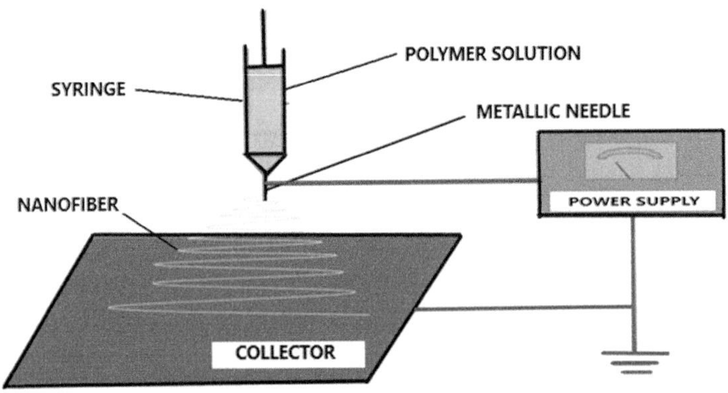

FIGURE 18.2 Electrospinning setup which contains pump, syringe, polymer solution, power supply, and collector.

The reason for this technology being used on a large scale is its one-step process, suitable for the preparation of a wide range of molecules and natural polymers. The nanofibers generated by this technique are polycrystalline, porous, amorphous, dense, hollow, and partially aligned when the rotating collector is used and unaligned randomly oriented when the static collector is used.[16] The length and diameter of the nanofiber can be adjusted only by changing the spinning parameters such as polymer concentration, flow rate of solution, and applied electric field.[17] Some of the energy applications of nanofibers will be briefly discussed below.

18.2 APPLICATIONS

18.2.1 TRIBOELECTRIC NANOGENERATORS

For small-scale physical change conversion into electricity done either by thermal or mechanical energy, for this energy conversion nanogenerators come into the picture, which is basically of three types: piezoelectric, triboelectric, and pyroelectric nanogenerators,[18] but discussion focused on the application of nanofibers in triboelectric nanogenerators. Humans do lots of mechanical work in daily life by their actions such as arm movement, breathing, pulse movement, stepping, and so on for which an amount of energy is used,[19] which is an alternative energy source that can be converted into electrical energy via promising power generation technology known as triboelectric nanogenerator (TENG) and then used on micro-level like wearable and portable electronic devices.[20] This type of nanogenerator was first introduced by professor Zhong Lin Wang's group in 2012,[21] before that triboelectrification was seen as negative as it also causes accidents or troubles because it exposes extremely high voltages. According to a study, approximately 35,000 V triboelectricity is generated from everyday activities.[22] After the invention of TENG, this huge amount of voltage is used for energy harvesting. It is generally seen that less conductive materials or insulators have a strong triboelectrification effect. Although almost every material has this effect especially wood, metals, nanofibers, and polymers but here our discussion will be a focus on nanofibers in which polyvinylidene fluoride (PVdF), polyvinyl chloride (PVC), polyimide (PI), polydimethylsiloxane (PDMS), fluorinated ethylene propylene (FEP), and polytetrafluoroethylene (PTFE) are widely used tribonegative(TN) nanofibers.[23] While tribopositive (TP) materials are limited in comparison to TN materials such as polyamide

(PA), metals, cellulose, and metal oxides are widely used TP materials in which polymers can be flexible and easy to process, and offer higher flexibility compared to metals and metal oxides, but TP effect of polymer is less.[24,25] To improve or increase thc TP cffect, we can change the surface of polymer materials by physical and chemical action with the help of nanoparticles, graphene oxide, and nanowires, which will change the contact characteristics of the polymer,[26–28] when researchers embed the nanostructure in the polymer matrix, the electrification and permittivity of both the surfaces of the materials, will change and making the electrostatic induction and coupling electrification more effective.[29]

According to the working principle of TENG, the selection of dissimilar materials and microstructure are two important factors that affect the performance of TENG.[30] In terms of microstructure, on a large scale, it is observed that structured surfaces have better charge generation in comparison to smooth surfaces it is due to the increased contact and frictional areas.[31] The use of polyamide nanofiber and polyimide film in TENG as TP and TN material, respectively, which is prepared by an electrospinning method that theoretically leads to higher triboelectric output.[32] TENG operates entirely on the triboelectric effect, triboelectrification occurs in almost all material combinations: solid–gas, solid–solid, and gas–liquid,[33] which is contact-induced electrification, in which a material becomes electrically charged, when it encounters another type of material (TP and TN materials) through friction so there is the transfer of charges from one side to the other to equalize the electrochemical potential. These charges are either electrons or ions/molecules. Now after charge transfer, when they are separated, some bonded atoms tend to retain extra electrons and release some of them, which probably produces triboelectric charges on surfaces, connected with electrodes.[34,35]

Requisition to improve the TENGs performance is the selection of materials that have a mass propensity to gain and lose triboelectric charges during the frictional process between triboelectric layers.[36] Figure 18.3 depicts the triboelectric material series which is according to the ability to gain and lose charges. The material at bottom of the series has a loud tendency to get electrons and become negatively charged, whereas materials at top of the table readily lose electrons so they bear positive charges. So, the materials listed in the table at long last avail triboelectric charges after they meet the material listed at top of the table, this implies that the topmost material in the series and bottom material should exchange more triboelectric charges that increase the triboelectric effect which results in maximum output power.

Thus, the selection of triboelectric materials plays an important role in the generation of electric responses in nanogenerators.

Obtained output voltage after triboelectrification can be written as:

$$V = \frac{\sigma d}{\varepsilon_0} \quad (18.1)$$

This depends on distance (d) among the triboelectric layers, charge density (σ) produced during friction between two layers, and permittivity constant (ε_0), so from eq 18.1 we conclude that the output of TENG increases when charging density and separation between the oppositely charged layer increases.

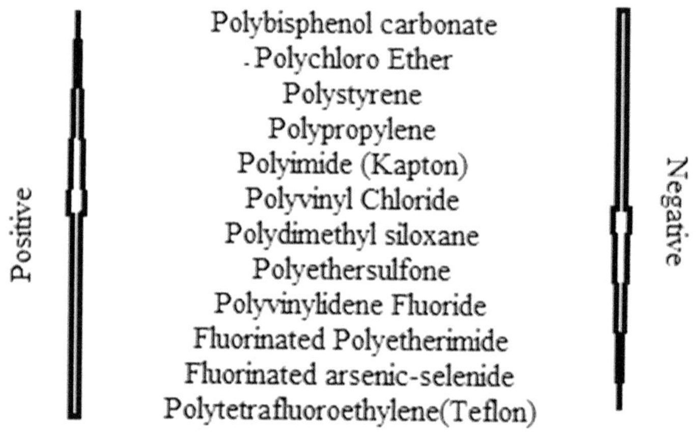

FIGURE 18.3 Tribopositive and tribonegative materials.

TENG mainly works in three ways such as contact sliding, free-standing, and vertical contact separation, and all these three types are either of the single electrode or double electrode type.[19]

18.2.2 DYE-SENSITIZED SOLAR CELLS

Attraction behind the energy harvesting through third-generation solar cells known as dye-sensitized solar cells are due to their low cost, environment friendly, semi-flexible, semi-transparent, and good energy efficiency and it is also used as an alternative to silicon-based solar cells. Sunlight is the primary source of energy and provides a huge amount of energy in a single day more than the entire world consumes in a year, to harvest this energy with high efficiency, economical, and environmentally friendly way third-generation solar cells also known as a dye-sensitized solar cell (DSSCs) is used.[37] Nowadays DSSCs are used in so many applications like building-integrated photovoltaics, power windows, designed as decorative elements in indoors also used in portable devices like calculators and mobiles.[38–40] A DSSC works on light absorption, electron injection, transportation of carrier, and current collection and it consists of an electrolyte, counter electrodes, a conductive glass substrate, dye sensitizer attached to photoanode which absorbs sunlight and electrons become excited, and photoanodes which provides a surface area to dye adsorption to generate power, from Glass substrate solar radiation passes so, it must be transparent in nature. A schematic band diagram is shown in Figure 18.4.[41–43] Near about 12% power conversion efficiency was achieved by DSSCs.

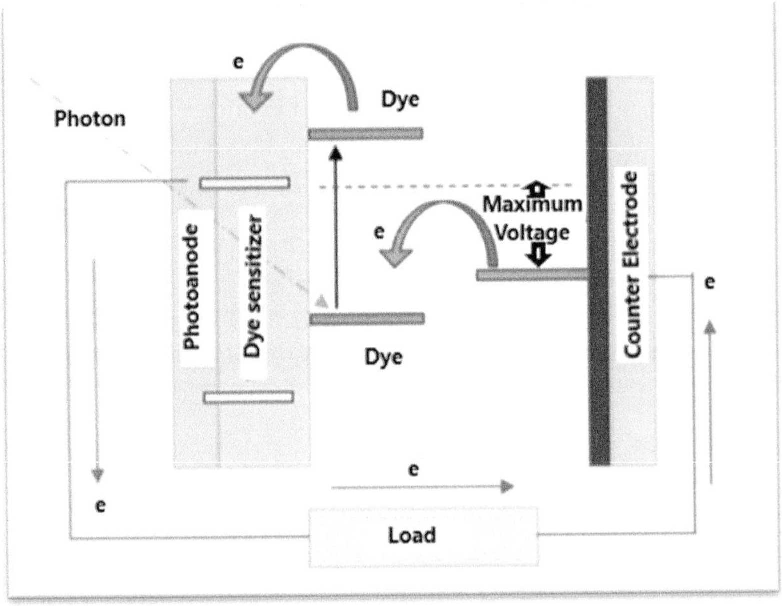

FIGURE 18.4 Working of DSSC.[69]

Source: Reprintedwith permissionfrom Ref. [69] © 2000. Elsevier Science S.A.

A high specific area is the nature of nanofibers, which offers fine dye loading, good electron transport, and light-harvesting by NF-based photoanode that lifts the power conversion efficiency.[45] Some of them are bilayer

TiO_2 nanofiber.[46] Some research group fabricates nanoparticle composite photoelectrode which has high flexibility[47] and by using CNT/TiO_2 nanofiber-based photoelectrode energy conversion efficiency reached 3.39%.[48] Recently TiO_2/Ag nanofiber-based photoanode is used in DSSCs with an energy conversion efficiency of 5.13%.[49] For counter electrode to replace the use of Pt because economically it is not good and also reduces the catalytic activity,[50] after so many investigation researchers find carbon nanofibers are a good alternative to replace platinum[51] and saw that the transfer of ions and electrons is promoted by activated carbon nanofibers, which enhances the association between electrode and electrolyte because of an increase in the contact area between them and slowed down the charge transfer but the gross efficiency is alike Pt-based DSSCs.[52] In this series, transition metal carbides/CNFs[53] and transition metals/CNFs[9] are very good for increasing electrocatalytic activity. Subashini Gnanasekar and group prepared VO_2(M)/CNF as counter electrode for DSSC which have low charge transfer resistance in comparison to Pt counter electrode and good electrocatalytic activity,[54] it instates around 6.53% power conversion efficiency which is close to platinum counter electrode conversion efficiency (7.39%). These are some examples of nanofiber used in DSSCs as photoanode and counter electrode. The next important part of DSSCs is an electrolyte that plays a crucial role in charge transfer from electrode to photoanode and reduces the oxidized dye,[55] but some problems associated with conventional ionic and organic electrolytes they are not stable for a long time because of leakage and volatile nature.[56] So this urge is completed by using polymer electrolytes and electrospun nanofibers like PVdF-HEP,[57] PVdF-HFP,[58] SiO_2NF/PEO-(PVdF-HFP),[59] and many more which have high thermal stability, good ionic conductivity, and permeability.

18.2.3 SUPERCAPACITORS

Energy storage devices play a vital role in the field of renewable energy (like wind and solar) to store energy and then supply it to usufructuary. Therewith to increase the vehicle efficiency that demands storing kinetic energy and release whenever the vehicle stops or slows down and many more uses. All these operations are also performed by the rechargeable battery on a low power scale so we need an enhancement in the efficiency so there is a need for energy storage technology that provides power on a large scale this requirement should be done by another energy storage device known as supercapacitor,[60] just like rechargeable batteries SCs also consists of an electrode, separator, electrolyte, and current collector which collects the current as shown in Figure 18.5.[61] Evaluation of supercapacitor is from 1950 through the design of rechargeable batteries and fuel cells.[62] Supercapacitor is also based on electrochemical energy conversion principle, but it comes into the limelight when a low resistance supercapacitor was developed by a pinnacle research institute for the US military purposes in 1982 and known as PRI ultra-capacitor.[63] The fascination behind this energy storage device is there long charging-discharging cycle life, high specific capacitance, free of memory effect, and low maintenance but lower energy density than rechargeable batteries.[64] So, the researcher is interested in the enhancement of SCs energy density comparable to rechargeable batteries. Like normal capacitors, SCs consist of thin dielectric layers and electrolytes inside them which separate electrodes having high surface area,[65] so the huge number of charges are stored that's why it stocks more energy than a normal capacitor. Mainly supercapacitor is of three types: pseudo capacitor, electrochemical double-layer type capacitor (EDLC),[66] and the third one is a combination of a pseudo capacitor and EDLC known as hybrid type.[66]

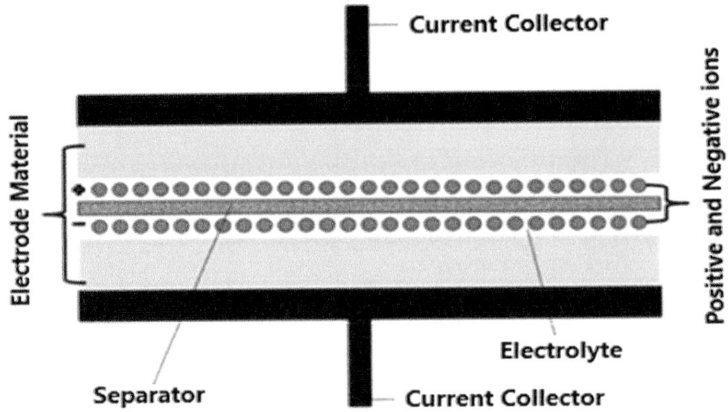

FIGURE 18.5 Supercapacitor consisting of electrode material separator, electrolyte, and current collector.[61]

Source: Reprinted from R ef. [61] © 2018. E lsevier Ltd.

In EDLC type, two different nature of charges are impulsively generated at the electrolyte and electrode interface when submersion of the electrode is done into the electrolyte, where either electrode atom interfuse into electrolyte or electrolyte ions are getting adsorbed on the electrode surface,[61] that is, there is no faradic charge storage process, and no redox reaction are involved here. Two different natures of charges generated on the interface are characterized as double layers of electric charges. When the electrode is connected to a power supply, then a double-layer capacitor gets charged and if the power supply is removed then there is no flow of electrons, so the charge is stored at the interface of electrode/electrolyte until the external is endowed;[67] while a pseudocapacitor is slightly different from the previous one because it works on electrochemical redox reactions and storage of energy is through electron transfer between electrolyte ions and electrodes.[65] Due to its faradic origin, it involves metal-doped nanofiber, conductive polymer, or metal-doped carbon-based high-energy electrode materials which enable SCs with a high energy density that's why pseudo capacitors provide high energy density than EDLCs.[68] A hybrid supercapacitor is the combination of a pseudo capacitor and EDLC, HSCs having enhanced features than other two, it's one half works like Pseudo capacitor and other half acts like an EDLC, so HSCs wins the race of maximum power densities and energy densities with other supercapacitors, which favors the use as compared to other two in the energy storage system as shown in Ragone plot (Figure 18.6) for different energy storage systems.[61] Different aspects of supercapacitor and capacitor in terms of power density as shown in the plot is due to dielectric plates used in capacitors for charge storage.[64]

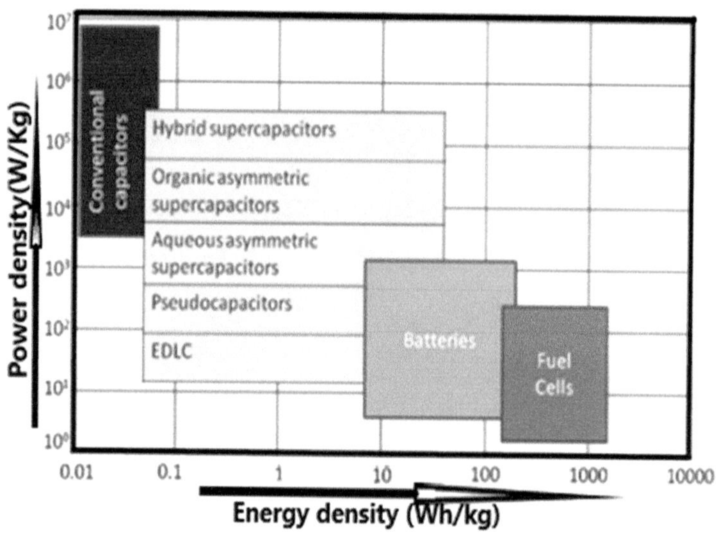

FIGURE 18.6 Ragone plot between energy density and power density.[61]

Source: Reprinted from R ef. [61] © 2018 E lsevier Ltd.

Since EDLCs stores energy through the electric charge compilation on the layer formed on the interface of electrode/electrolyte[70] and capacitance depends on the available surface area to the electrolyte. So for maximum surface area, porous carbon, graphene nanosheet, and lightweight carbon nanofiber which is fabricated in fiber form through electrospinning method and another nanofiber fabrication method (listed in Figure 18.1) is used.

Good temperature stability and conductivity, corrosion resistance, long-term chemical stability, cheap cost, are the fundamental requirements of an electrode material used in SCs. Since accessibility of the surface area to the electrolyte is an important factor in the EDLCs, that is, energy stored per unit voltage in supercapacitors is the function of electrode surface area. So the urge is electrodes which are made up of spongy and porous materials having high surface area, this urge is completed by use of porous carbon-based nanofibers like graphene, carbon nanotube, carbon fiber cloth, and nitrogen-enriched carbon nanofiber which is highly porous and provides high specific surface area.[71–73] Also, transition metal oxide encapsulated on electrospun carbon nanofibers like Co_3O_4/CNFs,[74] MnO_2/CNFs,[75,76] Fe_3O_4/CNFs,[77] MnOx/CNFs,[78] RuO_2/ACNFs,[79] ZnO/ACNFs,[80] NiO/RuO_2/CNFs,[81] V_2O_5/CNFs,[82] and Fe/CeO_2/CNFs[83] exhibit excellent electrochemical performance and electrical conductivity. Nanograin structure and porous fiber with α-Fe_2O_3 prepared by the electrospun method used as electrode material which displays excellent capacitance retentions above 3000 cycles.[84] In this series, some group makes NiO NFs improve by using citric acid which gives improved capacitive performance (336 F/g),[85] and very good cyclic performance. RuO_2 NF shows a specific capacitance of 886.8 F/g at a rate of 10 mV/s and capacity loss of only 30% from 10 to 2000 mV/s.[86] $La_xSr_{1-x}Cu_{0.1}Mn_{0.3}O_{3-x}$ ($0.1 \leq x \leq 1$) NFs, $La_xSr_{1-x}CoO_{3-x}$ ($0.3 \leq x \leq 1$) NFs, and $La_xSr_{1-x}NiO_{3-x}$ ($0.3 \leq x \leq 1$) NFs,[87,88] one dimension bimetallic $ZnCo_2O_4$ NTs shows a much higher specific capacitance of 770 F/g at 10 A/g and excellent cyclic performance only 10% loss after 3000 cycles.[89]

These examples show that nanofibers used in supercapacitors have high specific capacities and excellent electrochemical performances.

18.2.4 RECHARGEABLE BATTERIES

Discussion centralized on the use of nanofiber in lithium-ion batteries in which either lithium ions flow from cathode to anode through an electrolyte during charging and reverse when discharging.[90] In lithium-ion batteries, intercalated lithium compound is used as a cathode, and as an anode generally, graphite is used. In comparison to other prevailing energy storage devices, Li ion-based rechargeable batteries are more effectively fulfills the requirement of high electrochemical energy sources. Because it offers power and energy densities more than two-three times than traditional Ni–metal hydride and Ni–Cd-based batteries, and other energy storage devices are shown in Ragone plot (Figure 18.6). Lithium-ion batteries have high coulombic efficiency, high operating voltage, low self-discharge, and are free from any memory effect.[91] Supercapacitor which is evolved version of batteries having lower energy density than batteries. Li-ion batteries also consist of the cathode, anode, and electrolyte that lies in between these two electrodes which segregate the physical contact between them, which supports lithium-ion transfer during charging and discharging. Existing materials used in lithium-ion batteries are active powder materials like $LiFePO_4$[92] and graphite powder used in cathode and anode, respectively, to store energy but there is drawn back related with active powder materials. They provide a long migration path for the lithium-ion which results in large volume expansion during cycling that hinders the performance rate and affects cycling.[93] To remove these drawbacks there is a need for advanced material-based anode–cathode and electrolyte. Nanofibers fulfill material requirements because they provide a higher area mass ratio, high mechanical strength offers a large no. of lithium insertion sites, and shorter migration path for li-ion, due to which faster intercalation kinetics is expected. This shows that nanofibers are how much advantageous and promising for rechargeable batteries, in which some of them are briefly summarized here.

Carbon nanofibers (CNFs) are popular anode material for lithium-ion batteries because of their ease of availability, low cost, and long cycle life but it has a low rate capability and specific capacity, these drawbacks are reduced by highly flexible porous carbon nanofibers which deliver 1780 mAh/g at a current density of 50 mA/g after 40 cycles,[94] to increase electrochemical performance we use composites of nanofibers with transition metals and metal oxides, for example, Ag/CNFs in which Ag nanoparticles are embedded into CNFs which deliver 739 mAh/g at 50 mA/g with good capacity retention after 100 cycles[95] in this series carbon-coated Fe_3O_4 hollow nanofibers having excellent electrochemical performance with Li^+ diffusion coefficient (1.33×10^{-15} cm^2/s),[96] encapsulated GeO_x nanoparticles into hollow carbon shells by electrospinning[97] and other transition metal oxide like Co_3O_4/CNF core-shell,[98] γ-Fe_3O_4/CNFs,[99] and Mn_3O_4/CNF[100] are also used for the high-performance anode. Due to high lithium storage capacity (~4200 mAh/g),[101] silicon is good for anode with the format of membrane or nanowire but there is an issue of capacity loss during charging/discharging and electrode pulverization. So Si/CNFs are used as the anode which shows a fast discharge capacity of 1639 mAh/g. Tin (Sn) nanoparticle-loaded nanofibers are also used as a promising anode material due to their high capacity (990 mAh/g).[102]

High volumetric capacity, high discharge voltage, high specific heat capacity, and good cycling performance are requirements for any cathode material. The first commercial cathode is $LiCoO_2$ for lithium-ion batteries to achieve the performance as mentioned above some researchers synthesized the $LiCoO_2$ nanofibers which have higher first cycle discharge capacities (182 mAh/g) than powder-based cathode (140 mAh/g),[103,104] but poor cyclability. In this series, $LiCoO_2$-MgO NFs-based electrodes have better cyclability which still maintains a discharge capacity of around 90% of the initial value after 40 cycles and effectively avoids the increment in impedance generated during cycling by passive surface film formation.[105] Ternary layered material-based nanofibers like Li-Ni-Co-Mn-O, $LiNi_{0.5}Co_{0.2}Mn_{0.3}O_{2,}$ and $LiNi_{1/3}Co_{1/3}Mn_{1/3}O_2$[106–108] are also attractive for cathode material that has good thermal stability, high capacity, and good cyclability. Cathode materials with one electron reaction have limited practical capacity, which should be increased by developing a material that involves two or more electron reactions. Vanadium pentaoxide (V_2O_5) with the availability of two-electron reaction and also its low cost, rich abundance, high capacity, and ease of synthesis[109] makes them promising material in use as nanofiber for cathode material. The discharge capacity of V_2O_5 nanofibers for the first cycle in lithium-ion batteries is 350mAh/g and maintained capacity above 240 mAh/g even after 25 cycles. These are some of the applications of nanofibers in batteries.[110]

18.2.5 PHASE CHANGE MATERIAL (PCMS)

There is a different forms of energy storage methods like mechanical energy storage, electrical storage, thermal energy storage (sensible and latent heat storage), and thermochemical energy storage, and if we particularly talking about thermal heat storage techniques latent heat energy storage is attractive because it provides high-energy storage density.[111] To store this, thermal energy phase-change materials (PCMs) are promising materials, as the name suggests phase change materials are those materials, which change their phase during heating or cooling and release or absorb energy, this phase transition occurs between one state of matter to another state of matter it can be done from solid–solid, solid–liquid, solid–gas, liquid–gas, and vice versa.[112] PCMs easily respond to changes in environmental temperature and absorb store and release heat without any change in their temperature but there is a problem of leakage during this absorb and release process called the charging-discharging process which shortened the lifetime of PCM.[113] To reduce these drawbacks and for enhancement of thermal conductivity, phase change nanofibers come into picture which are prepared by electrospinning technology by encapsulating PCMs in support materials. PCM-based nanofibers are used in the field of energy storage, textiles, biotechnology, and many more because of their high surface area and high heat transfer rate. It was seen that in pulsed power operation requirement of solidification time is the same as that of solid-to-liquid transition duration. But diffusion is dominated over solidification and for fast transient power. The low thermal conductivity of the PCM is not a good choice so the thermal conductivity enhancers are used during the melt phase to amend the low diffusion rate, one of example is the addition of graphite nanofibers in n-tricosane (alkane),[114] which improves thermal performance of an organic PCM (n-tricosane) during melt phase. Generally, organic-based PCMs have been devised with the extraordinary properties of high thermal storage density, corrosion resistance, superior thermal stability, cost-effectiveness, and have room temperature transition which is necessary for textiles application also there is a future scope in dual-functional phase change material that holds energy storage capability. In this series, electrospun SiO_2 nanofiber and eutectic mixture (CA-LA-PA)-based PCM is used for thermal energy storage with high latent heat than those of individual fatty acids with lower phase transition temperature and good repeatability in thermal cycling also exhibits good thermal reliability.[115] Some of the polymer-based PCM nanofibers are as shown in the Table 18.1.

TABLE 18.1 Polymer-Based PCM Nanofibers.

Polymer	PCM	Cooling		Heating		Reference
		T_f (°C)	LHC (kJ/kg)	T_m (°C)	LHF (kJ/kg)	
PET	CA and LA	–	–	27.5	140	116
	CA and PA			31	152	
PS	LA	41.7	73.2	44.3	73.6	117

TABLE 18.1 *(Continued)*

Polymer	PCM	Cooling		Heating		Reference
		T_f (°C)	LHC (kJ/kg)	T_m (°C)	LHF (kJ/kg)	
PAN	LA-SA	–	–	42.9	80.3	118
SAN	LA	39.2	27.8	42.2	56.5	119
Silver-coated PU	CA-LA-MA-PA-SA	~12	97-106	~17	97-106	120
PAN	Methyl palmitate	18.7	110.8	31.9	110	121
	Ethyl stearate	19.8	95.6	29.7	93.4	
	Ethyl palmitate	12	95.9	29.4	95.4	
	Butyl stearate	12.6	55.1	21.2	55.1	
PMIA	Fatty acid ester of PA and SA	26.8	23.4	36.8	25.9	122
PMIA	LA and MA	10.9	29.7	20.5	31.1	123
PET	Lauric acid	38	32.1	44.6	57	121
PET	LA	45	51.9	40	54.9	124
	MA	55	55.9	51	61	
	PA	63	58.7	57	66.9	
	SA	69	61.2	64	67.9	
PET	Methyl stearate	28	88	39.9	90.4	125
CAc	CA	29.7	152.1	32.5	148	126
PA6	LA	40.1	62.5	44.7	74.12	127

18.3 CONCLUSIONS

Due to fascinating character of nanofibers having extraordinary chemical and physical properties which makes them a very good option for many energies application like nanogenerators, DSSCs, supercapacitors, PCMs, and rechargeable batteries, especially for high-density lithium-ion and sodium-ion batteries. Easily available and cheapest nanofiber making technique is used, known as electrospinning technique. Nanofibers are used as tribopositve and tribonegative electrode material in TENG to convert mechanical energy into electrical by triboelectrification and in DSSCs nanofiber-based photoelectrode increases the energy conversion efficiency. Nanofiber-based supercapacitor exhibits high specific capacities, and good electrochemical performance, and it is due to good thermal stabilities, high electrical conductivities, and high specific surface area of nanofibers. Whereas in lithium-ion batteries, NFs used as materials of electrodes that provide low diffusion pathways that increase the kinetic properties and high-power capabilities. In PCMs to increase the lifetime of material by avoiding leakage of PCM during charging and discharging cycle and to enhance the thermal conductivity encapsulation of PCM into nanofiber should be done or electrospun PCM had been developed. This is a short and quick overview of nanofiber in energy application but as earlier use a word "fascinating"is much meaningful for NFs. This fascinating fiber still has huge boundaries to explore in energy applications and many other applications.

ACKNOWLEDGMENT

Saurabh Pandey would like to thank Rajiv Gandhi Institute of Petroleum Technology (RGIPT) for research support.

KEYWORDS

- **energy storage devices**
- **nanofibers**
- **electrospinning**
- **phase change materials (PCMs)**

REFERENCES

1. Huang, Z. M.; Zhang, Y. Z.; Kotaki, M.; Ramakrishna, S. A Review on Polymer Nanofibers by Electrospinning and Their Applications in Nanocomposites. *Compos. Sci. Technol.* **2003,** *63*, 2223–2253. https://doi.org/10.1016/S0266-3538(03)00178-7.
2. Jiang, S.; Chen, Y.; Duan, G.; Mei, C.; Greiner, A.; Agarwal, S. Electrospun Nanofiber Reinforced Composites: A Review. *Polym. Chem.* **2018,** *9*, 2685–2720. https://doi.org/10.1039/c8py00378e.
3. Tiwari, J. N.; Tiwari, R. N.; Kim, K. S. Zero-Dimensional, One-Dimensional, Two-Dimensional and Three-Dimensional Nanostructured Materials for Advanced

Electrochemical Energy Devices. *Prog. Mater. Sci.* **2012,** *57*, 724–803. https://doi.org/10.1016/j.pmatsci.2011.08.003.
4. Xu, J.; Wang, K.; Zu, S.; Han, B.; Wei, Z. 17号 Nn1006539.Pdf, **2010,** *4*, 5019–5026.
5. Gao, W.; Razavi, R.; Fakhri, A. Preparation and Development of Fes2 Quantum Dots on Sio2 Nanostructures Immobilized in Biopolymers and Synthetic Polymers as Nanoparticles and Nanofibers Catalyst for Antibiotic Degradation. *Int. J. Biol. Macromol.* **2018,** *114*, 357–362. https://doi.org/10.1016/j.ijbiomac.2018.03.119.
6. Li, D.; McCann, J. T.; Xia, Y.; Marquez, M. Electrospinning: A Simple and Versatile Technique for Producing Ceramic Nanofibers and Nanotubes. *J. Am. Ceram. Soc.* **2006,** *89*, 1861–1869. https://doi.org/10.1111/j.1551-2916.2006.00989.x.
7. Huang, Y.; Song, J.; Yang, C.; Long, Y.; Wu, H. Scalable Manufacturing and Applications of Nanofibers, Mater. *Today.* **2019,** *28*, 98–113. https://doi.org/10.1016/j.mattod.2019.04.018.
8. Thavasi, V.; Singh, G.; Ramakrishna, S. Electrospun Nanofibers in Energy and Environmental Applications. *Energy Environ. Sci.* **2008,** *1*, 205. https://doi.org/10.1039/b809074m.
9. Sun, G.; Sun, L.; Xie, H.; Liu, J. Electrospinning of Nanofibers for Energy Applications. *Nanomaterials.* **2016,** 6. https://doi.org/10.3390/nano6070129.
10. Gibson, P.; H. Schreuder-Gibson, Pentheny, C. Electrospinning Technology: Direct Application of Tailorable Ultrathin Membranes. *J. Ind. Text.* **1998,** *28*, 63–72. https://doi.org/10.1177/152808379802800106.
11. Raghavan, P.; Lim, D. H.; Ahn, J. H.; Nah, C.; Sherrington, D. C.; Ryu, H. S.; Ahn, H. J. Electrospun Polymer Nanofibers: The Booming Cutting Edge Technology. *React. Funct. Polym.* **2012,** *72*, 915–930. https://doi.org/10.1016/j.reactfunctpolym.2012.08.018.
12. Shi, X.; Zhou, W.; Ma, D.; Ma, Q.; Bridges, D.; Review Article Electrospinning of Nanofibers and Their Applications for Electrospinning of Nanofibers and Their Applications for. **2015,** 2015. https://doi.org/10.1155/2015/140716.
13. Bhardwaj, N.; Kundu, S. C.; Electrospinning: A Fascinating Fiber Fabrication Technique. *Biotechnol. Adv.* **2010,** *28*, 325–347. https://doi.org/10.1016/j.biotechadv.2010.01.004.
14. Kohse, S.; Grabow, N.; Schmitz, K. P.; Eickner, T. Electrospinning of Polyimide Nanofibres–Effects of Working Parameters on Morphology. *Curr. Dir. Biomed. Eng.* **2017,** *3*, 687–690. https://doi.org/10.1515/cdbme-2017-0145.
15. Gugulothu, D.; Barhoum, A.; Nerella, R.; Ajmer, R.; Bechelany, M. Fabrication of Nanofibers: Electrospinning and Non-electrospinning Techniques. In *Handb, Nanofibers*, Springer International Publishing, 2019; pp. 45–77. https://doi.org/10.1007/978-3-319-53655-2_6.
16. Lan, Y.; Wang, Y.; Ren, Z. F. Physics and Applications of Aligned Carbon Nanotubes. *Adv. Phys.* **2011,** *60*, 553–678. https://doi.org/10.1080/00018732.2011.599963.
17. Ogueri, K. S.; Laurencin, C. T. Nanofiber Technology for Regenerative Engineering. *ACS Nano*. **2020,** *14*, 9347–9363. https://doi.org/10.1021/acsnano.0c03981.
18. Wang, Z. L.; Zhu, G.; Yang, Y.; Wang, S.; Pan, C. Progress in Nanogenerators for Portable Electronics. *Mater. Today* **2012,** *15*, 532–543. https://doi.org/10.1016/S1369-7021(13)70011-7.
19. Bera, B. Literature Review on Triboelectric Nanogenerator. *Imp. J. Interdiscip. Res. (IJIR).* **2016,** *2*, 1263–1271.

20. Wang, Y.; Yang, Y.; Wang, Z. L. Triboelectric Nanogenerators as Flexible Power Sources. *Npj Flex. Electron.* **2017,** *1*, 1–9. https://doi.org/10.1038/s41528-017-0007-8.
21. Cheng, T.; Gao, Q.; Wang, Z. L. The Current Development and Future Outlook of Triboelectric Nanogenerators: A Survey of Literature. *Adv. Mater. Technol.* **2019,** *4*, 1–7. https://doi.org/10.1002/admt.201800588.
22. Kim, W. G.; Kim, D. W.; Tcho, I. W.; Kim, J. K.; Kim, M. S.; Choi, Y. K. Triboelectric Nanogenerator: Structure, Mechanism, and Applications. *ACS Nano.* **2021,** *15*, 258–287. https://doi.org/10.1021/acsnano.0c09803.
23. Zhang, R.; Olin, H.; Material Choices for Triboelectric Nanogenerators: A Critical Review. *EcoMat.* **2020,** *2*, 1–13. https://doi.org/10.1002/eom2.12062.
24. Lee, B. Y.; Kim, S. U.; Kang, S.; Lee, S. D. Transparent and Flexible High Power Triboelectric Nanogenerator with Metallic Nanowire-Embedded Tribonegative Conducting Polymer. *Nano. Energy* **2018,** *53*, 152–159. https://doi.org/10.1016/j.nanoen.2018.08.048.
25. Mahapatra, B.; Kumar Patel, K.; Vidya, Patel, P. K. A Review on Recent Advancement in Materials for Piezoelectric/Triboelectric Nanogenerators. *Mater. Today Proc.* **2020,** *46*, 5523–5529. https://doi.org/10.1016/j.matpr.2020.09.261.
26. R. u. Ahmad, S.; Haleem, A.; Haider, Z.; Claver, U. P.; Fareed, A.; Khan, I.; Mbogba, M. K.; Memon, K.; Ali, W.; He, W.; Hu, P.; Zhao, G. Realizing the Capability of Negatively Charged Graphene Oxide in the Presence of Conducting Polyaniline for Performance Enhancement of Tribopositive Material of Triboelectric Nanogenerator. *Adv. Electron. Mater.* **2020,** *6*, 1–14. https://doi.org/10.1002/aelm.202000034.
27. Yar, A.; Karabiber, A.; Ozen, A.; Ozel, F.; Coskun, S. Flexible Nanofiber Based Triboelectric Nanogenerators with High Power Conversion. *Renew. Energy* **2020,** *162*, 1428–1437. https://doi.org/10.1016/j.renene.2020.08.030.
28. Choi, Y. S.; Jing, Q.; Datta, A.; Boughey, C.; Kar-Narayan, S. A Triboelectric Generator based on Self-Poled Nylon-11 Nanowires Fabricated by Gas-Flow Assisted Template Wetting. *Energy Environ. Sci.* **2017,** *10*, 2180–2189. https://doi.org/10.1039/c7ee01292f.
29. Jing, Q.; Kar-Narayan S. Nanostructured Polymer-Based Piezoelectric and Triboelectric Materials and Devices for Energy Harvesting Applications. *J. Phys. D. Appl. Phys.* **2018,** 51. https://doi.org/10.1088/1361-6463/aac827.
30. Shao, J. J.; Jiang, T.; Wang, Z. L. Theoretical Foundations of Triboelectric Nanogenerators (TENGs). *Sci. China Technol. Sci.* **2020,** *63*, 1087–1109. https://doi.org/10.1007/s11431-020-1604-9.
31. Zhou, L.; Liu, D.; Wang, J.; Wang, Z. L. Triboelectric Nanogenerators: Fundamental Physics and Potential Applications. *Friction* **2020,** *8*, 481–506. https://doi.org/10.1007/s40544-020-0390-3.
32. Chen, A.; Zhang, C.; Zhu, G.; Wang, Z. L. Polymer Materials for High-Performance Triboelectric Nanogenerators. *Adv. Sci.* **2020,** *7*, 1–25. https://doi.org/10.1002/advs.202000186.
33. Pan, S.; Zhang, Z.; Fundamental Theories and Basic Principles of Triboelectric Effect: A Review. *Friction* **2019,** *7*, 2–17. https://doi.org/10.1007/s40544-018-0217-7.
34. Zhu, G.; Peng, B.; Chen, J.; Jing, Q.; Lin Wang, Z. Triboelectric Nanogenerators as a New Energy Technology: From Fundamentals, Devices, to Applications. *Nano Energy* **2014,** *14*, 126–138. https://doi.org/10.1016/j.nanoen.2014.11.050.

35. Vasandani, P.; Mao, Z. H.; Jia, W.; Sun, M. Relationship Between Triboelectric Charge and Contact Force for Two Triboelectric Layers. *J. Electrostat.* **2017,** *90*, 147–152. https://doi.org/10.1016/j.elstat.2017.11.001.
36. Wu, C.; Wang, A. C.; Ding, W.; Guo, H.; Wang, Z. L. Triboelectric Nanogenerator: A Foundation of the Energy for the New Era. *Adv. Energy Mater.* **2019,** *9*, 1–25. https://doi.org/10.1002/aenm.201802906.
37. Gong, J.; Sumathy, K.; Qiao, Q.; Zhou, Z. Review on Dye-Sensitized Solar Cells (DSSCs): Advanced Techniques and Research Trends. *Renew. Sustain. Energy Rev.* **2017,** *68*, 234–246. https://doi.org/10.1016/j.rser.2016.09.097.
38. Li, G.; Sheng, L.; Li, T.; Hu, J.; Li, P.; Wang, K. Engineering Flexible Dye-Sensitized Solar Cells for Portable Electronics. *Sol. Energy* **2019,** *177*, 80–98. https://doi.org/10.1016/j.solener.2018.11.017.
39. Pandey, S.; Karakoti, M.; Surana, K.; Dhapola, P. S.; SanthiBhushan, B.; Ganguly, S.; Singh, P. K.; Abbas, A.; Srivastava, A.; Sahoo, N. G. Graphene Nanosheets Derived from Plastic Waste for the Application of DSSCs and Supercapacitors. *Sci. Rep.* **2021,** *11*, 1–17. https://doi.org/10.1038/s41598-021-83483-8.
40. Yoon, S.; Tak, S.; Kim, J.; Jun, Y.; Kang, K.; Park, J. Application of Transparent Dye-Sensitized Solar Cells to Building Integrated Photovoltaic Systems. *Build. Environ.* **2011,** *46*, 1899–1904. https://doi.org/10.1016/j.buildenv.2011.03.010.
41. Gong, J.; Liang, J.; Sumathy, K. Review on Dye-Sensitized Solar Cells (DSSCs): Fundamental Concepts and Novel Materials. *Renew. Sustain. Energy Rev.* **2012,** *16*, 5848–5860. https://doi.org/10.1016/j.rser.2012.04.044.
42. Tripathi, B.; Yadav, P.; Kumar, M. Charge Transfer and Recombination Kinetics in Dye-Sensitized Solar Cell using Static and Dynamic Electrical Characterization Techniques. *Sol. Energy* **2014,** *108*, 107–116. https://doi.org/10.1016/j.solener.2014.06.037.
43. Andualem, A.; Demiss, S. Review on Dye-Sensitized Solar Cells (DSSCs). *J. Heterocycl.* **2018,** *1*, 29–34. https://doi.org/10.33805/2639-6734.103.
44. Sugathan, V.; John, E.; Sudhakar, K. Recent Improvements in Dye Sensitized Solar Cells: A Review. *Renew. Sustain. Energy Rev.* **2015,** *52*, 54–64. https://doi.org/10.1016/j.rser.2015.07.076.
45. Cao, Y.; Dong, Y. J.; Feng, H. L.; Chen, H. Y.; Bin Kuang, D. Electrospun TiO2 Nanofiber Based Hierarchical Photoanode for Efficient Dye-Sensitized Solar Cells. *Electrochim. Acta* **2016,** *189*, 259–264. https://doi.org/10.1016/j.electacta.2015.12.073.
46. Lu, D.; Qin, L.; Liu, D.; Sun, P.; Liu, F.; Lu, G. High-Efficiency Dye-Sensitized Solar Cells Based on Bilayer Structured Photoanode Consisting of Carbon Nanofiber/TiO2 Composites and Ag@TiO2 Core-Shell Spheres. *Electrochim. Acta* **2018,** *292*, 180–189. https://doi.org/10.1016/j.electacta.2018.09.146.
47. Earis, P.; Nugent, N.; Martinez-fresno, M.; Wilkes, S.; Spring, S.; Dean, J.; Lewis, N.; Berkeley, U. C.; Bolt, H. Energy & Environmental Science Energy & Environmental Science. *Energy* **2009,** 3–10.
48. Hieu, N. T.; Baik, S. J.; Chung, O. H.; Park, J. S. Fabrication and Characterization of Electrospun Carbon Nanotubes/Titanium Dioxide Nanofibers used in Anodes of Dye-Sensitized Solar Cells. *Synth. Met.* **2014,** *193*, 125–131. https://doi.org/10.1016/j.synthmet.2014.04.010.

49. Nien, Y. H.; Chen, H. H.; Hsu, H. H.; Rangasamy, M.; Hu, G. M.; Yong, Z. R.; Kuo, P. Y.; Chou, J. C.; Lai, C. H.; Ko, C. C.; Chang, J. X. Study of How Photoelectrodes Modified by TiO2/Ag Nanofibers in Various Structures Enhance the Efficiency of Dye-Sensitized Solar Cells Under Low Illumination. *Energies* **2020,** 13. https://doi.org/10.3390/en13092248.
50. Aboagye, A.; Elbohy, H.; Kelkar, A. D.; Qiao, Q.; Zai, J.; Qian, X.; Zhang, L. Electrospun Carbon Nanofibers with Surface-Attached Platinum Nanoparticles as Cost-Effective and Efficient Counter Electrode for Dye-Sensitized Solar Cells. *Nano Energy* **2015,** *11*, 550–556. https://doi.org/10.1016/j.nanoen.2014.10.033.
51. D. Sebastián, Baglio, V.; Girolamo, M.; Moliner, R.; Lázaro, M. J.; Aricò, A. S. Carbon Nanofiber-Based Counter Electrodes for Low Cost Dye-Sensitized Solar Cells. *J. Power Sources* **2014,** *250*, 242–249. https://doi.org/10.1016/j.jpowsour.2013.10.142.
52. Park, S. H.; Kim, B. K.; Lee, W. J.; Electrospun Activated Carbon Nanofibers with Hollow Core/Highly Mesoporous Shell Structure as Counter Electrodes for Dye-Sensitized Solar Cells. *J. Power Sources* **2013,** *239*, 122–127. https://doi.org/10.1016/j.jpowsour.2013.03.079.
53. Gao, C.; Han, Q.; Wu, M. Review on Transition Metal Compounds based Counter Electrode for Dye-Sensitized Solar Cells. *J. Energy Chem.* **2018,** *27*, 703–712. https://doi.org/10.1016/j.jechem.2017.09.003.
54. Gnanasekar, S.; Kollu, P.; Jeong, S. K.; Grace, A. N. Pt-Free, Low-Cost and Efficient Counter Electrode with Carbon Wrapped Vo 2 (M) Nanofiber for Dye-Sensitized Solar Cells. *Sci. Rep.* **2019,** *9*, 1–12. https://doi.org/10.1038/s41598-019-41693-1.
55. Wu, J.; Lan, Z.; Hao, S.; Li, P.; Lin, J.; Huang, M.; Fang, L.; Huang, Y. Progress on the Electrolytes for Dye-Sensitized Solar Cells. *Pure Appl. Chem.* **2008,** *80*, 2241–2258. https://doi.org/10.1351/pac200880112241.
56. M. S. Su'ait, Rahman, M.; Y. A.; Ahmad, A. Review on Polymer Electrolyte in Dye-Sensitized Solar Cells (DSSCs). *Sol. Energy* **2015,** *115*, 452–470. https://doi.org/10.1016/j.solener.2015.02.043.
57. Lai, Y. H.; Chiu, C. W.; Chen, J. G.; Wang, C. C.; Lin, J. J.; Lin, K. F.; Ho, K. C. Enhancing the Performance of Dye-Sensitized Solar Cells by Incorporating Nanosilicate Platelets in Gel Electrolyte. *Sol. Energy Mater. Sol. Cells* **2009,** *93*, 1860–1864. https://doi.org/10.1016/j.solmat.2009.06.027.
58. Bandara, T. M. W. J.; Weerasinghe, A. M. J. S.; Dissanayake, M. A. K. L.; Senadeera G. K. R.; Furlani, M.; Albinsson, I.; Mellander, B. E. Characterization of Poly (Vinylidene Fluoride-Co-Hexafluoropropylene) (PVdF-HFP) Nanofiber Membrane based Quasi Solid Electrolytes and Their Application in a Dye Sensitized Solar Cell. *Electrochim. Acta* **2018,** *266*, 276–283. https://doi.org/10.1016/j.electacta.2018.02.025.
59. Zhao, X. G.; Jin, E. M.; Park, J. Y.; Gu, H. B. Hybrid Polymer Electrolyte Composite with SiO2 Nanofiber Filler for Solid-State Dye-Sensitized Solar Cells. *Compos. Sci. Technol.* **2014,** *103*, 100–105. https://doi.org/10.1016/j.compscitech.2014.08.020.
60. González, Goikolea, E.; Barrena, J. A.; Mysyk, R. Review on Supercapacitors: Technologies and Materials. *Renew. Sustain. Energy Rev.* **2016,** *58*, 1189–1206. https://doi.org/10.1016/j.rser.2015.12.249.
61. Muzaffar, A.; Ahamed, M. B.; Deshmukh, K.; Thirumalai, J. A Review on Recent Advances in Hybrid Supercapacitors: Design, Fabrication and Applications. *Renew. Sustain. Energy Rev.* **2019,** *101*, 123–145. https://doi.org/10.1016/j.rser.2018.10.026.

62. 318457763_Application_of_Battery_Energy_Storage_System_to_Optimize_Renewable_Energy_Resources_to_the_Grid @ www.researchgate.net, (2017). https://www.researchgate.net/publication/318457763_Application_of_Battery_Energy_Storage_System_to_Optimize_Renewable_Energy_Resources_to_the_Grid.
63. Sharma, P.; Bhatti, T. S. A Review on Electrochemical Double-Layer Capacitors. *Energy Convers. Manag.* **2010,** *51*, 2901–2912. https://doi.org/10.1016/j.enconman.2010.06.031.
64. Kim, B. K.; Sy, S.; Yu, A.; Zhang, J. Electrochemical Supercapacitors for Energy Storage and Conversion, Handb. *Clean Energy Syst.* **2015,** 1–25. https://doi.org/10.1002/9781118991978.hces112.
65. Karthikeyan, S.; Narenthiran, B.; Sivanantham, A.; Bhatlu, L. D.; Maridurai, T. Supercapacitor: Evolution and Review. In *Mater. Today Proc.*, Elsevier Ltd, 2020; pp. 3984–3988. https://doi.org/10.1016/j.matpr.2021.02.526.
66. Vangari, M.; Pryor, T.; Jiang, L. Supercapacitors: Review of Materials and Fabrication Methods. *J. Energy Eng.* **2013,** *139*, 72–79. https://doi.org/10.1061/(asce)ey.1943-7897.0000102.
67. Libich, J.; Máca, J.; Vondrák, J.; Čech, O.; Sedlaříková, M. Supercapacitors: Properties and Applications. *J. Energy Storage.* **2018,** *17*, 224–227. https://doi.org/10.1016/j.est.2018.03.012.
68. Stoller, M. D.; Park, S.; Yanwu, Z.; An, J.; Ruoff, R. S. Graphene-Based Ultracapacitors. *Nano Lett.* **2008,** *8*, 3498–3502. https://doi.org/10.1021/nl802558y.
69. Christen, T.; Carlen, M. W. Theory of Ragone Plots. *J. Power Sources* **2000,** *91*, 210–216. https://doi.org/10.1016/S0378-7753(00)00474-2.
70. Huang, Y.; Liang, J.; Chen, Y. An Overview of the Applications of Graphene-Based Materials in Supercapacitors. *Small* **2012,** *8*, 1805–1834. https://doi.org/10.1002/smll.201102635.
71. Le, T.; Yang, Y.; Huang, Z.; Kang, F. Preparation of Microporous Carbon Nanofibers from Polyimide by using Polyvinyl Pyrrolidone as Template and Their Capacitive Performance. *J. Power Sources* **2015,** *278*, 683–692. https://doi.org/10.1016/j.jpowsour.2014.12.055.
72. Lawrence, D. W.; Tran, C.; Mallajoysula, A. T.; Doorn, S. K.; Mohite, A.; Gupta, G.; Kalra, V. High-Energy Density Nanofiber-Based Solid-State Supercapacitors. *J. Mater. Chem. A.* **2015,** *4*, 160–166. https://doi.org/10.1039/c5ta05552k.
73. Ma, C.; Song, Y.; Shi, J.; Zhang, D.; Zhai, X.; Zhong, M.; Guo, Q.; Liu, L. Preparation and One-Step Activation of Microporous Carbon Nanofibers for Use as Supercapacitor Electrodes. *Carbon N. Y.* **2013,** *51*, 290–300. https://doi.org/10.1016/j.carbon.2012.08.056.
74. Abouali, S.; Akbari Garakani, M.; Zhang, B.; Xu, Z. L.; Kamali Heidari E.; Huang, J. Q.; Huang, J.; Kim, J. K. Electrospun Carbon Nanofibers with In Situ Encapsulated Co3O4 Nanoparticles as Electrodes for High-Performance Supercapacitors. *ACS Appl. Mater. Interfaces* **2015,** *7*, 13503–13511. https://doi.org/10.1021/acsami.5b02787.
75. Zhi, M.; Manivannan, A.; Meng, F.; Wu, N. Highly Conductive Electrospun Carbon Nanofiber/MnO 2 Coaxial Nano-Cables for High Energy and Power Density Supercapacitors. *J. Power Sources* **2012,** *208*, 345–353. https://doi.org/10.1016/j.jpowsour.2012.02.048.

76. Wang, J. G.; Yang, Y.; Huang, Z. H.; Kang, F. Coaxial Carbon Nanofibers/MnO2 Nanocomposites as Freestanding Electrodes for High-Performance Electrochemical Capacitors. *Electrochim. Acta* **2011,** *56*, 9240–9247. https://doi.org/10.1016/j.electacta.2011.07.140.
77. Mu, J.; Chen, B.; Guo, Z.; Zhang, M.; Zhang, Z.; Zhang, P.; Shao, C.; Liu, Y. Highly Dispersed Fe3O4 Nanosheets on One-Dimensional Carbon Nanofibers: Synthesis, Formation Mechanism, and Electrochemical Performance as Supercapacitor Electrode Materials. *Nanoscale* **2011,** *3*, 5034–5040. https://doi.org/10.1039/c1nr10972c.
78. Zhao, X.; Du, Y.; Li, Y.; Zhang, Q. Encapsulation of Manganese Oxides Nanocrystals in Electrospun Carbon Nanofibers as Free-Standing Electrode for Supercapacitors. *Ceram. Int.* **2015,** *41*, 7402–7410. https://doi.org/10.1016/j.ceramint.2015.02.053.
79. Yang, K. S.; Kim, B. H. Highly Conductive, Porous RuO2/Activated Carbon Nanofiber Composites Containing Graphene for Electrochemical Capacitor Electrodes. *Electrochim. Acta* **2015,** *186*, 337–344. https://doi.org/10.1016/j.electacta.2015.11.003.
80. Kim, C. H.; Kim, B. H.; Zinc Oxide/Activated Carbon Nanofiber Composites for High-Performance Supercapacitor Electrodes. *J. Power Sources* **2015,** *274*, 512–520. https://doi.org/10.1016/j.jpowsour.2014.10.126.
81. Wu, Y.; Balakrishna, R.; Reddy, M. V.; Nair, A. S.; Chowdari B. V. R.; Ramakrishna, S. Functional Properties of Electrospun NiO/RuO 2 Composite Carbon Nanofibers. *J. Alloys Compd.* **2012,** *517*, 69–74. https://doi.org/10.1016/j.jallcom.2011.12.019.
82. Kim, B. H.; Kim, C. H.; Yang, K. S.; Rahy, A.; Yang, D. J. Electrospun Vanadium Pentoxide/Carbon Nanofiber Composites for Supercapacitor Electrodes. *Electrochim. Acta* **2012,** *83*, 335–340. https://doi.org/10.1016/j.electacta.2012.07.093.
83. Ji, S. M.; Ghouri, Z. K.; Elsaid, K.; Ko, Y. H.; S. Al-Meer, Ahmad, M. I.; Son, D. I.; Kim, H. Y. Capacitance of MnO2 Micro-Flowers Decorated CNFs in Alkaline Electrolyte and Its Bi-Functional Electrocatalytic Activity Toward Hydrazine Oxidation. *Int. J. Electrochem. Sci.* **2017,** *12*, 2583–2592. https://doi.org/10.20964/2017.03.73.
84. Binitha, G.; Soumya, M. S.; Madhavan, A. A.; Praveen, P.; Balakrishnan, A.; Subramanian K. R. V.; Reddy, M. V.; Nair, S. V.; Nair, A. S.; Sivakumar, N. Electrospun α-Fe2O3 Nanostructures for Supercapacitor Applications. *J. Mater. Chem. A.* **2013,** *1*, 11698–11704. https://doi.org/10.1039/c3ta12352a.
85. Ren, B.; Fan, M.; Liu, Q.; Wang, J.; Song, D.; Bai, X. Hollow NiO Nanofibers Modified by Citric Acid and the Performances as Supercapacitor Electrode. *Electrochim. Acta* **2013,** *92*, 197–204. https://doi.org/10.1016/j.electacta.2013.01.009.
86. Hyun, T. S.; Tuller, H. L.; Youn, D. Y.; Kim, H. G.; Kim, I. D. Facile Synthesis and Electrochemical Properties of RuO2 Nanofibers with Ionically Conducting Hydrous Layer. *J. Mater. Chem.* **2010,** *20*, 9172–9179. https://doi.org/10.1039/c0jm00494d.
87. Cao, Y.; Lin, B.; Sun, Y.; Yang, H.; Zhang, X. Synthesis, Structure and Electrochemical Properties of Lanthanum Manganese Nanofibers Doped with Sr and Cu. *J. Alloys Compd.* **2015,** *638*, 204–213. https://doi.org/10.1016/j.jallcom.2015.03.054.
88. Cao, Y.; Lin, B.; Sun, Y.; Yang, H.; Zhang, X.; Symmetric/Asymmetric Supercapacitor Based on the Perovskite-type Lanthanum Cobaltate Nanofibers with Sr-substitution. *Electrochim. Acta* **2015,** *178*, 398–406. https://doi.org/10.1016/j.electacta.2015.08.033.
89. KumarMarka, S.; Peruswamula V. V. H.; Srikanth Vadali V. S. S. Graphene and Related Materials as Anode Materials in Li Ion Batteries: Science and Practicality. *Nanocarbon Electrochem.* **2019,** 85–121. https://doi.org/10.1002/9781119468288.ch3.

90. Li, M.; Lu, J.; Chen, Z.; Amine, K. 30 Years of Lithium-Ion Batteries. *Adv. Mater.* **2018,** *30*, 1–24. https://doi.org/10.1002/adma.201800561.
91. Etacheri, V.; Marom, R.; Elazari, R.; Salitra, G.; Aurbach, D. Challenges in the Development of Advanced Li-Ion Batteries: A Review. *Energy Environ. Sci.* **2011,** *4*, 3243–3262. https://doi.org/10.1039/c1ee01598b.
92. Toprakci, O.; Toprakci, H. A. K.; Ji, L.; Zhang, X. 28_2010008, **2010,** *28*, 50–73.
93. Pampal, E. S.; Stojanovska, E.; Simon, B.; Kilic, A. A Review of Nanofibrous Structures in Lithium Ion Batteries. *J. Power Sources* **2015,** *300*, 199–215. https://doi.org/10.1016/j.jpowsour.2015.09.059.
94. Li, W.; Li, M.; Wang, M.; Zeng, L.; Yu, Y. Electrospinning with Partially Carbonization in Air: Highly Porous Carbon Nanofibers Optimized for High-Performance Flexible Lithium-Ion Batteries. *Nano Energy* **2015,** *13*, 693–701. https://doi.org/10.1016/j.nanoen.2015.03.027.
95. Yang, C.; Yao, Y.; He, S.; Xie, H.; Hitz, E.; Hu, L. Ultrafine Silver Nanoparticles for Seeded Lithium Deposition toward Stable Lithium Metal Anode. *Adv. Mater.* **2017,** *29*, 1–8. https://doi.org/10.1002/adma.201702714.
96. Wang, B.; Wang, G.; Zheng, Z.; Wang, H.; Bai, J.; Bai, J. Carbon Coated Fe3O4 Hybrid Material Prepared by Chemical Vapor Deposition for High Performance Lithium-Ion Batteries. *Electrochim. Acta* **2013,** *106*, 235–243. https://doi.org/10.1016/j.electacta.2013.05.085.
97. Li, M.; Zhou, D.; Song, W. L.; Li, X.; Fan, L. Z. Highly Stable GeOx@C Core-Shell Fibrous Anodes for Improved Capacity in Lithium-Ion Batteries. *J. Mater. Chem. A* **2015,** *3*, 19907–19912. https://doi.org/10.1039/c5ta05400a.
98. An, G. H.; Ahn, H. J. Carbon Nanofiber/Cobalt Oxide Nanopyramid Core-Shell Nanowires for High-Performance Lithium-Ion Batteries. *J. Power Sources* **2014,** *272*, 828–836. https://doi.org/10.1016/j.jpowsour.2014.09.032.
99. Wu, Y.; Zhu, P.; Reddy, M. V.; Chowdari B. V. R.; Ramakrishna, S. Maghemite Nanoparticles on Electrospun CNFs Template as Prospective Lithium-Ion Battery Anode. *ACS Appl. Mater. Interfaces* **2014,** *6*, 1951–1958. https://doi.org/10.1021/am404939q.
100. Park, S. H.; Lee, W. J. Hierarchically Mesoporous Carbon Nanofiber/Mn3O4 Coaxial Nanocables as Anodes in Lithium Ion Batteries. *J. Power Sources* **2015,** *281*, 301–309. https://doi.org/10.1016/j.jpowsour.2015.01.156.
101. Ashuri, M.; He, Q.; Shaw, L. L. Silicon as a Potential Anode Material for Li-Ion Batteries: Where Size, Geometry and Structure Matter. *Nanoscale* **2016,** *8*, 74–103. https://doi.org/10.1039/c5nr05116a.
102. Brousse, T.; Crosnier, O.; Devaux, X.; Fragnaud, P.; Paillard, P.; J. Santos-Pea, Schleich, D. M. Advanced Oxide and Metal Powders for Negative Electrodes in Lithium-Ion Batteries. *Powder Technol.* **2002,** *128*, 124–130. https://doi.org/10.1016/S0032-5910(02)00186-9.
103. Goodenough, J. B.; Park, K.; Goodenough, J. B.; K.-Park, S. J. Amer. Chem. Soc.. *J. Am. Chem. Soc.* **2012,** *135*, 1167.
104. Gu, Y.; Chen, D.; Jiao, X. Synthesis and Electrochemical Properties of Nanostructured LiCoO 2 Fibers as Cathode Materials for Lithium-Ion Batteries. *J. Phys. Chem. B* **2005,** *109*, 17901–17906. https://doi.org/10.1021/jp0521813.

105. Gu, Y.; Chen, D.; Jiao, X.; Liu, F. $LiCoO_2$-MgO Coaxial Fibers: Co-Electrospun Fabrication, Characterization and Electrochemical Properties. *J. Mater. Chem.* **2007,** *17*, 1769–1776. https://doi.org/10.1039/b614205b.
106. Shaju, K. M.; Bruce, P. G.; Macroporous $Li(Ni_{1/3}Co_{1/3}Mn_{1/3})O_2$: A High-Power and High-Energy Cathode for Rechargeable Lithium Batteries. *Adv. Mater.* **2006,** *18*, 2330–2334. https://doi.org/10.1002/adma.200600958.
107. Liang, C.; Liu, L.; Jia, Z.; Dai, C.; Xiong, Y. Synergy of Nyquist and Bode Electrochemical Impedance Spectroscopy Studies to Particle Size Effect on the Electrochemical Properties of $LiNi_{0.5}Co_{0.2}Mn_{0.3}O_2$. *Electrochim. Acta* **2015,** *186*, 413–419. https://doi.org/10.1016/j.electacta.2015.10.190.
108. Chen, Z.; Wang, J.; Chao, D.; Baikie, T.; Bai, L.; Chen, S.; Zhao, Y.; Sum, T. C.; Lin, J.; Shen, Z. Hierarchical Porous $LiNi_{1/3}Co_{1/3}Mn_{1/3}O_2$ Nano-/Micro Spherical Cathode Material: Minimized Cation Mixing and Improved li^+ mobility for Enhanced Electrochemical Performance. *Sci. Rep.* **2016,** *6*, 1–10. https://doi.org/10.1038/srep25771.
109. Murphy, D. W.; Christian, P. A.; Disalvo, F. J.; Waszczak, J. V. Lithium Incorporation by Vanadium Pentoxide. *Inorg. Chem.* **1979,** *18*, 2800–2803. https://doi.org/10.1021/ic50200a034.
110. Yan, B.; Li, X.; Bai, Z.; Li, M.; Dong, L.; Xiong, D.; Li, D. Superior Lithium Storage Performance of Hierarchical Porous Vanadium Pentoxide Nanofibers for Lithium Ion Battery Cathodes. *J. Alloys Compd.* **2015,** *634*, 50–57. https://doi.org/10.1016/j.jallcom.2015.01.292.
111. Sharma, A.; Tyagi, V. V.; Chen, C. R.; Buddhi, D. Review on Thermal Energy Storage with Phase Change Materials and Applications. *Renew. Sustain. Energy Rev.* **2009,** *13*, 318–345. https://doi.org/10.1016/j.rser.2007.10.005.
112. Farid, M. M.; Khudhair, A. M.; Razack S. A. K.; Al-Hallaj, S. A Review on Phase Change Energy Storage: Materials and Applications. *Energy Convers. Manag.* **2004,** *45*, 1597–1615. https://doi.org/10.1016/j.enconman.2003.09.015.
113. Zhou, L.; Shi, F.; Liu, G.; Ye, J. P.; Han, P. S.; Zhang, G. Fabrication and Characterization of In Situ Cross-Linked Electrospun Poly(vinyl alcohol)/Phase Change Material Nanofibers. *Sol. Energy* **2021,** *213*, 339–349. https://doi.org/10.1016/j.solener.2020.11.039.
114. Sanusi, O.; Warzoha, R.; Fleischer, A. S. Energy Storage and Solidification of Paraffin Phase Change Material Embedded with Graphite Nanofibers. *Int. J. Heat Mass Transf.* **2011,** *54*, 4429–4436. https://doi.org/10.1016/j.ijheatmasstransfer.2011.04.046.
115. Cai, Y.; Sun, G.; Liu, M.; Zhang, J.; Wang, Q.; Wei, Q. Fabrication and Characterization of Capric-Lauric-Palmitic Acid/Electrospun SiO_2 Nanofibers Composite as Form-Stable Phase Change Material for Thermal Energy Storage/Retrieval. *Sol. Energy* **2015,** *118*, 87–95. https://doi.org/10.1016/j.solener.2015.04.042.
116. Chen, C.; Wang, L.; Huang, Y. Morphology and Thermal Properties of Electrospun Fatty Acids/Polyethylene Terephthalate Composite Fibers as Novel Form-Stable Phase Change Materials. *Sol. Energy Mater. Sol. Cells.* **2008,** *92*, 1382–1387. https://doi.org/10.1016/j.solmat.2008.05.013.
117. Lu, P.; Chen, W.; Zhu, M.; Murray, S. Embedding Lauric Acid into Polystyrene Nanofibers to Make High-Capacity Membranes for Efficient Thermal Energy Storage. *ACS Sustain. Chem. Eng.* **2017,** *5*, 7249–7259. https://doi.org/10.1021/acssuschemeng.7b01476.

118. Ke, G.; Wang, X.; Pei, J. Fabrication and Properties of Electrospun PAN/LA–SA/TiO2 Composite Phase Change Fiber. *Polym. - Plast. Technol. Eng.* **2018,** *57*, 958–964. https://doi.org/10.1080/03602559.2017.1370101.
119. Mu, S.; Guo, J.; Zhang, B.; Qi, S.; Yang, L.; Wang, D.; Zhang, S.; Yu, Y. On Preparation and Characterization of the Phase Change Nanofibers from the Copolymer of Poly(Styrene-Co-Acrylonitrile) and Lauric Acid. *J. Macromol. Sci. Part A Pure Appl. Chem.* **2015,** *52*, 699–706. https://doi.org/10.1080/10601325.2015.1063863.
120. Ke, H.; uddin Ghulam M. H.; Li, Y.; Wang, J.; Peng, B.; Cai, Y.; Wei, Q. Ag-Coated Polyurethane Fibers Membranes Absorbed with Quinary Fatty Acid Eutectics Solid-Liquid Phase Change Materials for Storage and Retrieval of Thermal Energy. *Renew. Energy* **2016,** *99*, 1–9. https://doi.org/10.1016/j.renene.2016.06.033.
121. Ke, H.; Li, Y. A Series of Electrospun Fatty Acid Ester/Polyacrylonitrile Phase Change Composite Nanofibers as Novel Form-Stable Phase Change Materials for Storage and Retrieval of Thermal Energy. *Text. Res. J.* **2017,** *87*, 2314–2322. https://doi.org/10.1177/0040517516669078.
122. Chen, W.; Ni, S.; Weng, W.; Fu, M. The Preparation and Characterization of Ultrafine Fatty Acid Ester/Poly(Meta-Phenylene Isophthalamide) Phase Change Fibers Designed for Thermo-regulating Protective Clothing. *Fibers Polym.* **2018,** *19*, 498–506. https://doi.org/10.1007/s12221-018-7180-9.
123. Chen, W.; Weng, W. Ultrafine Lauric-Myristic Acid Eutectic/Poly (Meta-Phenylene Isophthalamide) Form-Stable Phase Change Fibers for Thermal Energy Storage by Electrospinning. *Appl. Energy* **2016,** *173*, 168–176. https://doi.org/10.1016/j.apenergy.2016.04.061.
124. Golestaneh, S. I.; Karimi, G.; Babapoor, A.; Torabi, F. Thermal Performance of Co-Electrospun Fatty Acid Nanofiber Composites in the Presence of Nanoparticles. *Appl. Energy* **2018,** *212*, 552–564. https://doi.org/10.1016/j.apenergy.2017.12.055.
125. Ke, H. Electrospun Methyl Stearate/PET Form-Stable Phase Change Composite Nanofibres for Storage and Retrieval of Thermal Energy. *Mater. Res. Innov.* **2018,** *22*, 150–158. https://doi.org/10.1080/14328917.2016.1266203.
126. Zhang, J.; Yang, Q.; Cai, Y.; Song, X.; Qiao, H.; Zhang, J.; Wei, Q. Fabrication and Characterization of Electrospun Porous Cellulose Acetate Nanofibrous Mats Incorporated with Capric Acid as Form-Stable Phase Change Materials for Storing/Retrieving Thermal Energy. *Int. J. Green Energy* **2017,** *14*, 1011–1019. https://doi.org/10.1080/15435075.2017.1354298.
127. Cai, Y.; Gao, C.; Xu, X.; Fu, Z.; Fei, X.; Zhao, Y.; Chen, Q.; Liu, X.; Wei, Q.; He, G.; Fong, H. Electrospun Ultrafine Composite Fibers Consisting of Lauric Acid and Polyamide 6 as Form-Stable Phase Change Materials for Storage and Retrieval of Solar Thermal Energy. *Sol. Energy Mater. Sol. Cells* **2012,** *103*, 53–61. https://doi.org/10.1016/j.solmat.2012.04.031.

Index

C

D

E

F

G

H

I

L

Q

R

S

T

U

W

X